Circulatory and Developmental Aspects of Brain Metabolism

Circulatory and Developmental Aspects of Brain Metabolism

Edited by

Maria Spatz

U.S. Department of Health, Education, and Welfare
Laboratory of Neuropathology and Neuroanatomical Sciences
N.I.N.C.D.S., N.I.H., Bethesda, Maryland

B.B. Mršulja, Lj. M. Rakić

University of Belgrade, Belgrade, Yugoslavia

and

W.D. Lust

Laboratory of Neurochemistry
N.I.N.C.D.S., N.I.H., Bethesda, Maryland

PLENUM PRESS · NEW YORK AND LONDON

Library of Congress Cataloging in Publication Data

International Symposium on the Pathophysiology of Cerebral Energy Metabolism,
2d, Belgrad, 1979.
Circulatory and developmental aspects of brain metabolism.

"Proceedings of the Second International Symposium on the Pathophysiology of
Cerebral Energy Metabolism, organized and sponsored by the Serbian Academy of
Sciences and Art, in Belgrade, Yugoslavia, September 16–20, 1979."
Includes index.
1. Cerebral ischemia–Congresses. 2. Brain chemistry–Congresses. 3. Brain–Growth–
Congresses. 4. Energy metabolism–Congresses. I. Spatz, Maria. II. Srpska akademija
nauka i umetnosti, Belgrad. III. Title. [DNLM: 1. Brain–Metabolism–Congresses.
2. Energy metabolism–Cogresses. 3. Cerebral ischemia–Physiopathology–Congresses.
W3 IN924GE 2d 1979c / WL300 I6127 1979c]
RC388.5.I516 1979   616.8'1   80-18746
ISBN-13: 978-1-4684-3838-3   e-ISBN-13: 978-1-4684-3836-9
DOI: 10.1007/978-1-4684-3836-9

Proceedings of the Second International Symposium on the Pathophysiology of Cerebral
Energy Metabolism, organized and sponsored by the Serbian Academy of Sciences and
Art, and held in Belgrade, Yugoslavia, September 16–20, 1979.

© 1980 Plenum Press, New York
Softcover reprint of the hardcover 1st edition 1980

A Division of Plenum Publishing Corporation
227 West 17th Street, New York, N.Y. 10011

PREFACE

This monograph contains the Proceedings of the Second International Symposium on the Pathophysiology of Cerebral Energy Metabolism, organized and sponsored by the Serbian Academy of Sciences and Art, in Belgrade, September 16-20, 1979.

The purpose of the Symposium was to promote an interdisciplinary discussion on Circulatory and Developmental Aspects of Brain Metabolism. Doctors G. Buznikov (Moscow), I. Klatzo (Bethesda), B.B. Mrsulja (Belgrade). O.Z. Selinger (Ann Arbor), and M. Spatz (Bethesda) served as members of the Advisory Board, chaired by Dr. Lj.M. Rakić (Belgrade), and helped to plan the scientific program.

CONTENTS

INTRODUCTION

S. Kanazir

Serbian Academy of Sciences and Arts
Belgrade, Yugoslavia

Ladies and Gentlemen: It is a privilege and an
honor to welcome all of you on behalf of the Serbian
Academy of Sciences and Arts to the Symposium on
Circulatory and Developmental Aspects of Brain Metabolism.

I am fully aware that an appropriate welcome for
an audience of extremely competent and prominent scientists such as this one is not a very easy task. The
fact that I am not a neuroscientist makes it an even
more difficult assignment. In any case, let me proceed
in a way characteristic of a molecular biologist who
began his career by investigating bacteria (E. coli and
salmonella), but has considered switching over to brain
research. The brain is a most fascinating, challenging
and attractive enigma. The problems of brain and mind
are not new; they have been known since ancient times
but now with a greater understanding of molecular
biology and neuropharmacology the undertaking seems
more reasonable.

The human brain represents a most complex and
highly integrated system of molecules. Its structure
along with its bioelectrical, biophysical or biochemical
functions, forms an organic unit. The advances in
neural sciences and pathophysiology of the brain led to
a remarkable understanding of some facets of brain
functions. This knowledge has improved the treatment
of some neurological and mental disorders. The spectacular progress made in the fields of brain anatomy,
physiology, molecular biology, neuroendocrinology and

psychopharmacology strongly suggests a relationship
between the macromolecular interactions of cells in the
brain and behavior, thought, perception, emotion and
sleep. These cellular reactions are interrelated and
genetically controlled: the regulated gene expression
results in the orderly appearance of specific proteins.
Many factors such as hormones, neurotransmitters, drugs
and other substances modulate the rate of gene expression in the brain cell, as well as in other cells, but
these changes in the brain may be particularly important
in the regulation of thoughts and behavior of human
beings. Recent studies have also shown that the metabolism not only proceeds with great rapidity and intensity
in the brain cells, but that the turnover of most substrates also occurs at a fairly high rate. Therefore,
a question can be raised: what is the molecular basis
for the stability of brain structure? Consequently,
does the gene expression in the brain control the
structure and perhaps the higher mental functions of
the brain? The developing brain provides a challenging
biological system for the study of the regulation of
gene expression (i.e., the dynamic molecular processes
underlying the structure and function of the brain).
There is hope that these studies will relate gene
expression in the brain to the changing patterns of
brain biochemistry, as well as to the physiological and
mental (associative) activities.

In spite of the spectacular achievements in neural
sciences, the brain is so imperfectly understood that
we simply do not know enough about the molecular biology
of brain in order to deduce any facts related to its
integral performance. The present research provides a
description and a comprehensive picture of the brain
metabolism, but it does not explain the higher mental
functions in molecular terms.

We have therefore reached a stage of knowledge
where, as in other fields of molecular biology, "the
more we know, the better we understand, the more we have
to learn". As we learn more about the brain, the more
we realize how little we know about its integral performance. There are some fragmentary principles known
about the nervous system including the nature of the
nerve impulse and its transmission, but we have no
scientific explanation for the working brain. With
respect to this matter, several questions can be raised:
Is the brain as a biological system comprehensible to
the human mind? Are our physiological concepts of the
brain's functioning too simplistic? Are they a reality?

What is the nature of the brain's coding of environmental experiences and the section of appropriate behavioral responses? Is the brain simply an instrument
made for reading peptide sequences and converting them
into behavior? These questions not only have scientific
significance, but are also extremely important from the
philosophical and psycho-sociological point of view.
This is due to the fact that the whole of human behavior
emerges from the working brain almost entirely without
relevant scientific description and understanding.
Since human behavior is the cause for most of the dif-
ficulties arising in contemporary civilization, the lack
of understanding of the brain performance in a science-
conscious society is a very serious matter and it can
affect the progress of research on mankind. These facts
urge more emphasis on research whose goal would include
the decoding of brain performance and mind. It is, at
the present moment, difficult to predict further evolution
of brain research, but it is clear that present approaches are analytically minded and too deductive. We
need therefore more intellectual and imaginative approaches with the aim of increasing not only data but
our knowledge of brain function.

Many efforts have already been made in this respect,
and with the rapid advancement of our knowledge in
molecular and submolecular biology (biology on the level
of electrons) in general, and in particular in the
brain, it is not too difficult to be an optimist. My
opening remarks did not intend to discourage the audience,
since I am aware of the fact that the existing, presently
unapproachable problems of brain research will not escape
the scientists' attention forever. This hope impels a
most rapid growth of our knowledge about the brain. Your
contributions will promote the achievement of this goal.
The present symposium will increase the level of our knowl-
ledge about the brain and promote international scientific
cooperation.

ISCHEMIA RELATED CHANGES IN ADENINE NUCLEOTIDE

METABOLISM

N. Murakami, W.D. Lust, F.A.M. de Azeredo and
J. V. Passonneau

Laboratory of Neurochemistry
National Institute of Neurological and
Communicative Disorders and Stroke
National Institutes of Health
Bethesda, Maryland 20205, U.S.A.

INTRODUCTION

Derangements in the concentrations of the adenine nucleotides are a common occurrence during and after an ischemic episode. ATP levels decrease rapidly at the onset of ischemia and remain depressed for the entire period of ischemia (12, 24). Upon recirculation of the tissue, the extent of ATP restoration depends on the duration of the ischemic insult. For example, the ATP levels were completely regenerated at 60 min of recirculation following 1 and 5 min of bilateral occlusion, but not after 20 and 60 min (16). The total adenylates (sum of ATP + ADP + AMP) decrease somewhat more slowly during ischemia; in decapitated mouse forebrains, the half-time for the disappearance of total adenylates was approximately 32 min. The total adenylates were also very slow to recover. Ljunggren et al. (10) reported that the levels were significantly less than control at 3 hours after a 3 min ischemic interval. Since the hydrolysis of ATP is the major source of energy for a host of biochemical processes, alterations in adenine nucleotide metabolism resulting from ischemia could have pronounced effects on brain function and therefore on the likelihood of survival after ischemia.

Another adenine nucleotide, cyclic AMP, is also extremely responsive to an ischemic insult (12, 23, 24). The levels of cyclic AMP increase 9-fold to a maximum

at 1 min after bilateral ischemia and thereafter the
levels decrease to about 4-fold greater than control at
5 min (16). During recovery, there is an additional
increase in cyclic AMP after most periods of ischemia.
The patho-physiological significance of the cyclic AMP
changes during and after ischemia may be related to its
role as an inhibitory neuroeffector.

In this paper, we describe experiments concerned
with the role of adenine nucleotides both during and
after an ischemic episode. In the first part, the
cyclic AMP response was measured in brain slices after
decapitation of the gerbil. The results indicate that
brain slices may be useful as a model for determining
the stimulus to the accumulation of cyclic AMP during
and after ischemia. In the second part, the cyclic AMP
accumulation was examined in vivo in gerbils pretreated
with theophylline, a potent anti-adenosine compound (22).
Pretreatment with the drug had a pronounced effect on
the ischemia-induced increase in cyclic AMP. In the final
section, the effects of ischemia and recovery on the
metabolite levels in different brain regions were determined. The brains were fixed by in situ freezing to
minimize the problems with fixation artefact.

MATERIALS AND METHODS

Mongolian gerbils were purchased from Tumblebrook
Farm (West Brookfield, MA). The animals, weighing 50-60 g,
were starved 24 hours prior to experimentation.

Brain Slices: Gerbils were decapitated, the brains
were removed and placed in a glucose-free phosphate-buffered saline (PBS) solution maintained at 37°C. The
slices, approximately 0.35 mm in thickness, were cut with
a McIlwain chopper. To study the effects of ischemia
alone, the slices were frozen in liquid nitrogen at 1.3,
2 and 10 min after decapitation. In the recovery studies,
the slices were transferred to another PBS medium which
was oxygenated with 95% O_2-5% CO_2 and contained 10 mM
glucose. These slices were frozen after 5, 10, 20, or
30 min of incubation at 37°C. The time between decapitation and being either frozen or placed in an oxygenated
medium represented the minutes of ischemia and the time
in the oxygenated medium the duration of recovery.

Theophylline Treatment: Gerbils were pretreated
either with 100 mg/kg ip of theophylline or with an equivalent volume of saline. After 20 min, the gerbils were

lightly anesthetized with diethyl ether, the scalp was
removed and the two common carotid arteries were exposed.
As the animals emerged from the anesthesia, the arteries
were occluded with Heifitz aneurysm clips for a period of
5 min. Sham-operated animals served as controls. At the
end of the 5 min period of bilateral occlusion the gerbils
were either frozen intact or the clips were removed and
the gerbils were frozen at 1, 5 or 30 min of recirculation.
The cerebral cortex was removed in a cryostat maintained
at -20°C and extracted as previously described (15).

   Regional Studies: The surgical procedures were the
same as described above except that only the right common
carotid artery was exposed and subsequently occluded with
an aneurysm clip. In those animals exhibiting positive
neurological signs of ischemia, the gerbil brains were
either frozen in situ with liquid nitrogen or the clips
were removed and these animals were frozen in situ 30 min
later. The in situ fixation is a modification of the
funnel-freezing technique first described by Kerr (7) in
1935 and later improved by other laboratories (18).
Conscious gerbils were restrained in a device which places
a funnel directly over the calvarium. While the liquid
nitrogen is being poured into the funnel, the gerbils
continue to breath spontaneously for up to 90 sec. The
two hemospheres were removed in a cryostat (-20°C) and the
cerebral cortex, cerebellum, hippocampus, caudate-putamen,
thalamus and hypothalamus were dissected out of each side
of the brain.

   Metabolite Measurements: The cyclic nucleotides were
assayed according to the method of Steiner et al. (23) as
modified by Harper and Brooker (4). The intermediary
metabolites were determined enzymatically using the fluorometric procedures described by Lowry and Passonneau (11).
The ATP in the regional studies was measured using a
standardized luciferin-luciferase technique. In the analysis of the total adenylates, the ADP and AMP were enzymatically converted to ATP in a single incubation using
adenylate kinase and pyruvate kinase and the total ATP
determined with luciferin-luciferase.

   Reagents: The reagents were purchased from the following companies: theophylline from K & K Laboratories
(Plainview, NY), luciferin-luciferase from Sigma Chemical
Co. (St. Louis, MO), materials for the cyclic nucleotide
assays from Schwarz-Mann (Orangeburg, NY), Dulbecco's
phosphate-buffered saline from Grand Island Biological Co.
(Grand Island, NY), and the remaining substrates, cofactors
and enzymes from Boehringer Mannheim (Indianapolis, IN).

RESULTS AND DISCUSSION

Brain Slices: There is a vast literature on cyclic
AMP metabolism in brain slices; however, the primary purpose of those studies was to examine the receptor-cyclic
AMP interaction (for review, see Daly (1)). To measure
the hormone-sensitive production of cyclic AMP in brain
slices, the basal levels of cyclic AMP should be constant
as well as relatively low. In freshly cut brains, these
conditions were not met and a 40 min preincubation was
required to stabilize the concentration. Because in
previous investigations the earliest time the cyclic AMP
levels were examined was almost an hour after decapitation,
many of the more interesting aspects of cyclic AMP and
ischemia may have been overlooked in brain slices. In this
study, the levels of high-energy phosphates and cyclic AMP
were measured in brain slices at 1.3, 2 and 10 min after
decapitation (ischemic duration). The metabolites were
also measured in brain slices that were incubated in oxygenated medium containing 10 mM glucose (recovery period).

The high-energy phosphate levels in the brain slices
decreased to about the same extent as observed in bilaterally ischemic brains in vivo (16) (Fig. 1). Incubation of
the slices in oxygenated medium resulted in the rapid
restoration of both ATP and P-creatine. However, the
recovery of these metabolites never exceeded 50% of the
control values determined in vivo. The rate and extent
of recovery was essentially the same after all periods of
ischemia.

The concentrations of cyclic AMP were maximal at 1.3
and 2 min after decapitation and thereafter decreased to
about 45% of peak values at 10 min (Fig. 2). The changes
in brain slices were somewhat greater than the 80 pmole/mg
protein observed in brains from gerbils made ischemic in
vivo (16, 24). During the recovery period in the oxygenate
medium, the concentrations of cyclic AMP in the 1.3 and
2 min groups decreased with increasing periods of incubation; whereas, in slices that were ischemic for 10 min the
levels essentially doubled after 5 min of incubation. The
elevation persisted for 5 min and then the levels dropped
at 20 and 30 min of incubation. At 30 min, all values were
substantially higher than those of controls measured in vivo
(6.4 pmole/mg protein) (16). The cyclic AMP profile during
ischemia and recovery in brain slices was reminiscent of
that observed in brains from bilaterally ischemic gerbils.
It would appear that the brain slices may be a useful model
for the investigation of the molecular mechanism involved
in the fluctuations of cyclic AMP both during and after

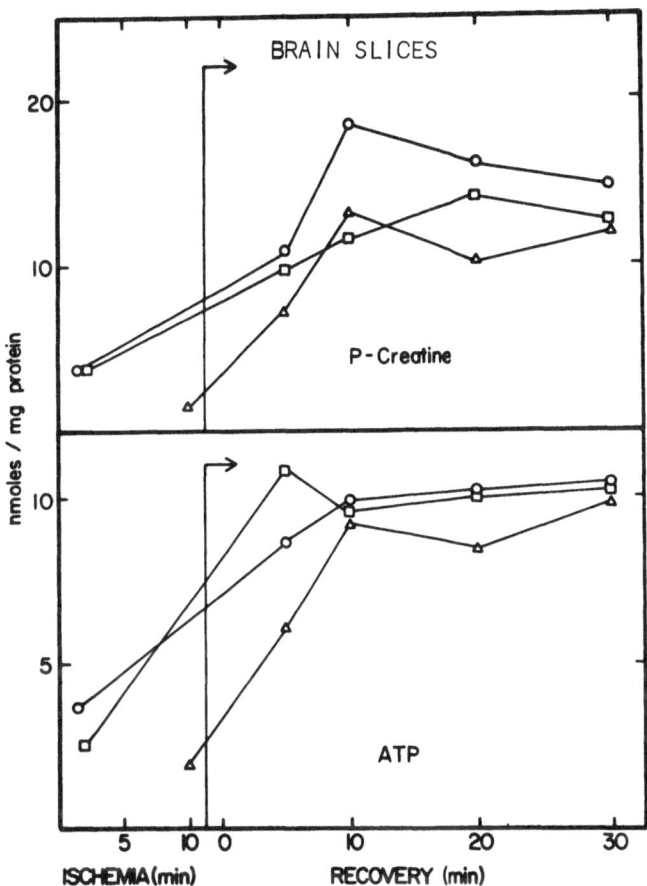

Fig. 1. Concentrations of high-energy phosphates in brain slices. The brain slices from decapitated gerbil brains were prepared and the experiments performed as described in the Materials and Methods. The brain slices were kept at 37°C in a medium free of glucose and oxygen for 1.3 min O , 2 min □ or 10 min Δ and were then frozen in liquid nitrogen (ischemia). Another set of slices following the 1.3, 2 or 10 min of ischemia were incubated at 37°C in oxygenated medium containing 10 mM glucose for an additional 5, 10, 20 or 30 min and then were frozen (recovery). Each point represents the mean of 4 to 5 determinations measured in duplicate.

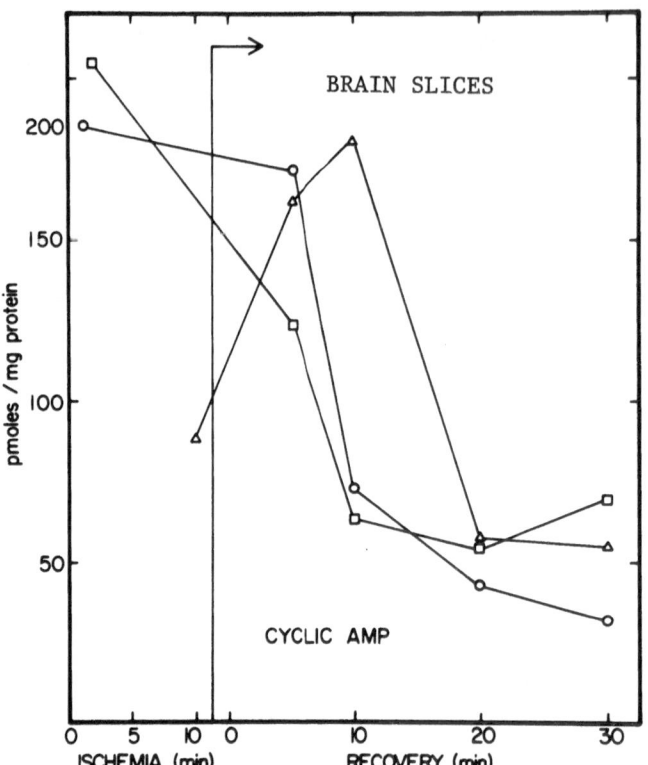

Fig. 2.   Concentrations of cyclic AMP in brain slices.
For details, see legend for Fig. 1.

ischemia. This approach would circumvent many of the
problems that would arise if such a study were pursued
in vivo. The difficulty of evaluating the regulation of
cyclic AMP in vivo will become more apparent in the next
section.

Theophylline: Several laboratories have reported
that adenosine increases dramatically in the brain during
ischemia (8, 20). Since adenosine is a potent stimulus
to the production of cyclic AMP in brain slices, the
possibility exists that it may be one of the agonists
responsible for the large increases in cyclic AMP associated with ischemia. Initially, theophylline was demonstrated to be a phosphodiesterase inhibitor; however, it

now appears that it is more potent as an anti-adenosine compound (22). Theophylline pretreatment of gerbils should therefore reduce the cyclic AMP accumulation during and after ischemia if adenosine is involved.

Gerbils were pretreated with 100 mg/kg ip of theophylline 20 min prior to bilateral occlusion of the common carotid arteries. Peak brain levels of the drug are achieved within 15 to 20 min after injection (21). After 5 min of bilateral occlusion, the levels of high-energy phosphates were significantly higher in the cerebral cortex of theophylline-treated gerbils ($p < 0.05$), albeit the differences were rather small (Fig. 3). During recovery, the rate of ATP restoration was not significantly different in the two groups, whereas the rate for P-creatine was somewhat slower in the cortex from the theophylline-treated gerbils. While the P-creatine levels completely recovered in the untreated group at 5 min of recirculation, those for the theophylline group were still significantly depressed ($p < 0.05$). The overshoot of P-creatine, a phenomenon which commonly occurs during recovery (9, 19), was evident in both groups; however, the effect was more pronounced with theophylline treatment.

In the sham-operated controls, the levels of both cyclic nucleotides were essentially doubled after theophylline treatment (Fig. 4). Cyclic GMP decreased to about the same level in both groups following bilateral ischemia. While the restoration of cyclic GMP was complete at 1 min of recirculation in the untreated group, full restoration in the theophylline-treated group was only evident after 5 min. The cyclic GMP levels in the theophylline group are significantly lower than the theophylline controls at 30 min of recirculation ($p < 0.05$). This may reflect a partial loss of the theophylline effect due to the metabolism and excretion of the drug at 60 min after administration. In the untreated group, the concentrations of cyclic AMP increased 12-fold during the 5 min of ischemia and were further elevated from 29- to 46-fold greater than controls during the early stages of the recovery period. In the theophylline-treated gerbils, the cyclic AMP levels after 5 min of ischemia were not significantly different from those of the sham-operated controls. During the first minute of recirculation, cyclic AMP did accumulate but to a much lesser degree than in the untreated group. By 5 min of recovery, the levels of cyclic AMP were not significantly different from those in the untreated group. The levels of cyclic AMP decreased between 5 and 30 min of recirculation, but remained significantly greater than controls in both groups ($p < 0.05$).

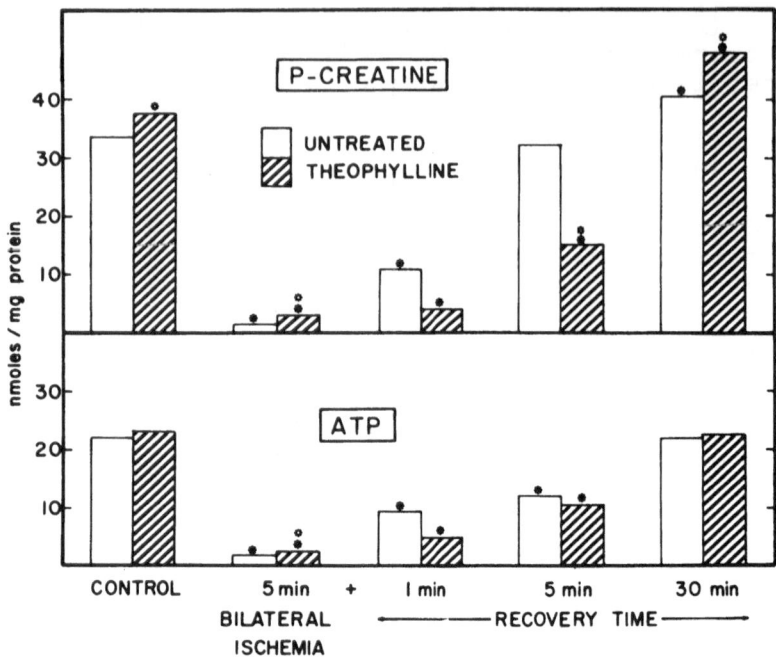

Fig. 3.   The effects of theophylline on the high-energy
          phosphate levels in the cerebral cortex both
          during and after bilateral ischemia.   The
          gerbils were treated and the experiments per-
          formed as described in the Material and Methods.
          The values for the gerbils receiving saline are
          shown in the clear bars and for those receiving
          100 mg/kg ip of theophylline in the hatched bars.
          Each experimental value represents the mean of
          5 determinations.   The asterisks are the values
          which are significantly different from the
          corresponding sham-operated controls and the
          stars are values from theophylline-treated
          animals which are significantly different from
          those of the untreated group (p < 0.05).

     The results for this experiment clearly indicate that
the effect of theophylline was not limited to cyclic AMP.
In previous studies, it has been shown that the P-creatine

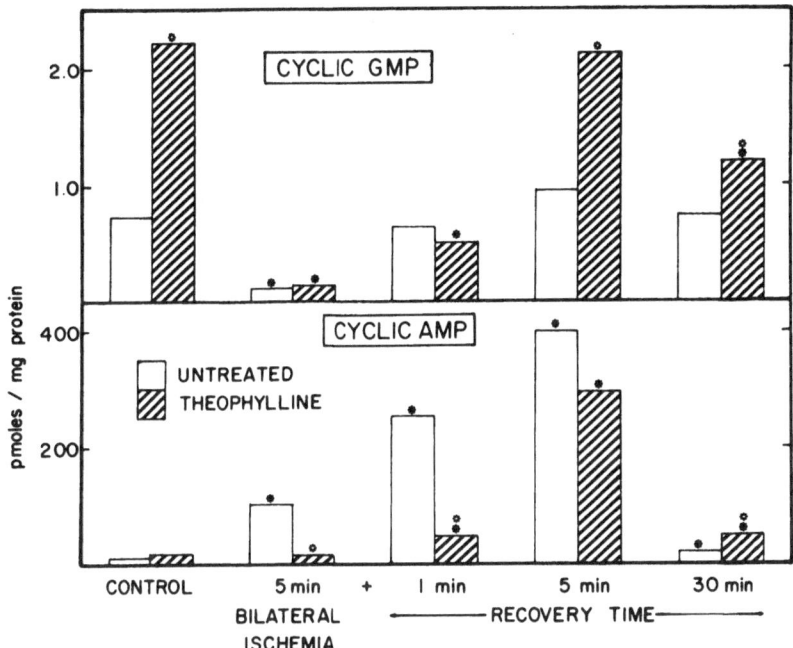

Fig. 4.   The effects of theophylline on the cyclic
          nucleotide levels in the cerebral cortex both
          during and after bilateral ischemia.   For
          details, see legend for Fig. 3.

overshoot was greater and the rate of restoration slower
following longer periods of ischemia (13, 16).   The
slower recovery and the greater overshoot of P-creatine
indicates that the ischemic insult may have been more
severe following theophylline treatment.   In support of
this is the report that aminophylline, also a methylxan-
thine, decreased cerebral blood flow (3).   While this may
not be critical at the onset of bilateral ischemia, it
may have an adverse effect during recirculation.   Another
effect of theophylline was the elevation of cyclic GMP in
the cerebral cortex of sham-operated controls.   Since a
number of other CNS stimulants increase cyclic GMP (2),
it may be the enhanced CNS excitability which accounts
for this effect of theophylline.   The multiple effects of
theophylline clearly demonstrate the hazards of using
drugs to determine regulatory mechanism in vivo.

If theophylline were exerting its effect on cyclic AMP metabolism by blocking the effect of adenosine, then adenosine is the primary stimulus to the production of the cyclic AMP during ischemia, but has little or no effect on the post-ischemic period. It is intriguing to speculate about the significance of this finding. Based on electrophysiological data, adenosine and cyclic AMP both have been shown to depress excitability in neurons of the cerebral cortex (17). The elevation of metabolites during ischemia would be compatible with the quiescent state of the brain which occurs under these conditions. Loss of the inhibitory influence of the adenosine-cyclic AMP system would leave the brain susceptible to excessive excitability at the onset of recirculation. Indirectly, the greater metabolic demands as a consequence of greater CNS excitability could account for the slower restoration of the energy status following recirculation in the theophylline-treated gerbils.

Regional Studies: There is histological evidence that certain regions of the brain are more susceptible to an ischemic episode than others. For example, Ito et al. (5) demonstrated that in the earlier stages of ischemia only the H2 neurons of the hippocampus were affected, but as the ischemia progressed, other regions including the basal ganglia and cerebral cortex were also affected. To determine if there was a biochemical correlation to these histological findings, the levels of P-creatine and adenine nucleotides were measured in 6 regions of the brain both during and after unilateral ischemia. Mrsulja and co-workers have previously examined the concentrations of selected glycolytic intermediates in 4 regions of the brain during an ischemic episode (14). While the overall effects on the metabolites were quite similar in the different regions, any differences may have been obscured by the method of fixation used.

In previous experiments on ischemia in this laboratory the measurements of brain metabolites have been limited to the cerebral cortex from gerbils frozen intact in liquid nitrogen. When the animals are frozen in this manner, respiration ceases as soon as they are plunged into the coolant and hypoxemia ensues. The resulting hypoxia in the outer 1 mm of the brain, where it takes less than 10 sec to freeze, would be relatively minor. However, the freezing time in the hypothalamus, a depth of about 6 mm, would be approximately a minute and the effects of hypoxia would be more severe. To avoid these problems with fixation artefact, the gerbils were frozen in situ with liquid nitrogen as described in Materials and Methods. In this study,

unanesthetized animals continued to breath spontaneously
for up to 90 sec during the freezing process. Since the
circulation remained intact, all areas of the brain not
yet frozen continued to receive oxygenated blood.

The concentration of P-creatine is a good criteria
on which to evaluate the energy status of the brain. As
shown at the bottom of Fig. 5, the levels were relatively
high even in the deep regions like the thalamus and hypothalamus. Thus, preservation of this labile metabolite
indicates that in situ freezing minimizes the fixation
artefact. Following 20 min of right common carotid artery
occlusion, the levels on the right side were depressed in
all regions of the forebrain, although the reduction was
not as great in the thalamus and hypothalamus. The cerebellar concentrations were unaffected as would be expected,
since the circulation to the cerebellum is supposedly not
compromised by ligating the common carotid arteries.
During 30 min of recirculation, the levels of P-creatine
were restored and were not significantly different from
controls. P-creatine in all regions on the left side was
essentially unaffected by the treatment.

The changes in ATP concentrations were very much like
those for P-creatine. The ATP decreased during ischemia in
all regions with the exception of cerebellum (Fig. 6). The
reduction in the thalamus and hypothalamus was not as great
as that observed in the other regions. After 30 min of
recirculation, the ATP values in the hippocampus were still
significantly lower than those of control ($p < 0.05$). While
the ATP in the left hemisphere is somewhat depressed, the
values are not significantly lower than those of controls.

The principal source of the adenine moiety for the
rapid regeneration of ATP is the adenylate pool. Since
the brain is dependent on the hydrolysis of ATP as the
major source of energy, the ability of the brain to regain
function depends to a large extent on the restoration of
ATP and therefore on the size of the adenylate pool. The
total adenylates on the right side decreased in all regions
except the hypothalamus and cerebellum (Fig. 7). As with
ATP, the recovery of the total adenylates was essentially
complete at 30 min of recirculation and the left side
involvement was minimal.

The regional distribution of cyclic AMP is shown in
Fig. 8. The levels are consistent, with the exception of
the caudate-putamen and hypothalamus, with those reported
by Jones et al. (6) for regions of the mouse brain which
was uniformly fixed by microwave irradiation. While the

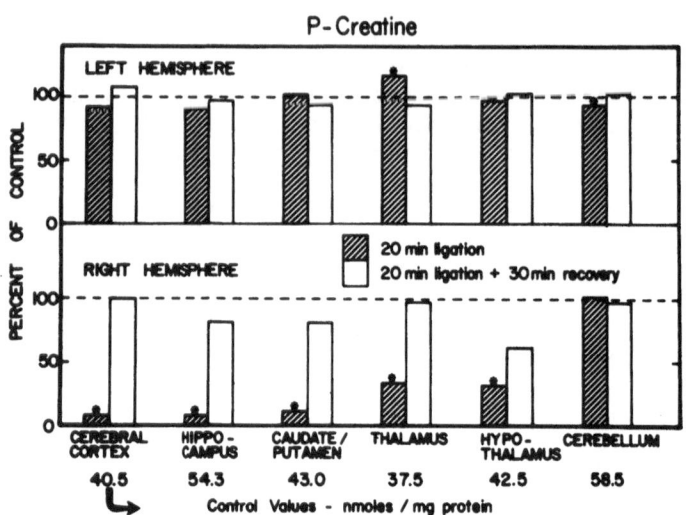

Fig. 5.   Regional distribution of P-creatine during and
           after unilateral ischemia. The gerbils were
           treated and the experiments performed as
           described in the Materials and Methods.   The
           right common carotid artery was occluded for
           20 min and only those animals exhibiting
           positive neurological signs of ischemia were
           used. The control values shown at the bottom
           of the figure are those for the corresponding
           brain region of sham-operated controls and are
           indicated by the dashed line. The experimental
           values are expressed as percent of control and
           represent the mean of at least 5 determinations.
           The open bars represent the metabolite levels
           in the brain regions from ischemic animals and
           the hatched bars those from ischemic gerbils
           30 min after recirculation. Asterisks represent
           values significantly different from sham-operated
           controls (p < 0.05).

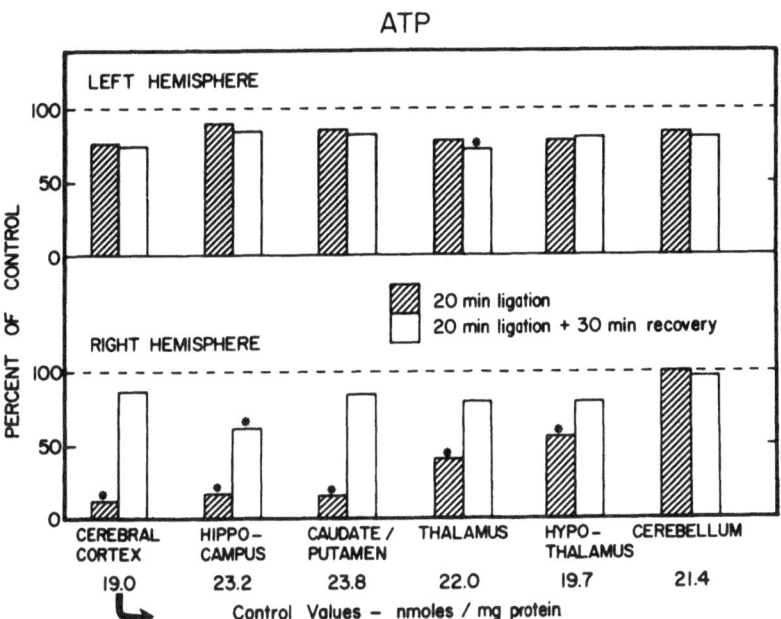

Fig. 6. Regional distribution of ATP during and after
unilateral ischemia. For details, see legend
for Fig. 5.

elevated cyclic AMP in the caudate-putamen and hypothalamus could be attributed to improper fixation, the high
values for P-creatine in these regions do not support
this explanation. There was a 12-fold elevation in cyclic
AMP in the cerebral cortex during ischemia, whereas in
the other regions it was only 5- to 6-fold. While the
cyclic AMP levels returned toward control in all groups
during recirculation, the levels remained significantly
elevated in the cerebral cortex, hippocampus and thalamus
30 min after release (p < 0.05). With the exception of
the thalamus, the effects on the left side were insignificant.

The metabolite data support the contention that
regional studies can be performed with in situ fixation.
P-creatine is perhaps the most labile of the metabolites
measured and the levels were as high, if not higher, in
the deeper regions than in the superficial areas represented

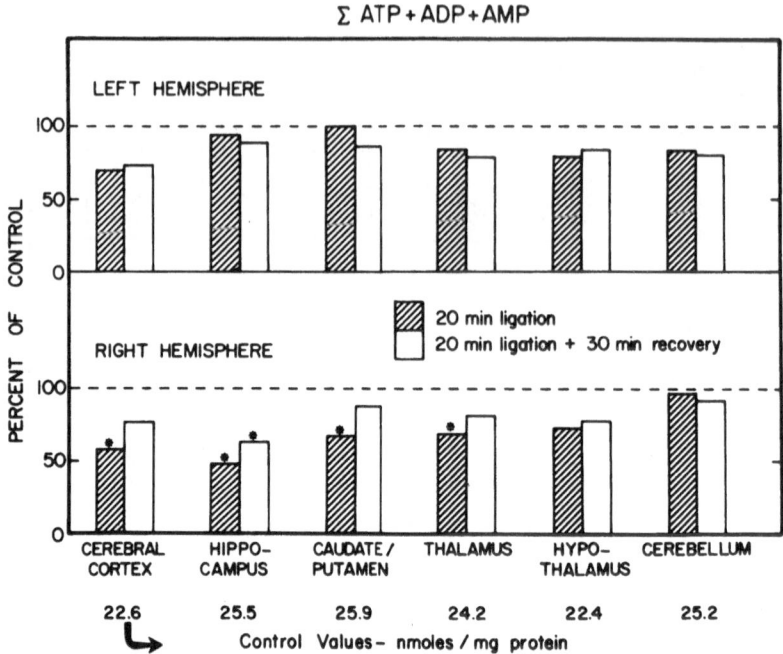

Fig. 7.   Regional distribution of total adenylates
           during and after unilateral ischemia. The
           concentration of the total adenylates refers
           to the sum of ATP, ADP and AMP. For additional
           details, see legend for Fig. 5.

by the cerebral cortex. Proper fixation of brain is not
only important in this type of study but also in pro-
jected studies on the discrete biochemical changes within
a given region. In addition, many of the pathophysiological events which are manifested during recirculation
involve subtle changes which could be masked if fixation
artefact is substantial.

    The regional changes in the levels of the adenine
nucleotides and P-creatine following unilateral ischemia
were marked more by their similarities than their differences. These findings support the work of Mrsulja et
al. (14) that ischemia has a similar effect independent
of the region examined. The only indication of a regional

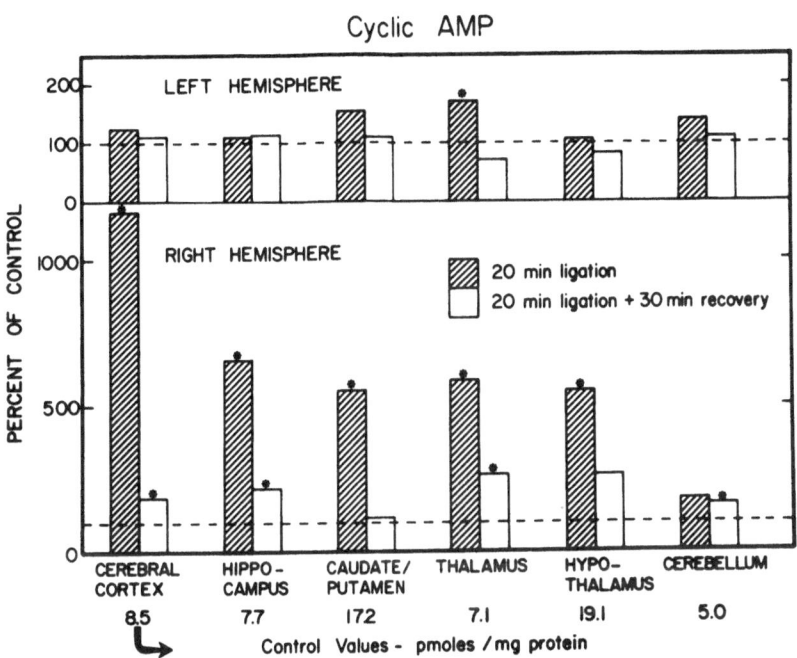

Fig. 8.   Regional distribution of cyclic AMP during and
          after unilateral ischemia. For details, see
          legend for Fig. 5.

difference was the attenuated response of P-creatine, ATP
and total adenylates in the hypothalamus during unilateral
ischemia. If the ischemia were not complete, owing to its
anatomical location, then it is the severity of the insult,
not the response of the hypothalamus to ischemia, that is
responsible for the differences in the metabolite levels.
Further sampling at more dynamic periods of ischemia,
however, may serve to distinguish actual biochemical differences between regions of the brain.

REFERENCES

1. Daly, J.W. (1976): The nature of receptors regulating
    the formation of cyclic AMP in brain slices. Life
    Sciences 18: 1349-1358.

2. Ferrendelli, J.A., Gross, R.A., Kinscherf, D.A. and
    Rubin, E.H. (1979): Effects of seizures and anti-
    convulsant drugs on cyclic nucleotide regulation in
    the CNS. In: "Neuropharmacology of Cyclic Nucleotides
    G.C. Palmer, ed., pp. 211-227, Urban & Schwarzenberg,
    Baltimore.

3. Gottstein, U. and Paulson, O.B. (1972): The effect
    of intracarotid aminophylline infusion on cerebral
    circulation. Stroke 3: 560-565.

4. Harper, J.F. and Brooker, G. (1975): Femtomole sensi-
    tive radioimmunoassay for cyclic AMP and cyclic GMP
    after 2'0 acetylation by acetic anhdride in aqueous
    solution. J. Cyclic Nucleot. Res. 1: 207-218.

5. Ito, U., Spatz, M., Walker, J.T. and Klatzo, I. (1975)
    Experimental cerebral ischemia in mongolian gerbils.
    Acta Neuropath. (Berl.) 32: 209-223.

6. Jones, D.J. and Stavinoha, W.B. (1977): Levels of
    cyclic nucleotides in mouse regional brain following
    300 msec microwave irradiation. J. Neurochem. 28:
    759-763.

7. Kerr, S.E. (1935): Studies on the phosphorus com-
    pounds of the brain. I. Phosphocreatine. J. Biol.
    Chem. 110: 625-635.

8. Kleihues, P., Kobayashi, K. and Hossmann, K.-A. (1974)
    Purine nucleotide metabolism in the cat brain after
    1 hour of complete ischemia. J. Neurochem. 23: 417-425

9. Kobayashi, M., Lust, W.D. and Passonneau, J.V. (1977):
    Concentration of energy metabolites and cyclic nucleo-
    tides during and after bilateral ischemia in the
    gerbil cerebral cortex. J. Neurochem. 29: 53-59.

10. Ljunggren, B., Ratcheson, R.A. and Siesjo, B.K. (1974)
    Cerebral metabolic state following complete compressio
    ischemia. Brain Res. 73: 291-307.

11. Lowry, O.H. and Passonneau, J.V. (1972): A Flexible
    System of Enzymatic Analysis. Academic Press, New Yor

12. Lust, W.D., Mrsulja, B.B., Mrsulja, B.J., Passonneau,
     J.V. and Klatzo, I. (1975): Putative neurotransmitters
     and cyclic nucleotides in prolonged ischemia of the
     cerebral cortex. Brain Res. 98: 394-399.

13. Mrsulja, B.B., Lust, W.D., Mrsulja, B.J., Passonneau,
     J.V. and Klatzo, I. (1976): Post-ischemic changes in
     certain metabolites following prolonged ischemia in
     the gerbil cerebral cortex. J. Neurochem. 26: 1099-1103.

14. Mrsulja, B.B., Mrsulja, B.J., Ito, U., Walker, J.T.,
     Spatz, M. and Klatzo, I. (1975): Experimental cerebral
     ischemia in mongolian gerbils. II. Changes in carbo-
     hydrates. Acta Neuropath. (Berl.) 33: 91-103.

15. Murakami, N., Lust, W.D., Wheaton, A.B. and Passonneau,
     J.V. (1979): Short-term unilateral ischemia in gerbils:
     a reevaluation. In: "Pathophysiology of Cerebral Energy
     Metabolism," B.B. Mrsulja, Lj. M. Rakic, I. Klatzo and
     M. Spatz, eds., pp. 33-46, Plenum Press, New York.

16. Passonneau, J.V., Kobayashi, M. and Lust, W.D. (1977):
     The effect of bilateral ischemia and recirculation on
     energy reserves and cyclic nucleotides in the cerebral
     cortex of gerbils. In: "Alcohol and Aldehyde Metabolizin
     Systems," R.G. Thurman, J.R. Williamson, H.R. Drott
     and B. Chance, eds., v 3: pp. 485-498, Academic Press,
     New York.

17. Phillis, J.W. and Kostopoulous, G.K. (1975): Adenosine
     as a putative neurotransmitter in the cerebral cortex.
     Studies with potentiators and antagonists. Life Sci-
     ences 17: 1085-1094.

18. Ponten, U., Ratcheson, R.S., Salford, L.G. and
     Siesjo, B.K. (1973): Optimal freezing conditions for
     cerebral metabolites in rats. J. Neurochem. 21: 1127-
     1138.

19. Nordstrom, D.-H, Rehncrona, S. and Siesjo, B.K. (1978):
     Restitution of cerebral energy state as well as of
     glycolytic metabolites, citric acid cycle intermediates
     and associated amino acids after 30 minutes of complete
     ischemia in rats anesthetized with nitrous oxide or
     phenobarbital. J. Neurochem. 30: 479-486.

20. Nordstom, C.-H., Rehncrona, S., Siesjo, B.K. and
     Westerberg, E. (1077): Adenosine in rat cerebral
     cortex: Its determination, normal values and correla-
     tion to AMP and cyclic AMP during shortlasting ischemia.
     Acta Physiol. Scand. 101: 63-71.

21.   Sattin, A. (1971): Increase in the content of
       adenosine 3',5'-monophosphate in mouse forebrain
       during seizures and prevention of the increase by
       methylxanthines. J. Neurochem. 18: 1087-1096.

22.   Sattin, A. and Rall, T.W. (1970): The effect of
       adenosine and adenine nucleotides on the cyclic
       adenosine 3',5'-phosphate content of guinea pig
       cerebral cortex slices. Mol. Pharmacol. 6: 13-23.

23.   Steiner, A.L., Wehmann, R.E., Parker, C.W. and
       Kipnis, D.M. (1972): Radioimmunoassay for cyclic
       nucleotides. J. Biol. Chem. 247: 1121-1124.

24.   Watanabe, H. and Ishii, S. (1976): The effect of
       brain ischemia on the levels of cyclic AMP and
       glycogen metabolism in gerbil brain in vivo.
       Brain Res. 102: 385-389.

REGIONAL GLUCOSE METABOLISM AND NERVE CELL DAMAGE AFTER CEREBRAL ISCHEMIA IN NORMO- AND HYPOGLYCEMIC RATS

N. H. Diemer and E. Siemkowicz

Institutes of Neuropathology and Medical
Physiology
University of Copenhagen, and Department of
Anesthesia
Hvidovre University Hospital, Denmark

Cerebral ischemia was induced in rats by lowering the blood pressure to about 50 mm Hg and inflating a pneumatic neck cuff to a pressure of one atmosphere for 10 minutes. Clinically, all rats survived this treatment but with minor neurological deficits (13). The induction of hypoglycemia prior to ischemia exacerbated the symptoms and reduced the recovery of the rats when compared to ischemia per se.

Histologically, the brain of the normoglycemic animals showed a marked loss of Purkinje cells (up to 35%) and hippocampal H-1 neurons (70%) (2) and a slight loss of neurons in the globus pallidus and substantia nigra. Hypoglycemia protected the H-1 and Purkinje cells to some degree, but caused small infarcts in the brain stem, which could have been responsible for the greater neurological deficits after ischemia.

Previous studies of the regional cerebral glucose metabolism had shown areas of both high and low metabolism (5, 6, 7, 8).

The purpose of the present study was to investigate the relationship of the regional nerve cell damage to the local cerebral glucose uptake/metabolism.

METHODS

Male Wistar rats, weighing about 350 g, were deprived of food 24 hours before the onset of the experiments. In one group of animals, hypoglycemia was produced by an intraperitoneal injection of 2 IE/kg insulin
2 1/2 hours prior to ischemia, while in the other group
the rats were exposed to ischemia only. The procedures
of animal preparation will be described only briefly
since they were reported in detail elsewhere (13).

Initially the animals were anaesthetized with ether,
and intubated with a metal tube through which they were
ventilated with gas mixture of 1 vol. % halothane and
35% oxygen in air. Catheters were inserted in the
femoral artery and vein for the periodical measurement
of arterial blood gases, glucose concentration and pH
and for the continuous recording of blood pressure and
temperature. EEG was monitored by means of two-lead
parietal electrodes.

After heparin administration, arterial blood was
slowly removed during a period of 1 minute until the
mean blood pressure was 50 mm Hg, and cerebral ischemia
was then established by inflating a pressure cuff around
the neck. Ten minutes later the cuff pressure was rapidly
released and the blood reinfused within 1 minute. Five
μg adrenalin was added to the last ml of blood, assuring
that the blood pressure was above 150 mm Hg. The rats
treated with insulin received an intraperitoneal injection of 50% glucose during ischemia, in order to
normalize the blood glucose concentration.

In order to investigate the local cerebral glucose
metabolism, 3 normoglycemic and 2 hypoglycemic rats were
injected with 50 μCi ^{14}C-2-deoxyglucose (Amersham,
specific activity 53 Ci/mol) 5 minutes after ischemia
and were decapitated 10 minutes later. The control group
consisted of 2 rats. Two additional animals were treated
similarly but with the nonmetabolizable glucose analogue
^{14}C-3-0-methylglucose (Amersham, specific activity 58.9
Ci/mol) for the evaluation of the carrier mediated glucose transport. Plasma glucose concentration was determined in blood samples obtained at 15, 30, 45 and 90
seconds and 2, 3, 5 and 10 minutes after injection (14).

^{14}C-mannitol (Amersham, specific activity 50 Ci/mol)
was used for the evaluation of passive blood-brain barrier (BBB) permeability in two rats.

All brains were quickly removed and cooled to -
60°C in isopentane. Later they were cut into μm frozen
sections and autoradiograms were developed as described
by Sokoloff and associates (14;9) using Mamoray RP-3
(Agfa-Gevaert) or Kodak MIN-R film exposed for 10 days
(in case of mannitol 20 days).

From the autoradiograms, the local cerebral ^{14}C-2-
deoxyglucose content was determined using calibrated
^{14}C-methylacrylate standards (Amersham) and a Leitz-TAS
image analyzer with densitometer.

RESULTS

Both groups of rats remained unconscious after
ischemia and required respiratory assistance. The EEG
activity reappeared in about 17 minutes after ischemia
in these animals (13). The ischemic cerebral 2-deoxy-
glucose (2-DG) autoradiograms showed striking differences
from the normal appearance (10), as may be seen in Figs.
1A and 5, respectively.

In the normoglycemic group, the increased 2-DG
uptake was noticed in the globus pallidus, substantia
nigra and in the hippocampal H-1 region. On the other
hand, the normally seen high uptake in the 4th layer of
the cortex, in the medial geniculate body, and in the
inferior colliculus was absent. In the brain stem, the
high uptake of 2-DG seen under normal conditions in the
cerebellar nuclei and inferior olives was absent too
(Fig. 3).

The hypoglycemic animals showed predominantly the
same changes in the forebrain as the normoglycemic rats
subjected to ischemia. They were manifested by the
increased uptake of 2-DG in globus pallidus, the substantia nigra, and in the H-1 zone of hippocampus.
However, the brain stem 2-DG uptake was different, as
most of the animals displayed small pale areas indicative
of a reduced 2-DG uptake (Fig. 4) as compared to 2-DG
uptake in hypoglycemia only (Fig. 6).

A closer inspection of the hippocampus in the 2-DG
autoradiograms (Fig. 2) revealed various densities in the
H-1 zone. Normally, the lacunosum layer has the highest
2-DG uptake and this was also the case in both ischemic
groups. However, the stratum radiatum showed a higher
2-DG uptake than the stratum oriens, the two layers in
which the processes of the pyramidal cells run.

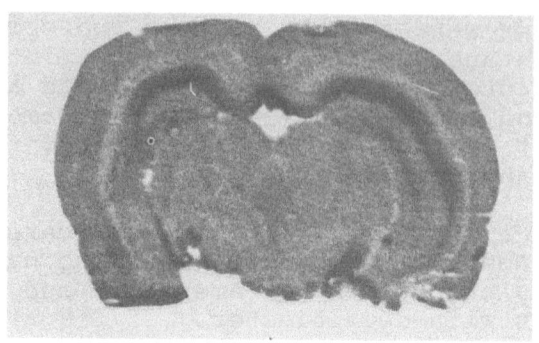

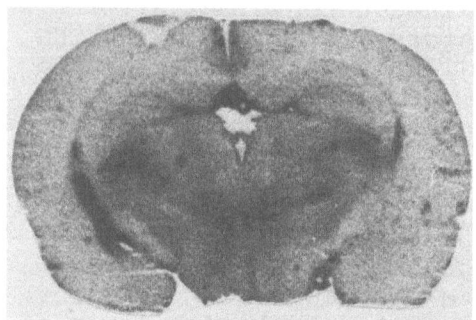 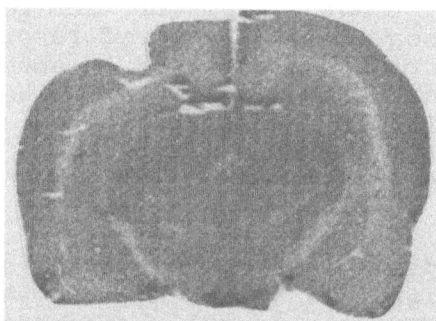

Fig. 1.   A. Autoradiogram from a normoglycemic rat
injected with ^{14}C-2-deoxyglucose 5 minutes
after ischemia, circulation time 10 minutes.
Increased uptake is seen in the hippocampal
H-1 region and in the right substantia nigra
(due to oblique sectioning the right substantia
nigra is not seen). Cortex lacks the high
activity usually seen in the 4th layer.

B. Autoradiogram from a normoglycemic rat
injected with ^{14}C-mannitol 5 minutes after
ischemia. Circulation time 10 minutes. Increased density is seen in the choroid plexus
and in some of the thalamic and hypothalamic
areas, whereas the hippocampal region shows
normal uptake.

C. Autoradiogram from a normoglycemic rat injected with ^{14}C-3-0 methylglucose 5 minutes
after ischemia. Circulation time 10 minutes.
All gray structures show homogeneous and almost
equal tracer uptake.

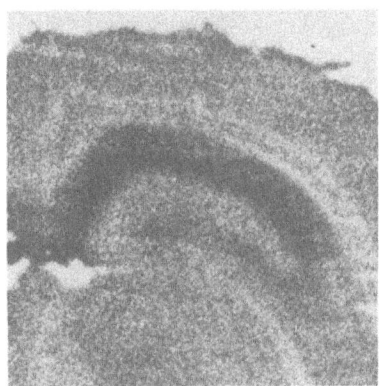

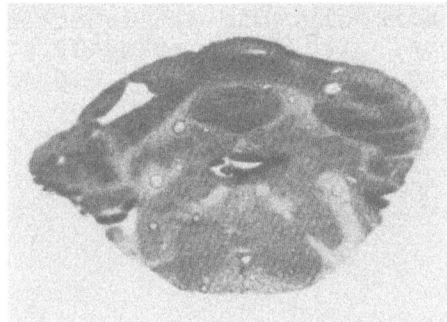
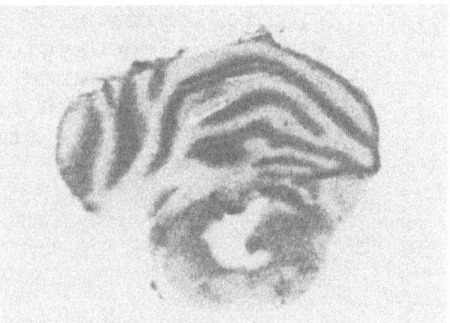

Fig. 2.   Higher magnification of the hippocampal region
          from a 2-deoxyglucose autoradiogram. The upper
          half of the dark band represents stratum oriens,
          the lower one stratum radiatum and lacunosum.
          In between the two layers the pyramidal nerve
          cell bodies are situated.

Fig. 3.   Autoradiogram of cerebellum and brain stem from
          a normoglycemic rat injected with 2-deoxyglucose
          5 minutes after ischemia. The usual high activity
          seen in central cerebellar and brain stem nuclei
          is absent.

Fig. 4.   Autoradiogram of cerebellum and brain stem from
          a hypoglycemic rat injected with 2-deoxyglucose
          5 minutes after ischemia. In the central brain
          stem is seen an area without glucosemetabolism.

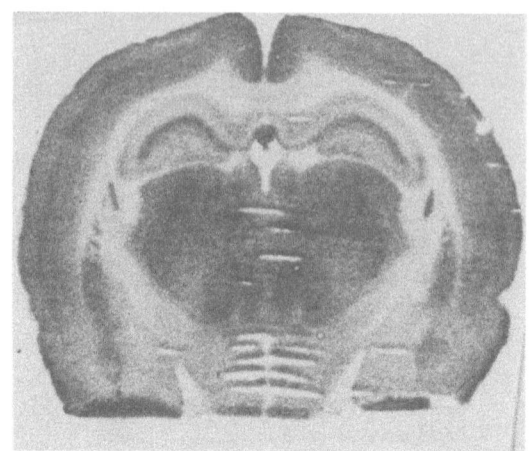

Fig. 5.   Autoradiogram from a normoglycemic rat injected
           with ^{14}C-2-deoxyglucose, circulation time 10
           minutes.   Compared with Fig. 1A there is higher
           uptake in the 4th cortical layer, and in the
           hippocampus only stratum lacunosum shows high
           uptake.

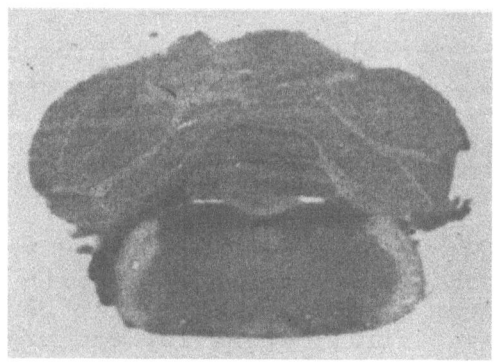

Fig. 6.   Autoradiogram of cerebellum and brain stem from
           a hypoglycemic rat injected with ^{14}C-2-deoxy-
           glucose.   Compared with Fig. 4 there is high
           density in vestibular and cochlear nuclei.   In
           the cerebellar cortex the uptake is highest in
           the granular layer in contrast to the ischemic
           animal (Fig. 3) which has high activity in the
           molecular layer.

Densitometric comparison between normoglycemic ischemia animals and controls showed an overall slight increase in 2-DG uptake (3) and a pronounced increase in the globus pallidus, substantia nigra, and hippocampus.

No attempts were made to calculate the absolute glucose metabolism as described by Sokoloff et al. (14), because a) the isotope circulation time was too short, b) the rate constants (K^*_1, K^*_2 and K^*_3) were not known for the damaged regions, for which furthermore, c) the 2-DG/2-DG-6-phosphate ratio was not known. ^{14}C-mannitol uptake was found to be low in most regions except for some central areas (Fig. 1B) and it was especially noted that none of the structurally damaged regions showed increased mannitol permeability of the BBB.

As may be seen in Fig. 1C, all regions of the gray matter showed a homogeneous and almost equal uptake of ^{14}C-3-0-methylglucose but a somewhat lower uptake in the white matter. Thus, no increased glucose transport took place in the regions with irreversible nerve cell damage.

DISCUSSION

The most conspicuous sequalae of the total ischemia in the rat are the loss of some hippocampal neurons H-1 and Purkinje cells and a slight cellular change of globus pallidus and substantia nigra. While preischemic hypoglycemia protects to some extent these cells, it leads to the infarction of brain stem and lowers the neurological restitution of the animals (2, 13).

The increased 2-DG uptake demonstrable in the hippocampus, globus pallidus and substantia nigra by autoradiography correlated fairly well with the histologically observed areas of ischemic injury, although the most affected area was in the hippocampus. In the hypoglycemic animals, the "white spots" in the brain stem corresponded well to the observed small infarcts in this part of the brain (11). These changes in the 2-DG uptake represent regionally altered glucose metabolism since the uptake of the nonmetabolizable 3-0-methyl glucose or an increased BBB passage of mannitol was not observed in the area of cell loss. In the same model, it had been found that the cerebral metabolic rate for glucose is increased after 3 minutes of ischemia (12), and this is consistent with our semiquantitative situation of increased glucose metabolism in the regions which eventually display nerve cell loss (especially hippocampus H-1).

As there normally is a coupling between cerebral glucose metabolism and flow (CBF) (14), then the H-1 zone should have an increased CBF shortly after ischemia, but experiments using ^{14}C-iodo-antipyrine to visualize the CSF showed no isolated flow increase in any specific region. Thus, there seems to exist an uncoupling of glucose metabolism and flow in globus pallidus, substantia nigra and hippocampus, and this phenomenon, although it disappears in 1 hour after ischemia (3) may represent the factor which leads to a partial loss of nerve cells.

In a previous study of the regional cerebral glucose metabolism after acute focal ischemia in the cat (6), the area of deprived blood flow showed centrally a zone of greatly suppressed glucose utilization surrounded by a narrow zone of increased utilization. The latter was ascribed to anaerobic glycolysis. Rabbits with reversible cerebral emboli (5) showed mostly areas with increased glucose metabolism, but also areas of decreased uptake. In these animals there was no apparent incompabiility between areas of hypermetabolism and recovery. Also gerbils subjected to unilateral clipping of the common carotid artery (7) or air embolism (8) showed regional foci of increased and decreased 2-DG uptake. However, to the best of our knowledge no comparison has been made between regional pathological changes and glucose metabolism after ischemia. The study strongly supports that a linkage between the regional cerebral metabolism and the nerve cell loss might be present in postischemia.

Hypermetabolism alone does not appear to be normally deleterious (1, 4) but when the increased regional glucose metabolism after ischemia is not accompanied by increased CBF, a relative hypoxia may exist resulting in anaerobic glycolysis and damage to the tissue.

REFERENCES

1. Collins, R.D. (1978): Use of cortical circuits during focal penicillin seizures: an autoradiographic study with (^{14}C) deoxyglucose. Brain Research 150: 487-501.

2. Diemer, N.H. and Siemkowicz, E. (1978): The influence of different blood-glucose levels on the brain damage in rats after cerebral ischemia. Neuropath. appl. Neurobiol. 4: 236.

3. Diemer, N.H. and Siemkowicz, E. (1979): Regional glucose metabolism in the rat brain after ischemia. Neuropath. appl. Neurobiol. 5: 83-84.

4. Divac, I. and Diemer, N.H. (1979): Prefrontal system in the rat visualized by means of labelled deoxy-glucose. Evidence for functional heterogenity of the neostriatum. J. Comp. Neurol. (in press).

5. Fieschi, C., Sakurada, O. and Sokoloff, L. (1978): Local cerebral glucose utilization during resolution of embolic experimental ischemia. In: Pathology of Cerebrospinal Microcirculation, J. Cervos-Navarro, E. Betz, G. Ebhardt, R. Ferszt and R. Wullenweber, (eds.), 20: pp. 223-230, Raven Press, NY.

6. Ginsberg, M.D., Reivich, M., Giandomenico, A. and Greenberg, J.H. (1977): Local glucose utilization in acute focal cerebral ischemia, local dysmetabolism and diaschisis. Neurology 27: 1042-1048.

7. Kakari, S., Nishimoto, K., Walker, Jr., J.T. and Spatz, M. (1977): Effects of ischemia on regional glucose metabolism in the gerbil. J. Neuropath. exp. Neurol. 36: 608.

8. Nishimoto, K., Wolman, M., Spatz, M. and Klatzo, I. (1978): Pathophysiologic correlations in the blood-brain barrier damage due to air metabolism. In: Pathology of Cerebrospinal Microcirculation. J. Cervos-Navarro, E. Betz, G. Ebhardt, R. Ferszt and R. Wullenweber (eds.), 20: pp. 237-444, Raven Press, NY.

9. Plum, F., Gjedde, A. and Samson F.E. (1976): Neuro-anatomical functional mapping by the radioactive 2-deoxy-D-glucose method. Neurosciences Research Program Bulletin, 14/4, pp. 457-518.

10. Schwartz, W.J. and Sharp, F.R. (1978): Autoradiographic maps of regional brain glucose consumption in resting, awake rats using (^{14}C)-2-deoxyglucose. J. Comp. Neurol. 177: 335-359.

11. Siemkowcz, E. and Diemer, N.H.: ^{3}H- and ^{14}C-2-deoxylglucose autoradiographical determination of the glucose metabolism after cerebral ischemia in normo-, hypo- and hyperglycemic rats. (To be published).

12.   Siemkowicz, E. and Gjedde, A. (1979): Post-ischemic
       coma in rat: Effect of different pre-ischemic blood
       glucose levels on cerebral metabolic recovery after
       ischemia.   (In press).

13.   Siemkowicz, E. and Hansen, A. J. (1978):   Clinical
       restitution following cerebral ischemia in hypo-,
       normo- and hyperglycemic rats.   Acta Neurol.
       Scandinav. 58: 1-8.

14.   Sokoloff, L., Reivich, M., Kennedy, C., Des Rosiers,
       M.H., Patlak, C.S., Pettigrew, K.D., Sakurada, O.
       and Shinohara, M. (1977): The ^{14}C-deoxyglucose
       method for the measurement of focal cerebral glucose
       utilization: Theory, procedure and normal values in
       the conscious and anesthetized albino rat.   J.
       Neurochem. 28: 897-916.

NUCLEAR CELL REGULATORY MECHANISM

IN CEREBRAL ISCHEMIA AND ANOXIA

Takehiko Yanagihara, M.D.

Department of Neurology
Mayo Clinic and Mayo Medical School
Rochester, N.Y. 55901

In cerebral ischemia and anoxia, the depletion of oxygen supply promptly results in a rapid decline of the high energy source, such as adenosine triphosphate or creatine phosphate. However, the effect of reduction in energy state on the transcription step within the nuclear cell regulatory mechanism and the translation step at the polyribosomal level has not been well investigated. Since prolonged cerebral ischemia and anoxia result in irreversible cellular damage leading to cerebral infarction, the disturbance of the macro-molecular regulatory mechanism may play an important role in the reversibility of these pathophysiological conditions. In the past several years, the effort of our laboratory has been focused on the effects of cerebral ischemia and anoxia on the transcription and translation steps, and on comparison of cerebral ischemia and anoxia. For these purposes, we have utilized gerbils as an experimental model for cerebral ischemia and rabbits for an in vitro model of cerebral anoxia. In this communication, protein synthesis with brain slices in vitro, polypeptide synthesis with isolated microsomes, DNA-dependent RNA polymerase activity and phosphorylation of chromatin protein will be compared in cerebral ischemia and anoxia.

METHODS

For production of cerebral anoxia, an in vitro model with brain slices has been used, as reported elsewhere (11). Briefly, cerebral hemispheres of

33

albino rabbits were sliced with a McIlwain tissue
chopper at 300 μm and were incubated under constant
flow of oxygen and constant shaking at 35°C in a
buffered-electrolyte medium with 10 mM glucose (11).
After equilibration for 15 min, the gas flow to the
experimental group was abruptly switched to nitrogen
gas, and incubation was continued for up to 30 min.
Cerebral ischemia was produced in gerbils by ligation
or clipping of the right common carotid artery in the
neck, as described elsewhere (14, 16). Symptomatic
animals were identified by torsion of neck and trunk,
circling movement, jumping, left hemiparesis and
generalized seizure of the cork-screw type.

For isolation of purified nuclei, brain tissue
was homogenized either in 0.32M sucrose (12) or
directly in 2.3M sucrose (15, 17) containing 1 mM $MgCl_2$.
In the former instance, the crude nuclear fraction after
centrifugation at 700g was suspended in 2.3M sucrose
solution. The suspension was centrifuged at 54,000g
for 50 min, and the purified nuclei were obtained as a
pellet. For preparation of the microsomal fraction,
the postmitochondrial supernatant after centrifugation
at 12,000g for 20 min was further centrifuged at
150,000g for 90 min (12, 17).

Protein synthesis with brain slices was carried out
by addition of L-leucine-4,5[^{3}H] to the buffered-
electrolyte medium containing 10 mM glucose. For
cerebral anoxia, brain slices were resuspended in the
freshly prepared medium. Incubation was carried out at
35°C under constant flow of oxygen and constant shaking
for 30 min. At the end of incubation, one or two brain
slices were removed for determination of the acid-
soluble radioactivity. The rest was homogenized,
precipitated with an equal volume of cold 10% trichloroacetic acid (TCA), washed, dried and taken for determination of protein content according to Lowry et al
(5) and for determination of radioactivity (11, 16).
Polypeptide synthesis with isolated microsomes and pH
5 factor was carried out in the presence of ATP, GTP,
ATP-generating system, 19 amino acids and L-leucine-4,
5-[^{3}H] (12, 17). Incubation was carried out at 37°C
for 20 min, and was terminated by addition of cold 10%
TCA. Each precipitate was washed with cold 5% TCA,
extracted, and dried. The dry sample was either dissolved in 1N NaOH (12) for determination of protein
and radioactivity, or directly solubilized in Soluene-
350 (17) for determination of radioactivity.

DNA-dependent RNA polymerase assay was carried out
with purified nuclei in the presence of three non-
radioactive nucleotide triphosphates, ATP-generating
system and guanosine triphosphate-8-[3H]. For cerebral
anoxia, stimulation by Mg^{2+} and Mn^{2+} was utilized to
differentiate ribosomal RNA and messenger-like RNA (12).
For cerebral ischemia, -amanitin was used to differentiate RNA polymerase I and II (17). Incubation was
carried out for 15 min at 37°C for cerebral anoxia and
at 22°C for cerebral ischemia. The reaction was
terminated by addition of cold 10% TCA. Each precipitate
was washed, extracted and dried. The sample was either
hydrolyzed with 0.3N KOH or solubilized in Soluene-350
for determination of radioactivity. The DNA content of
each sample was determined according to Burton (1).

Phosphorylation of nuclear protein was carried out
with two different methods. The first method has been
described briefly, (15) and is a modification for the
method of Kish and Kleinsmith (3). With this method, the
nuclear proteins to be phosphorylated with endogenous
protein kinases are 0.4N-NaCl extractable chromatin
proteins or proteins loosely bound to DNA. Briefly,
nuclei were suspended in 0.14N NaCl, and stirred, and
the resulting pellet was extracted with 1N NaCl after
a brief period of dispersion. The soluble fraction
was reduced to 0.4N for the concentration of NaCl, and
the final soluble fraction was used for phosphorylation
with endogenous protein kinases in the same fraction.
Phosphorylation was carried out for 6 min at 37°C in
the presence of $[\gamma -^{32}P]$ATP and $MgCl_2$ (3). The reaction was terminated by addition of 0.5N HCl. Protein
which remained soluble was taken as the "HCl-soluble"
protein. The precipitate was extracted with phenol
and further treated according to Teng et al, (8) and
was taken as the "phenol-soluble" protein. In the
second method, isolated nuclei were resuspended and
phosphorylation was carried out at 37°C for 2 min in
the presence of $[\gamma -^{32}]$ATP and MgCl2, (15) and then
fractionated into the NaCl-soluble, HCl-soluble and
phenol-soluble nuclear protein according to Teng et al
(8). For determination of specific radioactivity, each
fraction was precipitated with cold 10% TCA, washed,
and dried. Protein and radioactivity were measured for
each fraction (15). Further separation of various
nuclear protein fractions was carried out by 10%
polyacrylamide disc gel electrophoresis in the presence
of 0.1% sodium dodecyl sulfate, according to Weber and
Osborn (9), as described elsewhere (15). One gel was
frozen and sliced at 1 mm thick, and the radioactivity

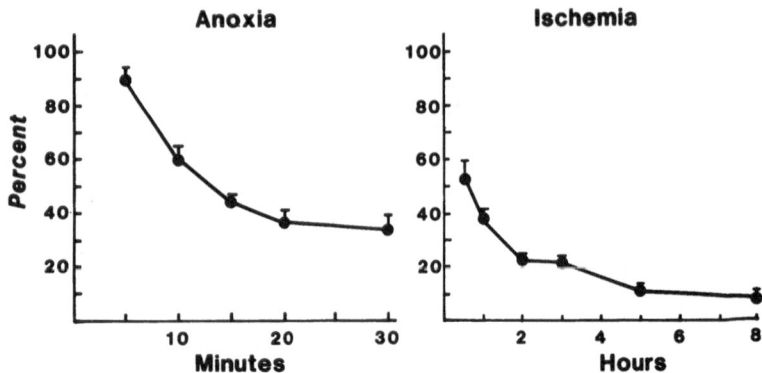

Fig. 1. Effects of cerebral anoxia and ischemia on brain
        slices in vitro. The figure was prepared from the
        data presented in references 11 and 16. The up-
        ward bars represent standard error of mean (SEM)
        bases on 4 to 8 experiments for cerebral anoxia
        and 4 to 6 experiments for cerebral ischemia.

of each gel slice was counted. Another gel was stained
with 0.25% Coomassie brilliant blue for evaluation of
protein pattern.

        The specific radioactivity was expressed as dis-
integration per minute per unit protein for protein
synthesis, and picomole phosphorus or count per minute
per unit protein for protein phosphorylation, while
DNA-dependent RNA polymerase activity was expressed as
disintegration per minute per unit DNA.

RESULTS

        The specific radioactivities for each assay system
and the optimal conditions have been given in the
original articles listed in the references. Fxcept
for analyses of gel electrophoresis, the data shown in
the Figures and Table are expressed as percent of the
specific radioactivities from the control samples.
The effects of cerebral anoxia and ischemia on protein
synthesis with brain slices in vitro are shown in
Fig. 1. There was a precipitous decline of protein

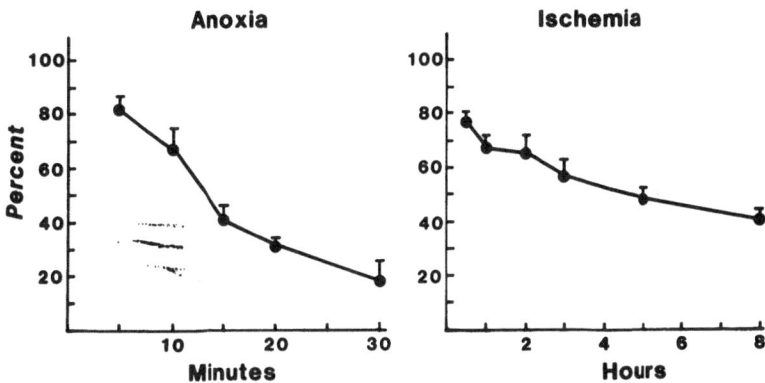

Fig. 2.   Effects of cerebral anoxia and ischemia on
           polypeptide synthesis with isolated microsomes.
           The figure was prepared from the data presented
           in references 12 and 17.   The upward bars
           represent SEM based on 6 to 8 experiments for
           cerebral anoxia and 4 to 6 experiments for
           cerebral ischemia.

synthesis in cerebral anoxia up to 20 min of anoxia.
For cerebral ischemia, a rapid decline of protein
synthesis was seen in ischemia of 2 hrs, and a steady
decline leading to 90% suppression was observed after
ischemia of 8 hrs.   In either case, there was no
significant alteration in the acid-soluble radioactivity
(data not shown).

    The results of polypeptide synthesis with isolated
microsomes in vitro are shown in Fig. 2.   While a rapid
decline was observed in cerebral anoxia to a similar
extent to that seen in protein synthesis with brain
slices in vitro, the decline was gradual and mild in
cerebral ischemia, showing marked contrast to that
seen in protein synthesis with brain slices in vitro.
Even after an ischemic period of 8 hrs, the suppression
only reached 60%.   The ability for RNA synthesis as
measured with DNA-dependent RNA polymerase activity is
shown in Fig. 3.   In contrast to polypeptide synthesis,
the decline of RNA polymerase activity was very slow
and slight in anoxia of 30 min, and there was no significant difference between Mg^{2+}- and Mn^{2+}-dependent RNA
polymerase activity.   In cerebral ischemia, both RNA

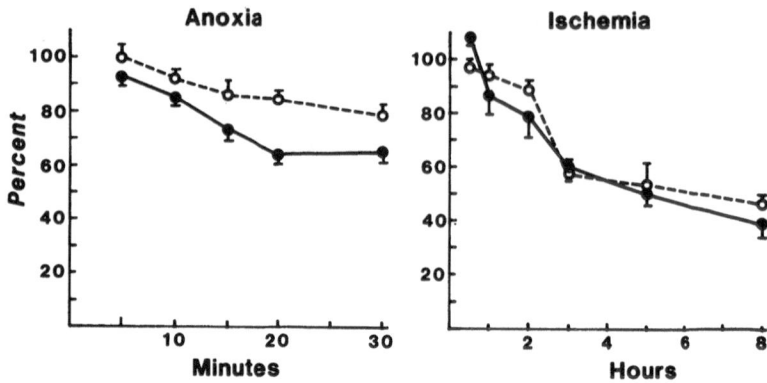

Fig. 3.   Effects of cerebral anoxia and ischemia on
           DNA-dependent RNA polymerase activity. The
           figure was prepared from the data presented
           in references 12 and 17. Closed and open
           circles indicate Mg^{2+}-and Mn^{2+}-dependent RNA
           polymerase activity, respectively, for
           cerebral anoxia, and RNA polymerase I and II,
           respectively, for cerebral ischemia. The
           bars represent SEM based on 6 experiments
           for cerebral anoxia and 4 to 6 experiments
           for cerebral ischemia.

polymerase I and II were relatively stable up to an
ischemic period of 2 hrs and then showed a significant
decrease. The extent of suppression after 8 hrs was
similar to that in polypeptide synthesis.

     Phosphorylation patterns of 0.4N NaCl-soluble
chromatin protein with endogenous protein kinases after
an anoxic period of 15 min are shown in Fig. 4.
There was no significant difference in the profile of
radioactivity after an anoxic period of 15 min. This
was also true even after anoxia for 30 min (data not
shown). The highest peak with molecular weight of
approximately 90,000 was higher in the anoxic sample
for the HCl-soluble fraction, and a similar tendency
was observed for the phenol-soluble fraction. Although
the protein distribution pattern was not shown in Fig.
4, no difference was found between the control and
the anoxic sample. The results from cerebral ischemia
are shown in Fig. 5.

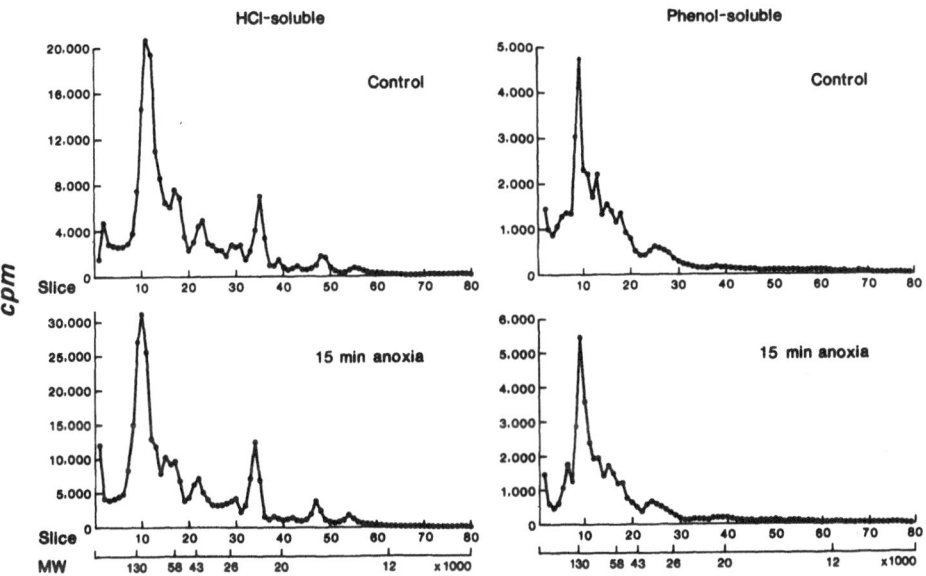

Fig. 4. Effects of cerebral anoxia on phosphorylation
pattern of 0.4N NaCl-soluble chromatin proteins
after an anoxic period of 15 min. The slice numbers are shown on the abscissa, and the location
of molecular weight as measured by marker proteins
is shown at the bottom.

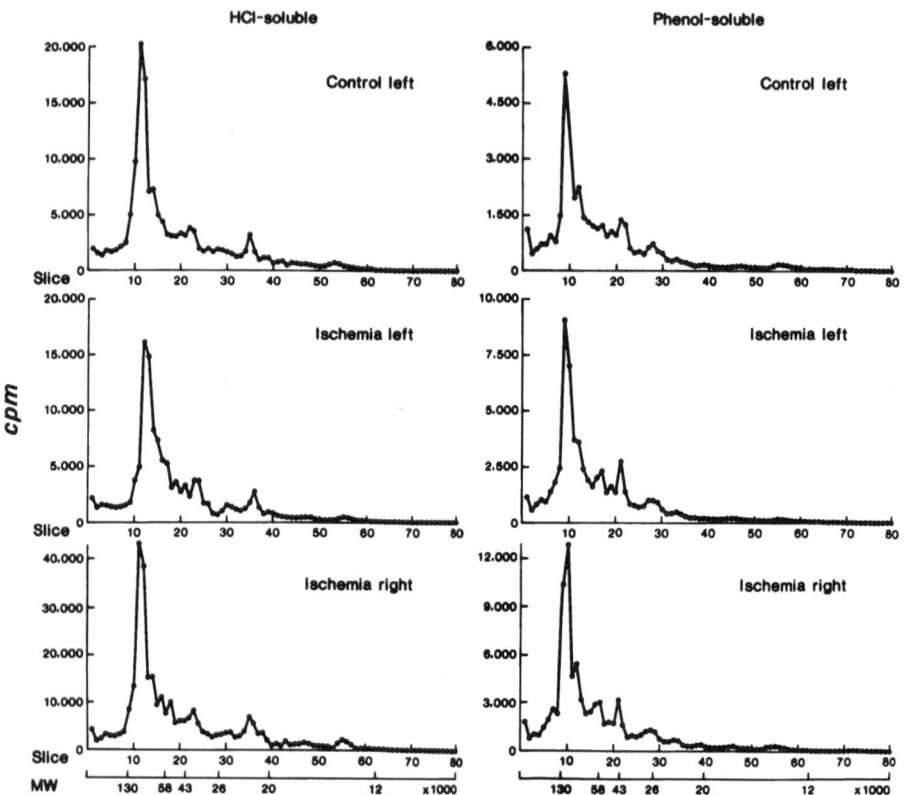

Fig. 5. Effects of cerebral ischemia on phosphorylation
pattern of 0.4N NaCl-soluble chromatin proteins
after cerebral ischemia for 3 hrs. The profiles
at the top were from the left cerebral hemispheres
of asymptomatic (control) animals.
The profiles in the middle were from the left
cerebral hemispheres of symptomatic (ischemic)
animals, while the profiles at the bottom were
from the right cerebral hemispheres of
symptomatic (ischemic) animals. See Fig. 4
for further details.

There was little difference between the left cerebral
hemispheres of asymptomatic (control) and symptomatic
(ischemic) animals. Even after an ischemic period
of 3 hrs, the profile of radioactivity remained
unchanged in either protein fraction, and there was
no alteration in protein patterns after staining with
Coomassie brilliant blue (photos not shown). As seen
in cerebral anoxia, the highest peak with molecular
weight of approximately 90,000 was higher in the
ischemic sample for the HCl-soluble fraction, and a
similar tendency was observed in the phenol-soluble
fraction. Phosphorylation of nuclear protein fractions
separated by step-wise extraction with 0.14N NaCl,
0.25N HCl and phenol in cerebral anoxia and ischemia
is shown in Table 1.

TABLE 1

PHOSPHORYLATION OF NUCLEAR PROTEIN FRACTIONS

Protein Fraction	Cerebral Anoxia (15 min)	Cerebral Ischemia (3 hrs)
Homogenate	87.5 + 1.3	113.9 + 23.4
NaCl-soluble	99.0 + 17.9	153.5 + 15.8
HCl-soluble	92.7 + 3.9	151.0 + 27.0
Phenol-soluble	72.7 + 3.3	97.6 + 12.5

The results are expressed as percent of the control
value (CPM/µg protein) + S.D. based on 5 experiments
for cerebral anoxia and 3 experiments for cerebral
ischemia. Nuclear protein fractionation was carried
out according to reference 8.

There was little change in the NaCl-soluble and HCl-
soluble fractions after an anoxic period of 15 min.
However, the phenol-soluble fraction showed a significant suppression of phosphorylation after the same
anoxic period. In cerebral ischemia, phosphorylation
of the NaCl-soluble and HCl-soluble fraction after 3
hrs was more than that of the control. Although the
extent of phosphorylation of the phenol-soluble
fraction was similar to the control value, it was
relatively decreased considering the much higher
values for the other fractions. Our preliminary

investigation indicated that there are qualitative alterations in phosphorylation pattern of the phenol-
soluble acidic chromatin proteins, as revealed by
electrophoresis (profiles not shown).

DISCUSSION

Although the depletion of high energy phosphates
and glucose in cerebral ischemia has been known for
a number of years, metabolism of macromolecules
which are essential for cellular function and integrity
has not drawn much attention. Kleihues and Hossman (4)
in 1971 observed disintegration of polyribosomal
function after re-establishment of cerebral circulation
following a prolonged period of complete ischemia.
An in vitro model for the investigation of cerebral
anoxia was reported from our laboratory in 1972, (10)
which demonstrated a rapid decline of protein synthesis
with irreversible damage beyond a certain period.
Deterioration of polyribosomal function actually does
occur during progression of cerebral anoxia, (12)
hypoxia (6, 13) or even ischemia (17). However, a
rapid deterioration occurs soon after re-establishment
of cerebral circulation in complete or incomplete
ischemia (2, 17). The results of protein synthesis
with brain slices and microsomes presented here indicate that the polyribosomal disaggregation may occur
continuously in cerebral anoxia, during both progression
and early phase of recovery, while deterioration of
polyribosomal function occurs more precipitously during
the early phase of the recovery process in cerebral
ischemia.

The deterioration of DNA-dependent RNA polymerase
activities occurred slowly in cerebral anoxia and up
to 2 hrs in cerebral ischemia. It is of interest to
note that the irreversibility of protein synthesis in
cerebral anoxia and the high mortality due to progressive stroke in cerebral ischemia occur when RNA polymerase activities decrease below 80% of the control
value. Since reaggregation of polyribosomes requires
fresh supply of messenger RNAs, it is quite possible
that interference of RNA synthesis is more closely
related to the irreversibility of cerebral anoxia or
ischemia than protein synthesis. Further evaluation
of phosphorylation of chromatin proteins revealed
heterogeneous responses. The negative results with
chromatin protein loosely bound to DNA suggest that
nuclear protein kinases are stable in cerebral anoxia
and ischemia, and that phosphorylation of at least

this group of chromatin protein is not affected by these pathophysiological conditions. On the other hand, the results with step-wise extraction of chromatin protein following phosphorylation with intact nuclei indicate that phosphorylation of the phenol-soluble acidic chromatin protein tightly bound to DNA is vulnerable to cerebral anoxia and ischemia, and does show significant quantitative and qualitative alterations at the time when these processes become irreversible. Thus, this series of presentations demonstrated significant alteration within the nuclear cell regulatory mechanism in cerebral anoxia and ischemia, which may play an important role in the reversibility of these pathophysiological processes. This presentation also demonstrated marked similarity between cerebral anoxia and ischemia regarding their effects on the nuclear cell regulatory mechanism.

SUMMARY

The effects of cerebral anoxia and ischemia on polyribosomal and nuclear cell regulatory mechanism were evaluated by reviewing the previously available information on protein and RNA synthesis, and by newly available information on phosphorylation of chromatin proteins. Even though the mechanism for suppression of protein synthesis might be different between these two pathophysiological conditions, the effects on RNA synthesis and phosphorylation of chromatin protein are very similar. There is a close temporal correlation among irreversibility of suppression of protein synthesis, significant decrease of RNA polymerase activity and significant alteration in phosphorylation of acidic chromatin protein tightly bound to DNA. This may be a very important determining factor for irreversibility of cerebral anoxia or ischemia.

ACKNOWLEDGMENT

The present series of investigations were supported by Research Grant NS-06663 from the National Institutes of Health, Public Health Service. The author wishes to thank Ms. Ruth Sargent and Patricia K. Olevson for their help in preparation of the manuscript.

REFERENCES

1.   Burton, K.   (1956): A study of the conditions and
      mechanism of the diphenylamine reaction for the
      colorimetric estimation of deoxyribonucleic acid.
      Biochem. J. 62: 315-323.

2.   Cooper, H.K., Zalewska, T., Kawakami, S.,
      Hossman, K. and Kleihues, P.   (1977):   The
      effect of ischemia and recirculation on protein
      synthesis in the rat brain.   J. Neurochem. 28:
      929-934.

3.   Kish, V.M. and Kleinsmith, L.J.   (1974):   Nuclear
      protein kinases.   Evidence for their heterogeneity,
      tissue specificity, substrate specificities, and
      differential responses to cyclic adenosine 3':5'-
      monophosphate.   J. Biol. Chem. 249:   750-760.

4.   Kleihues, P. and Hossman, K.A.   (1971):   Protein
      synthesis in cat brain after prolonged cerebral
      ischemia.   Brain Research 35:   409-418.

5.   Lowry, O.H., Rosebrough, N.J., Farr, A.L. and
      Randall, R.J.   (1951):   Protein measurement with
      Folin phenol reagent.   J. Biol. Chem. 193:   265-
      275.

6.   Metter, E.J. and Yanagihara, T.   (1979):   Protein
      synthesis in rat brain in hypoxia, anoxia, and
      hypoglycemia.   Brain Research 161:   481-492.

7.   Morimoto, K. and Yanagihara, T.   (1979):   Alter-
      ation of polyribosomes in cerebral ischemia in
      gerbils.   Trans. Amer. Soc. Neurochem. 10:   169.

8.   Teng, C.S., Teng, C.T. and Allfrey, V.G.   (1971):
      Studies of nuclear acidic proteins.   J. Biol.
      Chem. 246:   3597-3609.

9.   Weber, K. and Osborn, M.   (1969):   The reliability
      of molecular weight determination by dodecyl
      sulfate-polyacrylamide gel electrophoresis.
      J. Biol. Chem. 244:   4406-4412.

10.   Yanagihara, T.   (1972):   Cerebral anoxia: protein
      metabolism during recovery in vitro model.
      Stroke 3: 733-738.

11. Yanagihara, T.   (1973):   Cerebral anoxia: an improved in vitro model for biochemical study. Stroke 4:   409-411.

12. Yanagihara, T.   (1974):   Cerebral anoxia: effect on transcription and translation.   J. Neurochem. 22:   113-117.

13. Yanagihara, T.   (1974):   Protein metabolism in the neuronal and neuroglial fractions of rabbit brain during hypoxia.   Trans. Amer. Soc. Neurochem. 5: 108.

14. Yanagihara, T.   (1976):   Cerebral ischemia in gerbils: differential vulnerability of protein, RNA, and lipid syntheses.   Stroke 7:   260-263.

15. Yanagihara, T., Oh'hara, I., Arvidson, C. and Gintz, J.   (1978):   Phosphorylation of nuclear proteins from rabbit cerebrum, cerebellum and liver in vitro.   J. Neurochem. 31: 225-231.

16. Yanagihara, T.   (1978):   Experimental stroke in gerbils: correlation of clinical, pathological and electroencephalographic findings and protein synthesis.   Stroke 9:   155-159.

17. Yanagihara, T.   (1978):   Experimental stroke in gerbils: effect on translation and transcription. Brain Research 158:   435-444.

CORRELATION OF AMINO ACID CONCENTRATIONS, PERMEABILITY

AND STRUCTURAL CHANGES IN MONKEY BRAIN FOLLOWING BOTH

SUSTAINED AND TRANSIENT MIDDLE CEREBRAL ARTERY OCCLUSION

K.A. Conger and J.H. Garcia

Department of Pathology
University of Alabama Medical Center
Birmingham, Alabama   35294 USA

ABSTRACT

The effect of ischemia on the extravasation of inulin was determined following acute middle cerebral artery (MCA) occlusion by comparing amounts of inulin present in ischemic areas of rhesus monkey brain to values obtained from non-ischemic areas. The degree of ischemic damage present was evaluated by both biochemical and structural methods. Increased alanine to glutamate (A:G) ratios (a biochemical index of ischemia) correlated well with increased extravasation of inulin. In a separate study, transient ischemia was studied in monkeys whose MCA was occluded for one hour followed by one week's reperfusion. Throughout the study, local cerebral blood flow (CBF) was monitored. Following a week of reperfusion, no differences in sorbitol or inulin compartments were noted between hemispheres; in addition, A:G ratios had returned to preocclusive levels. These results are interpreted as evidence of blood-brain barrier (BBB) repair and brain metabolic recovery. Light microscopic demonstration of heterogeneous vacuolation correlated well with a biochemical permeability index (sorbitol:inulin ratio) calculated after equilibration of tissue slices in sorbitol and inulin. Equilibrium marker measurements may be useful adjuncts to light and electron microscopic evaluation of ischemic injury.

INTRODUCTION

Our knowledge of the mechanisms that lead to irreversible brain injury following vascular occlusion has
improved significantly through diverse analyses of animal
experiments that reproduce many of the features of human
brain infarction (12). Most biochemical evaluations of
ischemic brains prepared for structural analysis were
thought to be incompatible with the use of chemical fixa-
tives (aldehydes) that are customary in structural studies.
We have demonstrated that the effects of complete ischemia
on brain amino acid contents are reduced considerably by
rapid aldehyde fixation. We also found that ischemic
alterations on amino acid contents were both divergent and
time dependent (4). These features permit accurate quan-
titation of degree of ischemic injury which can be directly
correlated with structural alterations. Application of
this method to the study of a primate model of acute
cerebral infarction revealed a direct correlation between
structural damage and amino acid changes (2).

The biochemical definition of brain edema requires
demonstration of increased water content in brain tissues
(8). We have not measured water content in these studies;
instead we used a combination of marker solutes to evaluate
changes in permeability and compartmentation following
ischemic injury. Changes in marker compartments were
correlated to both structural and biochemical tissue
changes. Special emphasis has been placed on evaluating
(a) the utility of marker solutes in measuring change in
BBB permeability, and (b) the use of marker solutes to
characterize structural changes induced by edema.

We report the results of applying these techniques to
the evaluation of compartment changes following acute and
transient MCA occlusion. The main objective of this study
was to integrate information obtained in these two brain
edema models.

MATERIALS AND METHODS

Glutamate-pyruvate transaminase, sorbitol dehydrogenase and biochemical reagents were purchased from Sigma
Chemical Company, St. Louis, Mo. All other enzymes were
obtained from Boehringer Mannheim, Germany. Dextran T 70
(Registered Trademark) was purchased from Pharmacia,
Uppsala, Sweden. Other chemicals were of reagent grade
from standard sources.

Acute MCA Occlusion and Perfusion

Three adult male rhesus monkeys (M. mulatta) weighing 3.7 to 4.5 kg were maintained on standard monkey chow for 2 weeks before the experiments. One monkey served as a sham-operated control and two were used as experimental animals. Anesthesia was induced with 35 mg/kg ketamine hydrochloride; a second dose of ketamine (50 mg/kg) was administered immediately before perfusion with aldehyde fixative. Embolic occlusion of MCA was effected by injecting a latex cylinder (14 x 1.4 mm) into the internal carotid artery (13). The control animal was subjected to the same surgical procedure, except that 2 ml of Ringer's lactate solution was injected instead of the latex embolus. At 6 and 24 hours after embolization, or 7 hours after the sham-operation, animals were perfused with aldehydes. All perfusions were completed by inserting a cannula into the ascending aorta and opening the right cardiac atrium. Ringer's solution containing 28 mM lactate, 10 units/ml heparin and 0.1% procaine HCl, pH 7.4 (380 mOsmol) was circulated before perfusing with fixative solution, which consisted of 2% formaldehyde, 1% glutaraldehyde, 0.6% inulin in 0.15 M Na phosphate, pH 7.4 (1,000 mOsmol). All solutions were administered at 30-37°C and at a hydrostatic pressure of 180 cm of water. The time elapsed from chest opening to start of prewash was 1-2 minutes. Prewash perfusion lasted 30-50 seconds; perfusion with fixative lasted 5-12 minutes. The brain was removed within 10-15 minutes after fixation, cut coronally and chilled on ice. Adjacent samples of each brain region were cut for biochemical analysis and structural studies. Samples for biochemical analysis were frozen in liquid nitrogen and stored at -80°C. The areas sampled were: body of the caudate nucleus, putamen, corona radiata, internal capsule, insular cortex and superior temporal gyrus. Tissue samples designated ipsilateral were derived from the hemisphere whose MCA had been occluded by a silastic embolus or, in the control monkey, from the side of the sham-operation. Samples from the cerebral hemisphere opposite to the side of the arterial occlusion are designated contralateral.

Transient MCA Occlusion Followed by One Week's Reperfusion

Two unanesthetized monkeys (M. fasicularis) were subjected to MCA occlusion of 1 hour via the orbital approach followed by 1 week of reperfusion (14). Local CBF was monitored by hydrogen clearance measurements before, during and after MCA occlusion at multiple sites in the territory

of the occluded artery and in the contralateral hemisphere (14). In situ perfusion with aldehydes was performed
as described above; however, in addition to inulin, 10 mM
sorbitol was included in the aldehyde fixative. Cubes of
tissue measuring approximately 3 x 3 x 1 mm were taken from
the electrode sites. The tissue samples were divided; half
of the sample was processed for microscopy and the remaining sample was subdivided into 3 to 5 sections which were
either frozen immediately for analysis of amino acids and
perfusion marker extravasation measurements or placed in
fixative containing 10 mM sorbitol and 0.6% inulin for 2
days to determine equilibrative marker spaces.

Preparation of Tissue Extracts and Analysis of Substrates

Tissues were extracted with 0.3 M $HClO_4$ for analysis
of metabolites as described earlier (9). Alanine and
glutamate were assayed by direct fluorometric enzymatic
procedures (11). Inulin was measured as fructose following
hydrolysis (100°C for 20 minutes) directly in 0.3 M perchloric acid (PCA) (1). Sorbitol was measured by direct
fluorometric methods by a modification of Williams-Ashman's
(18) method as reported by Chan et al. (1). Dextran T 70
was measured as glucose following hydrolysis in 1 N HCl
for 2 hours at 100°C (11). Dextran and inulin spaces were
calculated by dividing the amount of hydrolysable glucose
and fructose, respectively, in tissues by the amount in
aldehyde fixative solution. Sorbitol space was calculated
in a similar manner. The sorbitol:inulin permeation index
was calculated by dividing the sorbitol tissue space by the
inulin tissue space.

Electron Microscopy

Cerebral tissue samples (selected from the same tissue used in biochemical assay) were immersed into the
fixative at 4°C for 1 hour. Tissues were then rinsed
overnight in a mixture of 0.2 m sucrose and 0.1 m sodium
cacodylate pH 7.4, post-fixed in 1% osmium tetroxide,
dehydrated in ethanol, en block stained with uranyl
acetate and embedded in epoxy resins. Randomly selected
specimens were sectioned at 1.0 µm and stained with tolu-
idine blue. Ultra-thin sections were stained with lead
citrate and examined in a JEM 100B electron microscope.

RESULTS

Inulin Fixative Space and A:G Ratios 6 Hours after MCA Occlusion

Comparison of A:G ratios obtained from the ipsilateral hemisphere to values obtained from the corresponding contralateral areas showed that the insular cortex and internal capsule had either modest increases or no change in this ratio 6 hours after occlusion of the MCA (Fig. 1). In contrast, the putamen had a marked increase in the A:G ratio. No changes were seen in the inulin fixative space in the insular cortex or internal capsule at this time; however, marked increases in this space were seen in the putamen.

Inulin Fixative Space and Alanine to Glutamate Ratios 24 Hours after MCA Occlusion

Although inulin fixative spaces increased 2- to 3-fold, A:G ratios determined in the insular cortex 24 hours after occlusion were not different from values obtained 6 hours after MCA occlusion (Fig. 2). However, further marked increases in both A:G ratios and inulin fixative spaces were seen in the putamen (note scale changes in illustrations), and the internal capsule values now closely paralleled values found in the putamen.

Correlation of A:G Ratios with Inulin Fixative Volumes

All individual determinations of the A:G ratio were plotted against measurements of the inulin fixative space in both contralateral and ischemic areas 6 and 24 hours after MCA occlusion (Fig. 3). A positive correlation ($r = .91$) was demonstrated between changes in the A:G ratio and the size of the inulin fixative space in samples obtained from ischemic areas, whereas no correlation was obtained in contralateral areas of the brain.

Measurement of A:G Ratios Following Transient MCA Occlusion

Table I lists values of the A:G ratio determined in the caudate nucleus, putamen and insular cortex of a control monkey and values found in both hemispheres of a monkey following 1 hour of MCA occlusion and 1 week of reperfusion. In contrast to the data obtained during acute MCA occlusion, A:G ratios did not differ markedly between

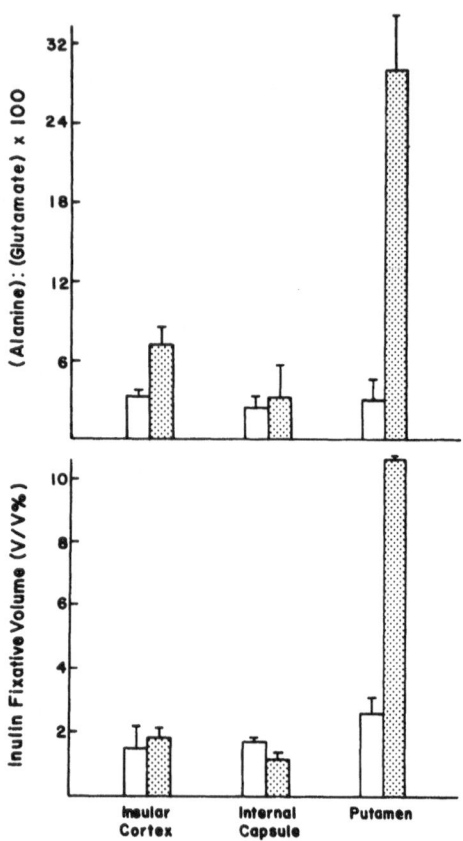

Fig. 1.   A̧:G Ratios and inulin spaces from a rhesus
          monkey brain perfused with fixative containing
          inulin 6 hours after occlusion of the MCA are
          given. Each value is the average of 2 samples
          from areas supplied by branches of the occluded
          (stippled) patent arteries (open bar). Each
          sample was extracted separately as described in
          Materials and Methods and assayed in duplicate.

hemispheres following transient MCA occlusion. Although
low A:G ratios were found in the ipsilateral caudate
nucleus, control values ranging from 0.6 to 4.0 determined
in both contralateral areas of occluded monkeys (n = 6)

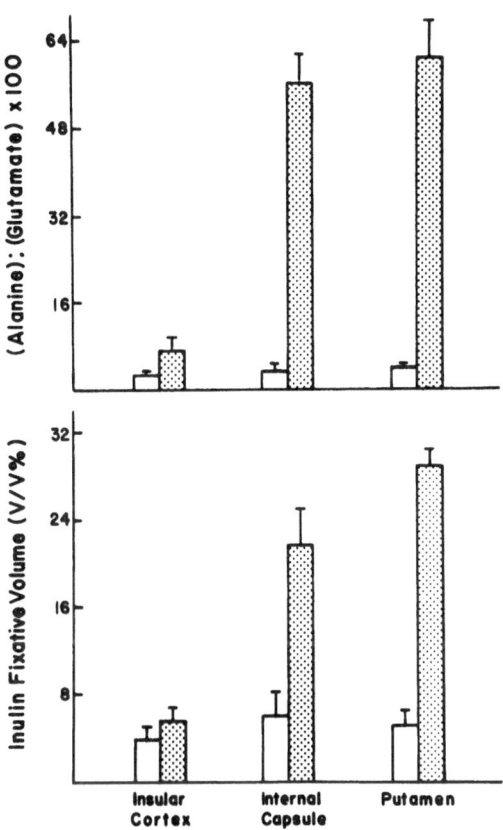

Fig. 2.   A:G ratios and inulin spaces from a rhesus
          monkey brain perfused with fixative containing
          inulin 24 hours after occlusion of the middle
          cerebral artery are given. Each value is the
          average of 3 samples obtained from ipsilateral
          (stippled) and contralateral (open bars) hemis-
          pheres. Each sample was extracted separately
          and assayed in duplicate as indicated in
          Materials and Methods.

and sham operated monkeys (n = 2) bracket the A:G values
reported here.

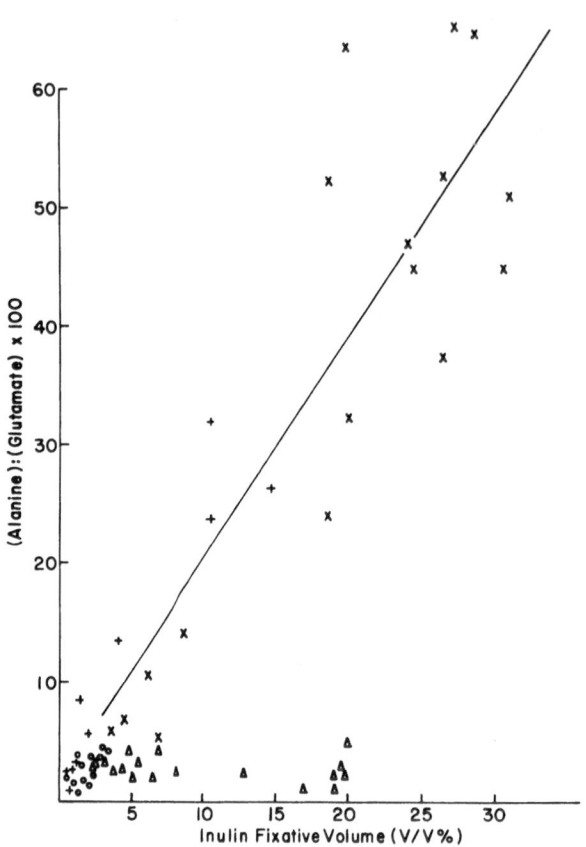

Fig. 3.   The relationship between changes in A:G ratios
          and changes in the inulin space in all areas of
          the 6- and 24-hour occluded monkeys is shown.
          All individual values on the ipsilateral sides
          of the 6-hour (+) and 24-hour (X) monkeys and
          values obtained from the areas in the contra-
          lateral hemispheres of the 6-hour (o) and 24-
          hour (Δ) monkeys were plotted. A positive
          correlation (r = 0.91) between changes in the
          A:G ratio and the size of the inulin space was
          found in ipsilateral areas.

Table I. Effect of Transient MCA Occlusion
on Alanine to Glutamate Ratios

Area	Sham Operated Control	Experimental Animal Contralateral	Ipsilateral
	$\frac{\text{(Alanine)}}{\text{(Glutamate)}}$ x 100		
Caudate Nucleus	2.8 ± 0.1	2.4 ± 0.1	0.6 ± 0.1
Putamen	3.7 ± 0.2	3.1 ± 0.1	2.7 ± 0.5
Insular Cortex	2.4 ± 0.9	1.8 ± 0.4	2.0 ± 0.2

Alanine and glutamate concentrations were measured in discrete areas of aldehyde fixed monkey brain following either a sham operation or occlusion of the MCA for 1.25 hour followed by 1 week's reperfusion. Each value is the mean of 3 samples. Each sample was extracted separately and assayed in duplicate. Values are means ± S.E.M.

Measurement of Marker Spaces in Aldehyde Fixed Rhesus Monkey Brain Cortex

Table II lists distribution volumes of dextran, inulin and sorbitol found in normal aldehyde fixed cortical gray matter. In vivo values obtained by Oppelt and Rall (16) from dog caudate nucleus are supplied for comparison to the values obtained in the aldehyde fixed tissue. Although dextran and inulin compartments are slightly higher than values determined in the caudate nucleus, the similarity to spaces determined in vivo is noteworthy. While the distribution of the high molecular weight solutes is very similar, the large sorbitol space found in the fixed tissue contrasts with the fairly modest sucrose space determined in vivo.

Correlation of the Sorbitol:Inulin Permeatility Index with a Microscopic Index of Vacuolation

Small tissue samples were obtained from areas immediately adjacent to electrodes used to monitor local CBF. Each of these samples was then subdivided for independent light microscopic and biochemical evaluation. Samples evaluated microscopically were assigned values ranging from 0 to 4, where 1 indicates minimal parenchymal changes, e.g. vacuolation only (Fig. 4a), and a value of 4 reflects marked vacuolation of the neuropil together with widened

Table II.   Dextran T 70, Inulin, and Sorbitol Spaces in
            Aldehyde Fixed Rhesus Monkey Brain Cortex

	Solute Molecular Weight	Aldehyde Fixed[a] Rhesus Cortex Gray Matter	In Vivo[b] Dog Caudate Nucleus
Marker Space (ml · 100 g wet tissue $^{-1}$)			
Dextran T 70	70,000	14.1 ± 1.4	10.2 ± 1.1
Inulin	5,000	18.2 ± 0.4	12.7 ± 3.9
Sucrose	342	----------	18.1 ± 5.9
Sorbitol	182	51.5 ± 2.1	----------

a.   Cubes of gray matter measuring 2 mm on a side were
     dissected from the cortex of a rhesus monkey brain
     which had been fixed by in vivo perfusion with
     aldehydes. The cubes of tissue were incubated at
     room temperature overnight in fixative containing
     10 mM sorbitol, 0.6% inulin and 1.0% dextran. After
     removal of surface fixative, the samples were frozen
     in liquid nitrogen. Dextran, inulin and sorbitol
     were measured in perchloric acid extracts of this
     tissue as indicated in Methods.

b.   Reference 16.

extracellular space (ECS), increased hypertrophic astrocytes, abundant lipid-laden macrophages, and so on
(Fig. 4b) (5). Tissues taken for equilibrium marker space
determinations were incubated in 10 mM sorbitol and 0.6%
inulin for 2 days. The penetration of sorbitol relative
to inulin was calculated and plotted against the microscopically derived vacuolation index. The marked correlation obtained is illustrated in Figure 5.

DISCUSSION

     Normally the BBB is relatively impermeable to inulin.
Inulin was included in the fixative in this study to quan-
titate changes in barrier permeability induced by acute
and transient occlusion of the MCA. Measurements of the
inulin space in the contralateral hemisphere following 6
hours of acute MCA occlusion suggest that inulin is largely

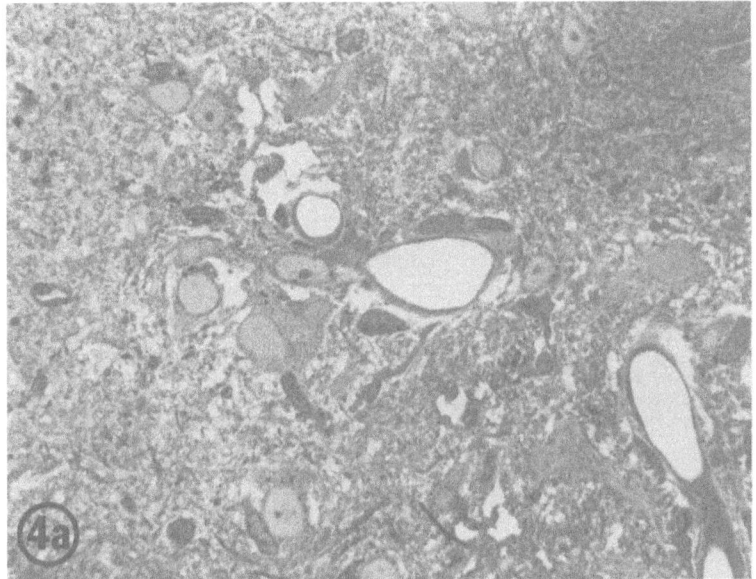

Fig. 4a. Effect of transient ischemia (1 hour) on the
          insular cortex. After 1 week of reperfusion,
          small amounts of perivascular and perineuronal
          edema were seen; however, much of the neuropil
          appeared relatively normal. This sample
          illustrates the pattern of swelling assigned
          a value of 1 (minor vacuolation).
          (Toluidine blue 400x)
          (Reduced 10% for reproduction)

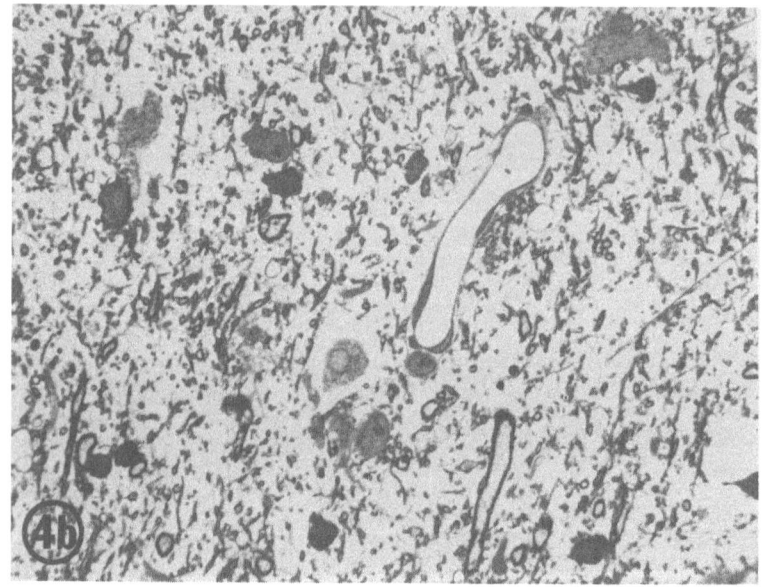

Fig. 4b.   Effect of transient ischemia (1 hour) on the
            putamen. One week after injury, extensive
            vacuolation is seen throughout the neuropil.
            This sample illustrates the typical pattern
            assigned a value of 4 (maximum vacuolation or
            edematous changes).
            (Toluidine blue 400x)
            (Reduced 10% for reproduction)

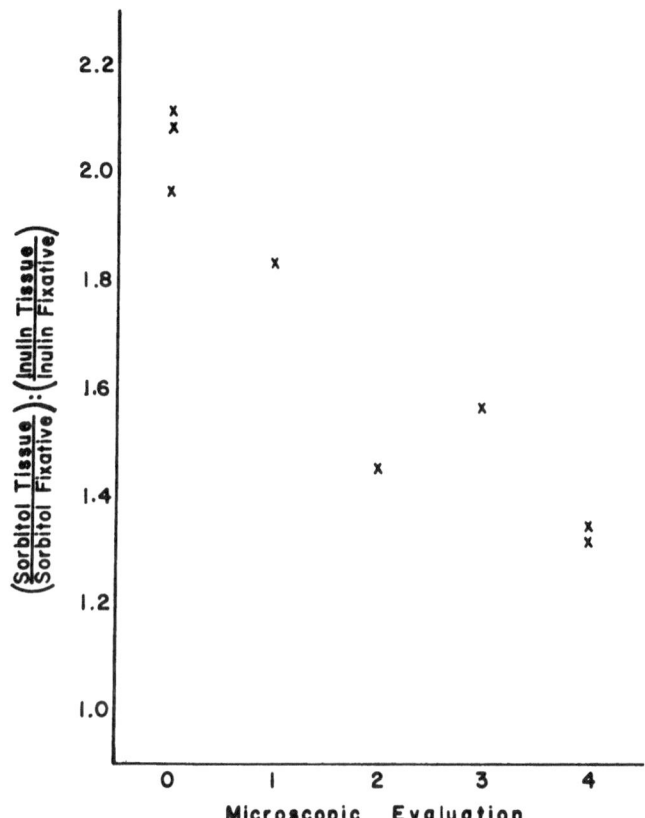

Fig. 5. The relationship between changes in the
sorbitol:inulin ratio and vacuolation in areas
of monkey brain following 1 hour of MCA occlusion and 1 week of post-occlusive recirculation
is shown. Compartment changes in discrete
areas of aldehyde fixed monkey brain were
evaluated by both light microscopy and by
equilibration with marker solutes as indicated
in Materials and Methods.

confined to the vascular compartment (Fig. 1), which
constitutes approximately 3% of total brain volume. When
injected in vivo, inulin readily enters the interstitial
space only after the BBB has been damaged (17). The
perfusion times used in our study were apparently too
short for inulin to equilibrate across areas where the
BBB was intact; however, in areas of compromised capillary integrity, rapid diffusion of inulin into the ECS,

as indicated by increased inulin space. The increased
inulin space in the ischemic putamen indicates altered
capillary permeability and a rapid accumulation of this
marker in either the ECS or astrocytic compartment.
Increased extravasation of inulin following 24 hours of
MCA occlusion in the putamen and internal capsule is
consistent with the progressive breakdown of BBB permeability in these areas (Fig. 2).

The maintenance of the integrity of the BBB until
marked tissue damage is present is in accord with data
obtained from in vivo permeability studies following
ischemic injury (6, 7, 15). Little et al. (10) included
fluorescein in the fixative and found similar extravasation patterns to those demonstrated in this study.

The marked correlation obtained between increases in
BBB permeability and increases in A:G ratios (Figs. 1 and
2) is in accord with the hypothesis that A:G ratios may
be used as a biochemical index of brain ischemic injury.
In order to elucidate the relationship between A:G ratios
and the inulin space, a scattergram (Fig. 3) was prepared.
The individual determinations of the A:G ratio were plotted against measurements of the inulin space in both
contralateral and ischemic areas of the 6- and 24-hour
animals. A positive correlation (r = .91) was demonstrated between changes in the A:G ratio and the size of
the inulin space.

A close correlation between the degree of ischemia,
recovery from ischemia and A:G ratios has been observed
in a study in gerbils (3). Gerbils' carotid arteries
were occluded bilaterally for 1, 5 and 20 minutes, after
which A:G ratios as well as a number of other metabolic
intermediates were measured. After reperfusing over a
half-hour period A:G ratios returned to normal in the
1-minute-occluded gerbils. Interestingly, the A:G ratio
remained high in the 20-minute-occluded animals (which
will not survive), despite return or even overshoot of
concentrations of high-energy phosphate intermediates.
It was proposed in this study that the sustained increases
in alanine might reflect impaired oxidation of pyruvate
in the post-ischemic recirculation period, reflecting
irreversible damage to mitochondria (3).

The data presented above indicate that, following a
transient occlusion of the MCA and a week of reperfusion,
A:G ratios no longer provide indices of the extent of
tissue damage demonstrated microscopically (Table I,
Figs. 4a,b). In addition, marker solutes included in

fixative solutions do not show extravasation patterns of the type seen during acute infarction (Figs. 1, 2 and 3). These data are consistent with the reestablishment of the BBB and the resumption of oxidative carbohydrate metabolism. The return of local CBF to pre-occlusive levels also supports the concept of reversibility (5).

Light and electron microscopic evaluation of this post-ischemic period revealed marked vacuolation of the neuropil, together with widened ECS, etc. (Figs. 4a,b) (5). The possibility that the marked compartmental changes seen microscopically might be quantitated by the use of marker solutes was investigated. Equilibration of small tissue samples directly in solutions containing marker solutes effectively bypasses the BBB. Values obtained with dextran and inulin were similar to compartment volumes determined for these markers in vivo (Table II) (16), and probably reflect the size of the ECS. In the fixed tissue, however, sorbitol penetrates into a much larger compartment than the larger marker solutes. It is proposed that the markedly different permeation characteristics of sorbitol and inulin be used to define a differential permeability scale. We suggest that the magnitude of the sorbitol to inulin ratio is a reflection of tissue permeability and compartment changes. If the ECS is enlarged or the tissue is structurally damaged in a manner which permits migration of the "extracellular" marker into a larger compartment, then the value of this ratio will drop and approach a limiting value of 1. The marked correlation obtained between measures of the sorbitol to inulin ratio and light microscopic evaluation of tissue vacuolation given in Figure 5 shows that these measurements are a useful supplement to the microscopic evaluation of ischemic tissue.

ACKNOWLEDGMENT

Financial support: USPHS Grant NS-06779 and NS-08802.

REFERENCES

1.   Chan, A.W.K., Burch, H.B., Alvey, T.R. and Lowry, O.H. (1975): A quantitative histochemical approach to renal transport. I. Aspartate and glutamate. Am. J. Physiol. 229: 1034-1044.

2. Conger, K.A., Garcia, J.H., Lossinsky, A.S.,
    Bielefeld, J.L., Fuld, R.A. and Kauffman, F.C. (1980):
    Alteration in selected amino acids in primate brain
    after cerebral infarction. Stroke (in preparation).

3. Conger, K.A., Garcia, J.H., Kauffman, F.C., Lust, W.D.
    and Passonneau, J.V. (1980): Cerebral ischemia in
    gerbils: Alanine to glutamate ratios as an index of
    reversibility. J. Neurochem. (in preparation).

4. Conger, K.A., Garcia, J.H., Lossinsky, A.S. and
    Kauffman, F.C. (1978): The effect of aldehyde fixa-
    tion on selected substrates for energy metabolism
    and amino acid in mouse brain. J. Histochem. Cytochem.
    26: 423-433.

5. Garcia, J.H., Conger, K.A., Morawetz, R. and Halsey,
    J.H., Jr. (1980): Post-ischemic brain edema: Quanti-
    tation and evolution. In: Brain Edema Symposium,
    Cervos-Navarro and Ferszt, (eds.), Raven Press, New
    York (in press).

6. Hossmann, K.A. and Olsson, Y. (1971): The effect of
    transient cerebral ischemia on the vascular permea-
    bility to protein tracers. Acta Neuropath. 18: 103-
    112.

7. Ito, U., Go, K.G., Walker, J.T.,Jr., Spatz, M. and
    Klatzo, I. (1976): Experimental cerebral ischemia
    in mongolian gerbils. III. Behavior of the blood-
    brain barrier. Acta Neuropath. 34: 1-6.

8. Katzman, R., Clasen R., Klatzo, I., Meyer, J.S.,
    Pappius, H.M. and Waltz, A.G. (1977): Report of Joint
    Committee for Stroke Resources. IV. Brain edema in
    stroke. Stroke 8: 512-540.

9. Kauffman, F.C., Brown, J.G., Passonneau, J.V. and
    Lowry, O.H. (1969): Effects of changes in brain
    metabolism on levels of pentose phosphate pathway
    intermediates. J. Biol. Chem. 244: 3647-3653.

10. Little, J.R., Sundt, T.M. and Kerr, F.W.L. (1974):
    Neuronal alterations in developing cortical infarction
    J. Neurosurg. 39: 186-197.

11. Lowry, O.H. and Passonneau, J.V. (1972): A Flexible
    System of Enzyme Analysis, pp. 146-218, Academic
    Press, New York.

12. Molinari, G.F. and Laurent, J.P. (1976): A classifi-
    cation of experimental models of brain ischemia.
    Stroke 7: 14-17.

13. Molinari, G.F., Moseley, J.I. and Laurent, J.P. (1974):
    Segmental middle cerebral artery occlusion in primates:
    An experimental method requiring minimal surgery and
    anesthesia. Stroke 5: 334-339.

14. Morawetz, R.B., DeGirolami, U., Ojemann, R.G.,
    Marcoux, F.W. and Crowell, R.M. (1978): Cerebral
    blood flow determined by hydrogen clearance during
    middle cerebral artery occlusion in unanesthetized
    monkeys. Stroke 9: 143-149.

15. Olsson, Y., Crowell, R.M. and Klatzo, I. (1971): The
    blood-brain barrier to protein tracers in focal
    cerebral ischemia and infarction caused by occlusion
    of the middle cerebral artery. Acta Neuropath. 18:
    89-102.

16. Oppelt, W.W. and Rall, D.P. (1967): Brain extra-
    cellular space as measured by diffusion of various
    molecules into brain. In: Brain Edema, Klatzo and
    Seitelberger (eds.), pp. 333-346, Springer-Verlag,
    New York.

17. Steinwall, O. and Klatzo, I. (1966): Selective vul-
    nerability of the blood-brain barrier in chemically
    induced lesions. J. Neuropath. Exp. Neurol. 25:
    542-559.

18. Williams-Ashman, G.H. (1965): D-Sorbitol. In:
    Methods of Enzymatic Analysis, H. U. Bergmeyer (ed.),
    pp. 167-170, Academic Press, New York.

HISTOCHEMICAL AND BIOCHEMICAL INVESTIGATIONS ON ATPASE ACTIVITY FOLLOWING TRANSIENT BRAIN ISCHEMIA IN GERBILS

Branka J. Mršulja and B. B. Mršulja

Division of Neurophysiology and Neurochemistry
Institute for Biological Research, Belgrade,

Laboratory of Neurochemistry, Institute of
Biochemistry, Faculty of Medicine, Belgrade,
Yugoslavia.

INTRODUCTION

Cerebral ischemia is a serious complication of many disease processes which involves a various number of structural and biochemical changes, including edema. The pathophysiology of edema associated with a cerebrovascular lesion and its developmental patterns are sufficiently specific to warrant a separate classification of ischemic brain injury (16). Since ion concentration gradients across cell membranes are maintained by an energy-dependent pump mechanism (15, 33), it was assumed for a long time that fluid accumulation was an expression of a disturbed cellular osmoregulation arising from an "acute energy crisis" which develops during ischemia (7). However, we have not found any relationship between the early post-ischemic brain edema and cerebral energy metabolism: in post-ischemic brain edema and cerebral energy metabolism: in post-ischemia ATP and energy charge are restored within minutes but brain edema progresses (26). Concurrent with the increase in water content, the Na-K-ATPase activity progressively decreased, and as the water content declined, the enzyme activity was restored (26). Therefore, it was tentatively concluded that, among the many factors which are involved in the development and/or persistence of post-ischemic brain edema, the functioning of the Na-K-ATPase is of the utmost importance.

The purpose of the study has been to measure the specific activity of the ouabain-sensitive Na-K-ATPase (E.C. 3.6.1.3) since the function of this intramembranous enzyme is to maintain the high potassium and the low sodium concentrations in the cell (32). In addition, the histochemical localization of the potassium-stimulated p-nitrophenyl phosphatase (K-NPPase) activity, a component of the Na-K-ATPase, was evaluated since this approach is more desirable for the follow-up of the regional enzyme activity. Experiments were performed to coincide with the appearance and disappearance of brain edema in the surviving animals (26).

MATERIALS AND METHODS

Ischemia was produced by bilateral common carotid artery occlusion for 15 min in the Mongolian gerbil. In the biochemical assay, the post-ischemic period was followed for 24 hours, while in the histochemical analyses, the times were extended for up to 1 week of recirculation.

Biochemical Procedures

For biochemical evaluation of Na-K-ATPase activity, fractions of brain capillaries and parenchyma have been obtained as described by Djuricic and Mrsulja (3). Na-K-ATPase activity was measured as follows:

Step 1. Protein samples (2-4 mg/ml) were added to the incubating medium containing 70 mM Imidazole-Cl buffer, pH 7.4, 3 mM ATP, 90 mM NaCl, 10 mM KCl, 3 mM $MgCl_2$, 0.02% BSA and 1 mM ouabain. Time of incubation for parenchymal and capillary preparations was 3 and 5 minutes, respectively. Incubation was stopped by boiling the samples for 1 minute at 100° C.

Step 2. The following reaction mixture (0.2 ml) was added to the tubes: 100 mM K-phosphate buffer, pH 7.1, 10 mM $MgCl_2$, 3 mM PEP, 0.1 mM NADH, 5 mg/ml lactate dehydrogenase and 5 mg/ml pyruvate kinase. Incubation was performed for 20-30 minutes at room temperature.

Step 3. Incubations were stopped with 0.03 ml of N HCl and after 10 minutes of standing at room temperature, 0.1 ml of 6N NaOH was added. Following a 15 minute reincubation at 60°C, the tubes were cooled.

Step 4.   The NAD$^+$ formed in the reaction was measured
fluorometrically and NAD$^+$ was used as the standard (21).
The difference between the samples measured with and
without ouabain represented the Na-K-ATPase activity.
Protein concentration was measured by the method of
Lowry et al. (22).

Histochemical Procedures.

Potassium-stimulated nitrophenyl phosphatase
activity was studied in cortex, caudate and thalamus.
The gerbil brain was fixed by perfusion for 20 minutes
with 3% paraformaldehyde in 0.1 M cacadylate buffer,
pH 7.4.   Frozen sections, 20 to 40 um thick, were cut
on a Reichert freezing microtome and were incubated in
the K-NPPase media according to the methods of Ernest
(4, 5) and Guth and Albers (9).   Nitrophenyl phosphate,
the disodium salt, was used as substrate for both
methods.   Since nonspecific alkaline phosphatase also
hydrolyzes nitrophenyl phosphate, levamisole (3 mM),
a specific inhibitor of alkaline phosphatase, was
added to the K-NPPase medium (1).   The following controls were carried out to substantiate the specificity
of the cytochemical procedure: a) omitting the substrate
or potassium, b) adding 5 mM ouabain to the K-NPPase
medium and c) heating the sections at 65° C for 30 minutes before incubation.

RESULTS

Biochemical analyses

In ischemia, the specific activity of the Na-K-
stimulated ATPase markedly fell by 80% in the vascular
endothelium, while the parenchymal activity only decreased by 20% (Fig. 1.).   With the onset of recirculation, vascular Na-K-ATPase activity slowly recovered
and reached normal values by the fourth hour of reflow,
while the parenchymal enzyme activity was further inhibited but eventually recovered by 24 hours of recirculation.

Histochemical analyses

The histochemical distribution of the reaction
product (K-NPPase activity) was the same irrespective
of the method used, although the enzymatic activity was
granular by the Ernst method and diffuse by the Guth-
Albers technique.   No reaction product was observed in

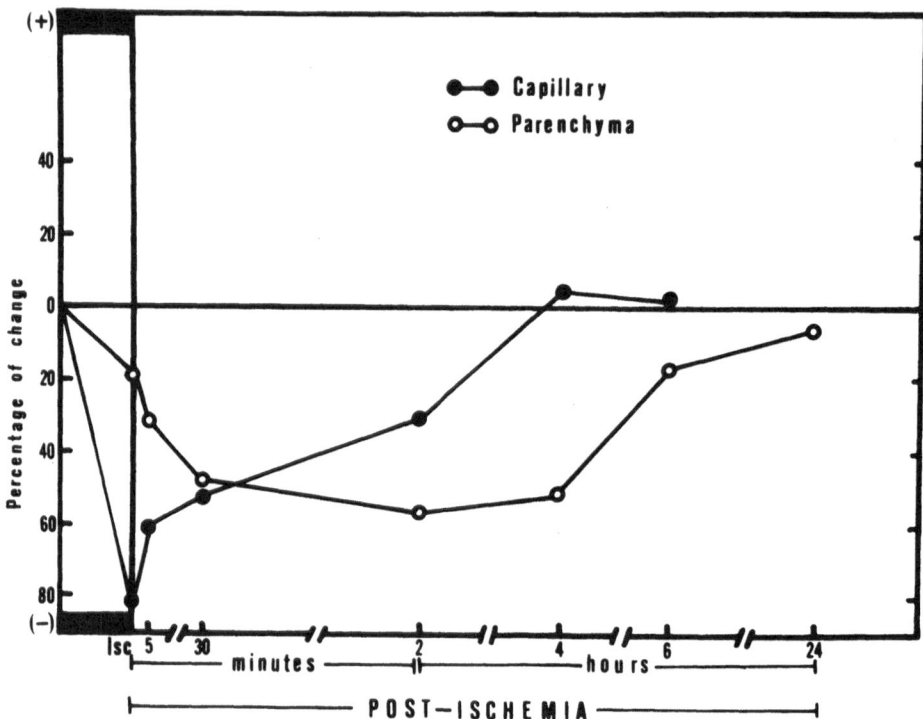

Fig. 1. Enzymatic activity of Na-K-ATPase in brain
        parenchyma and capillary fraction in 15-min
        ischemia and post-ischemic period.

sections incubated in medium without potassium. Addition of 5 mM ouabain to the K-NPPase medium markedly reduced the amount of the staining product.

In controls, the sites of K-NPPase activity as indicated by the reaction product were most obviously associated with the vascular endothelium and neuropil (Fig. 2, 5, 6, 10). Neuronal perikarya were shown to be relatively inert.

Animals sacrificed almost immediately after the release of the common carotid artery clips showed a slightly decreased staining product in the neuropil and blood vessels of the cerebral cortex. In the early post-ischemic periods, a patchy and/or horizontal con-fluent pattern of diminished K-NPPase activity within the cortex was observed (Fig. 3), while a markedly decreased reactivity was evident in the caudate and thalamus (Figs. 10, 11). The diminished amount of staining product within the lower cortical layers be-came apparent after 6 hours of reflow (Fig. 3); this laminar nature of K-NPPase activity in the cortex coexisted with the extensive foci of diminished enzyme activity in the caudate and thalamus of the same animal (Figs. 10, 11). Sometimes the cortical changes were characterized by a circumscribed reduction of enzyme activity involving the vascular endothelium and surrounding neuropil. Animals with the focal cortical changes frequently showed a normal pattern of the enzyme activity in the caudate and thalamus. The greatest decrease in the enzyme activity was in the central zone of the foci (Fig. 9). They were surrounded by a peripheral "reactive zone" (Fig. 11) showing a markedly dense reaction product in the neuropil and blood vessels (Fig. 11).

A prominent feature of the severely injured brain regions was the localization of the K-NPPase activity in the perivascular region (Figs. 4, 7, 8, 12-14). Beginning at 24 hours and continuing thereafter, numerous astrocytes were "overloaded" with the reaction product. The reaction product was easily visualized in the astrocytic processes which extended to the capillary wall (Figs. 12-14). These results are illustrated in parallel studies obtained with the Ernst and Guth-Albers methods (Figs. 7, 8).

COMMENTS

Ischemia led to a marked reduction in the Na-K-

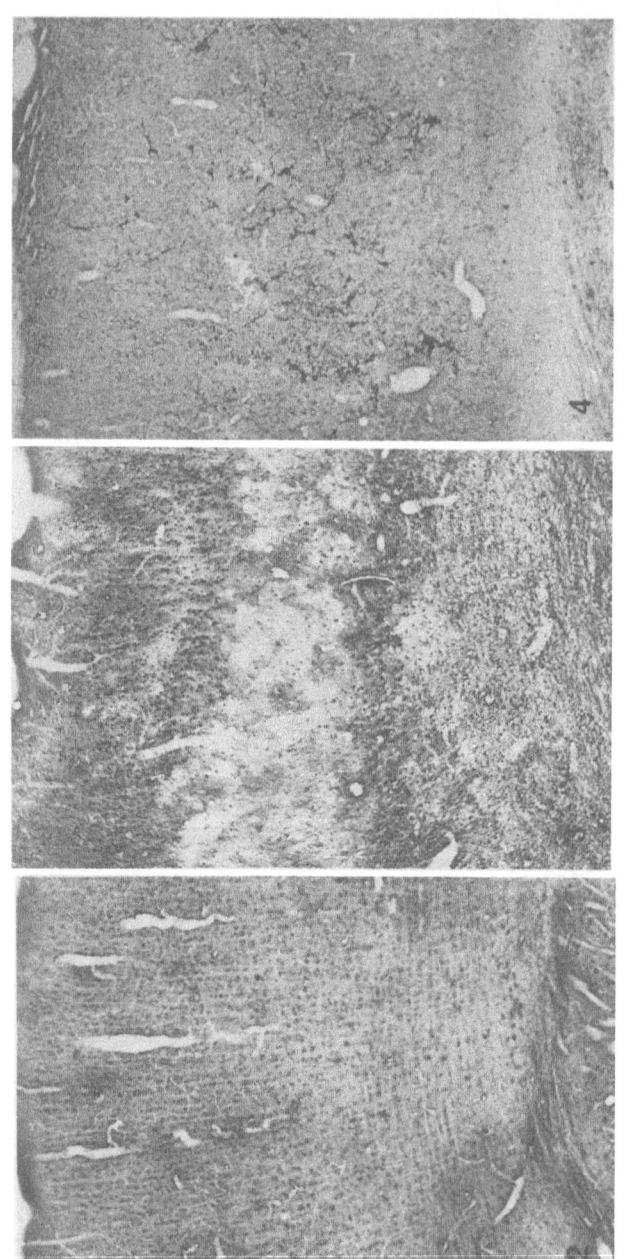

Fig. 2.   Control cerebral cortex

Fig. 3.   Strong decrease of K-NPPase activity in the lower
           cortical layers after 6 hr of postischemia.

Fig. 4.   Large number of reactive astrocytes are demonstrated
           in the same cortical region of 3-day-old infarction.
           X 28 (Reduced 10% for reproduction.)

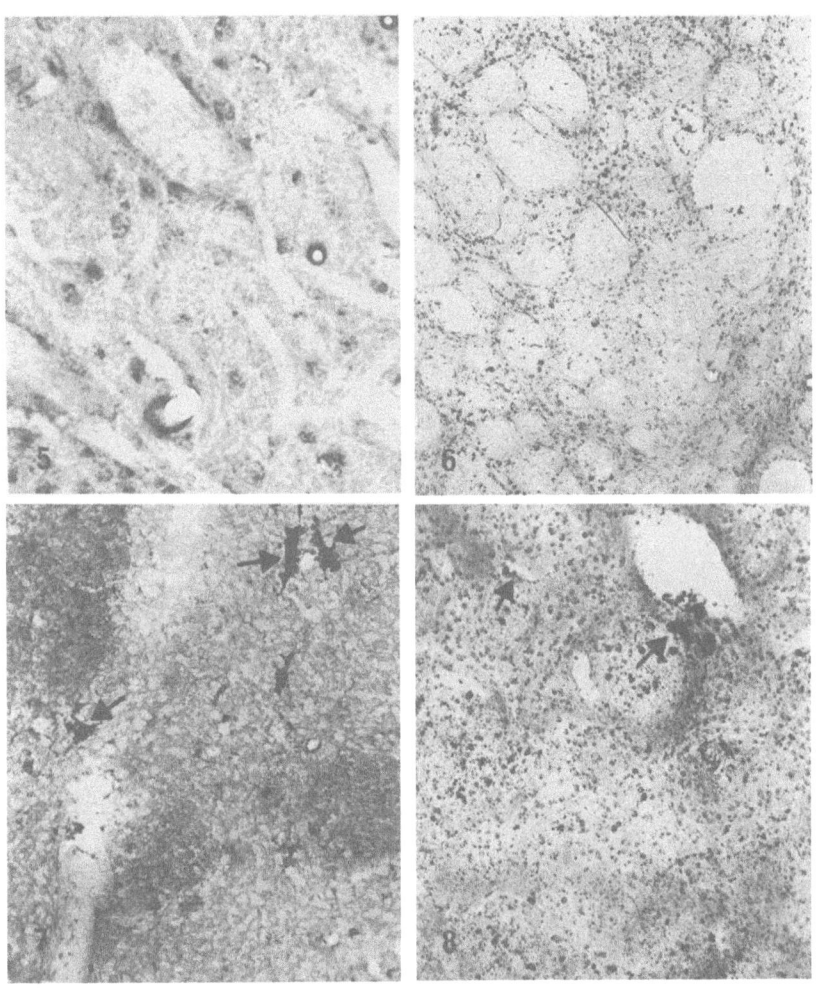

Figs. 5-8:   Caudate. Left side figures (Guth and Albers
              method); Right side figures (Ernst method).

Figs. 5-6:   K-NPPase activity in the vascular endothelium
              and neuropil of control caudate.   X109

Figs. 7-8:   The conspicuous staining product underlines
              the astrocytes, in the perivascular localiza-
              (arrows) and neuropil of 3-day-old infarct.
              X 175 (Reduced 10% for reproduction)

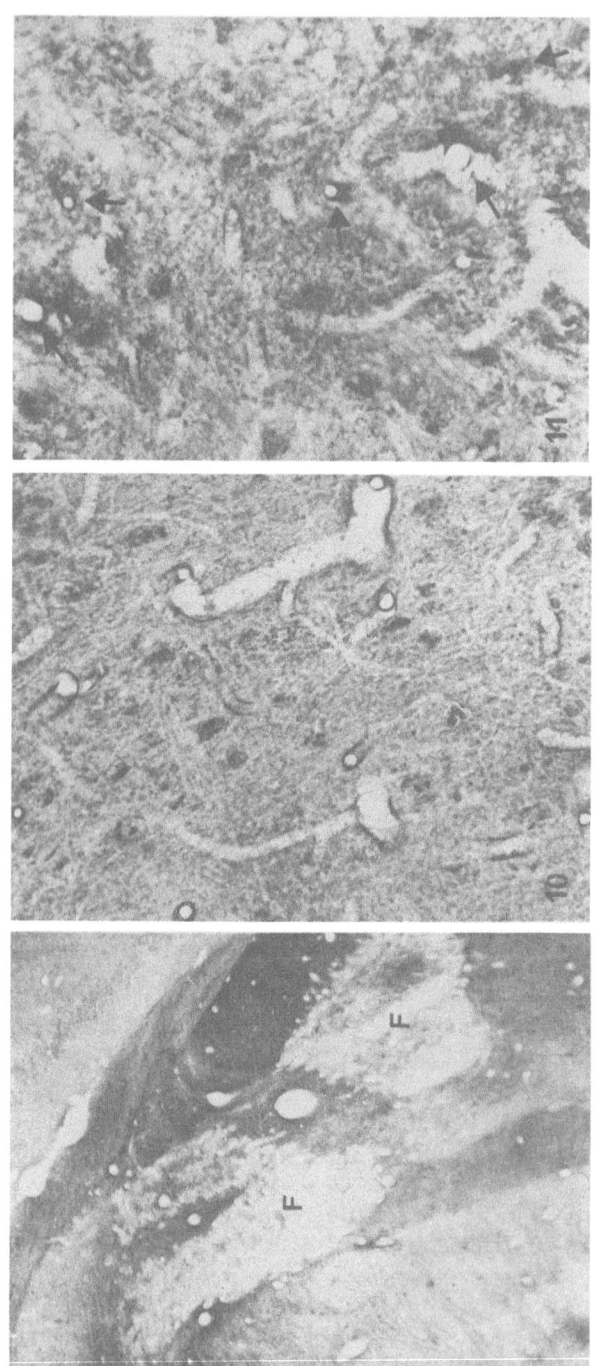

Fig. 9.   Extensive foci (F) of severe tissue injury in the
          dorsolateral thalamus of an animal sacrificed after
          6 hr of reflow. X 28

Fig. 10. K-NPPase activity in vascular endothelium and neuropil
          of the control thalamus. X 175

Fig. 11. Note intense granular staining product in the reactive
          zone of the thalamic infarction, after 12 hr occlusion
          of the artery. The blood vessel show increased K-NPPase
          activity as compared to control (Fig. 10). X 175
          (Reduced 10% for reproduction)

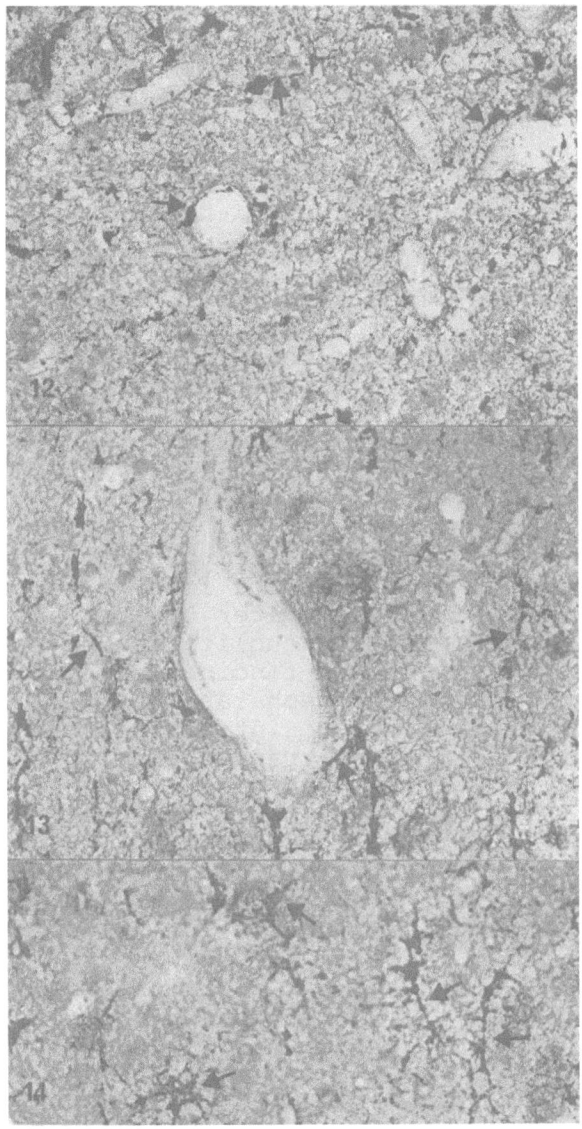

Figs. 12-14.   Numerous perivascular reactive astrocytes,
               loaded with staining product, are seen in
               the cerebral cortex of 7-day-old infarct
               (Fig. 12) and in the caudate of 3-day-old
               infarction (Figs. 13 & 14). The vascular
               processes (arrows) of the cells are clear-
               ly demonstrated by the enzymatic reaction.
               X 175 (Reduced 10% for reproduction.)

of rats previously treated with intraperitoneal administration of water (24).

In contrast with the marked biochemical changes,
the histochemical effect of ischemia on the activity
of K-NPPase is minimal, being only slightly reduced in
the blood vessel walls. While the Na-K-ATPase activity
in capillaries and parenchyma exhibited a pattern of
recovery, the response of the K-NPPase was heterogeneous
during post-ischemia and the injury was characterized by
a central decrease and a peripheral increase of the
reaction product. These results indicate that the
data on the enzymatic activity of the total homogenate
obtained by biochemical methods could lead to a false
conclusion about the normalization of the enzyme activity
in all regions of the affected tissue. Therefore, the
combination of the biochemical and histochemical methods
would appear to be the more desirable approach to such
an investigation.

The decreased K-NPPase activity in the cortex was
observed in the neuropil of the lower cortical layer
where the initial cellular water uptake was visualized
by light microscopy as a vacuolization of the neuropil
(16). Consistent with our findings, a reduced activity
of ATPase and alkaline phosphatase has also been reported
but at all postembolic time intervals studied, whereas
the activity of -glutamyl transpeptidase appeared to
be increased at 5 and 24 hours of recirculation (29).
Our results are in agreement with the marked decrease
of nucleoside phosphatase in the capillary walls
observed 15 minutes after circulatory hypoxia (30).

Normal astroglia have relatively low K-NPPase
activity as shown by other investigators (11, 19, 35, 36).
Hamberger (10) found similar activities in isolated
neuronal and glial cell fractions from beef brain;
conversely Medzihradsky and coworkers (23) found substantially higher Na-K-ATPase in so-called purified rat
glial fractions. An important finding in this study
is the increased K-NPPase activity of astroglia (Figs.
12-14). The nature of the changes in K-NPPase activity
in the perivascular astroglia and in the neuropil
could reflect the spatial progression of the ischemic
injury. The marked astrocytic swelling, an important
pathological feature of incomplete ischemia (8),
speaks in favor of a functional change. The extensive
glial reaction in the post-ischemic and post-hypoxic
periods are characterized by the prominent increases
in the activity of oxido-reductive enzymes (27) and

ATPase activity in the capillary fraction and to a much
lesser extent in the parenchyma (Fig. 1). With the
onset of post-ischemia, the recovery of the vascular
Na-K-ATPase activity was much faster than that of the
parenchyma. Moreover during the post-ischemic period,
the parenchymal activity continued to fall but thereafter
slowly reached control values by 24 hours of recirculation.
Our results led to the conclusion that vascular Na-K-
ATPase is more sensitive to ischemia and the parenchymal
enzyme to post-ischemia.

Both the changes in the vascular and parenchymal
Na-K-ATPase activities coincided with the onset and
development of the ischemic and of post-ischemic brain
edema (26). We have recently postulated that Na-K-
ATPase, serotonin and cyclic AMP are principle factors
involved in the development and/or persistence of
ischemic brain edema (26). It has previously been
demonstrated that the release of serotonin and the
accumulation of cyclic AMP were increased during ischemia
(25), and with the onset of recirculation the changes
were even more pronounced (26). The time course of the
post-ischemic changes in serotonin and cyclic AMP coincided with the postischemic inhibition of parenchymal
Na-K-ATPase and with the increased water content in the
brain. The activity of Na-K-ATPase in a preparation of
synaptosomal membranes has also been shown to be inhibited by increasing concentrations of serotonin (18).
In addition, the post-ischemic normalization of serotonin and cyclic AMP levels preceded the restoration of
the parenchymal Na-K-ATPase activity and the decline
of water content in the brain (26). The alterations in
cyclic AMP and serotonin and their reported effects on
Na-K-ATPase may be an important factor in the development of edema, since it appears that water movements
between the extra- and intracellular compartments are
directly related to the ionic gradients of sodium and
potassium which are maintained by an energy-dependent
pump (15).

The speculation that Na-K-ATPase functions as a
factor in the transport of sodium and potassium has
been checked in a number of different experimental
models. The cold injury affects the sensitivity of the
Na-K-ATPase in the rabbit cerebral hemisphere to the
K/Na ratio in the medium (31). On the other hand, no
significant changes in ATPase activity were observed
in white matter of rats in which the edematous process
was induced by the administration of triethyltin (14).
This activity was, however, decreased in glial cells

nucleoside phosphatase (30). In addition, ischemia of
long duration revealed an increased thiamine pyrophos-
phatase activity in astroglia (28). One morphological
manifestation of astrocytes is the ability to change
their shape and contents in response to a varieyt of
stimuli (2). Edema represents one rather universal
phenomenon under which reactive enzymatic changes in
astrocytes occur (6). However, the factors which trigger
the enzymatic response of astrocytes are unknown (6).
Recently the hypothesis of an active control by astrocytes of extracellular potassium concentrations in
nervous tissue has been postulated. The activity of
Na-K-ATPase in many tissues was found to increase in
situations requiring a higher rate of sodium and potassium transport (12, 13, 20, 23). The inhibition of
Na-K-ATPase activity is followed·by the disturbances
in ionic equilibrium between the intracellular and
extracellular compartments; the loss of potassium from
the cell is followed by the entering of sodium and water
into the cell during ischemia and early post-ischemia (26).
Swelling of astrocytes is especially prominent in hypoxic
and ischemic conditions (16, 17). The post-ischemic
decline of the water content coincides with the increase
of K-NPPase activity within the astrocytes. It has been
postulated that the redistribution of potassium ions
could be regulated by the glial cells (34). However,
the significance of the increased K-NPPase activity
within the astrocytes to the pathomechanism of post-
ischemic osmoregulation awaits clarification.

ACKNOWLEDGEMENTS

The authors are thankful to Mrs. Milana Stanisic
for her skilled technical assistance. The work has
been supported by the research grant No. 214 from
AMNU of Serbia.

REFERENCES

1.   Borgers, M.   (1973): The cytochemical application
      of new potenti inhibitors of alkaline phosphatases.
      J. Histochem. Cytochem. 21:   812-824.

2.   Brightman, M.W., Anders, J.J., Schmechel, D., and
      Rosenstein, J.M.   (1978): The lability of the shape
      and content of glial cells.   In: Dinamic properties
      of glia cells.   Eds: E. Schoffeniels, G. Franck,
      L. Hertz, and D.B. Tower, Pergamon Press, Oxford,
      New York, pp. 21-44.

3. Djuricic, B.M., and Mrsulja, B.B.   (1977): Enzymic activity
    of the brain: microvessels vs. total forebrain
    homogenate. Brain Res. 138: 561-564.

4. Ernst, S.A.   (1972): Transport adenosine triphos-
    phatase cytochemistry: II. Cytochemical localization
    of ouabain-sensitive, potassium dependent phosphatase
    activity in the secretory epithelium of the avian salt
    gland. J. Histochem. Cytochem. 20: 23-38.

5. Ernst, S.A.   (1973): Cytochemical localization of
    phosphatase activity in rat kidney cortex. J.
    Cell Biol. 59: (2, Pt. 2) 93a (abstr.)

6. Friede, R.L.   (1965): Enzyme histochemistry of
    neuroglia. In: Biology of Neuroglia,Progress in
    Brain Res., Vol. 15, Eds. E.D.P. de Robertis and
    R. Carrea, Elsevier Publishing Company., Amsterdam,
    London, New York, pp. 35-47.

7. Fujimoto, T., Walker, J.T., Jr., Spatz, M., and
    Klatzo, I.   (1976): Pathophysiologic aspects of
    ischemic edema. In: Dynamics of Brain Edema,
    eds. H.M. Pappius and W. Feindel, Springer Verlag,
    New York, pp. 171-180.

8. Garcia, J.H., Lossinsky, A.S., Nishimoto, K., Klatzo,
    I., and Lightfoote, W. Jr.   (1978): Cerebral mic-
    rovasculature in ischemia. In: Pathology of
    Cerebrispinal Microcirculation Advance in Neurology,
    Vol. 20, Eds. J. Cervos-Navarro, E. Betz, G. Ebhardt,
    R. Ferszt and R. Wullenweber, Raven Press, New York,
    pp. 141-149.

9. Guth, L., and Albers, R.W.   (1974): Histochemical
    demonstration of $(Na^+ - K^+)$-activated adenosine
    triphosphatase. J. Histochem. Cytochem. 22: 320-
    326.

10. Hamberger, A., Blomstrand, C., and Lehninger, A.L.
    (1970): Comparative studies on mitochondria isolated
    from neuron-enriched and glia-enriched fractions of
    rabbit and beef brain. J. Cell Biol. 45: 221-229.

11. Hess, H.H., Embree, L.J., and Shein, H.M.   (1972):
    Enzyme control of Na- and K- active transport in
    normal and neoplastic rodent astroglia. Prog. Exp.
    Tumor Res. 17: 308-315.

12. Huttenlocher, P.R., and Rawson, M.D.   (1968):
     Neuronal activity and adenosine triphosphatase in
     immature cerebral cortex. Exp. Neurol. 22: 118-
     129.

13. Katz, A.I., and Epstein, F.H.   (1967): Role of
     sodium potassium-activated adenosine triphosphatase
     in absorption of sodium in kidney. J. Clin. Invest.
     46: 1999-2011.

14. Katzman, R., Alen, F., and Wilson, C.   (1963):
     Further observations on triethyltin edema.
     Archs Neurol. (Paris) 9: 178-187.

15. Katzman, R., and Pappius, H.M.   (1973): Brain
     electrolytes and fluid metabolism. Williams and
     Wilkins, Baltimore.

16. Klatzo, I.   (1975): Pathophysiologic aspects of
     cerebral ischemia. In: The Nervous System,
     edited by D.B. Tower, Vol. 1, Raven Press, New
     York, pp. 313-322.

17. Klatzo, I.   (1979): Cerebral edema and ischemia.
     In: Recent Advances in Neuropathology, Eds. W.T.
     Smith and J.B. Cavanagh, Vol. 1, Churchill Livin-
     stone, Edinburgh-London-New York, pp. 27-39.

18. Kometiani, P., Kometiani, Z., and Mikeladze, D.
     (1978): 3',5'-AMP-dependent protein kinase and
     membrane ATPase of the nerve cell. Progress in
     Neurobiology 11: 223-247.

19. Lewin, E., and Hess, H.H.   (1964): Intralaminar
     distribution of Na^+, K^+-ATPase in rat cortex. J.
     Neurochem. 11: 473-482.

20. Lewin, E., and McCrimmon, A.   (1967): ATPase
     activity of discharging cirtical lesions induced
     by freezing. Archs Neurol. 16: 321-325.

21. Lowry, O.H., and Passonneau, J.V.   (1972): A
     Flexible System of Enzymatic Analyses. Academic
     Press, New York.

22. Lowry, O.H., Rosenbrough, N.J., Farr, A.L., and
     Randall, J.R.   (1951): Protein measurements with
     the Folin phenol reagent. J. Biol. Chem. 193:
     265-275.

23. Medzihradsky, F., Sellinger, O.Z. Nandhasri, P.S.,
     and Santiago, J.C. (1972): ATPase activity in
     glial cells and in neuronal perikarya of rat
     cerebral cortex during early postnatal development.
     J. Neurochem. 19: 543-545.

24. Medzihradsky, F., Sellinger, O.Z., Nandhasri, P.S.,
     and Santiago, J.C. (1974): Adenosine triphosphatase
     activity in glial cells and in neuronal perikarya of
     edematous rat brain. 67: 133-139.

25. Mrsulja, B.B., Mrsulja, B.J., Cvejic, V., Djuricic,
     B.M., and Rogac, Lj. (1978): Alterations of
     putative neurotransmitters and enzymes during
     ischemia in gerbil cerebral cortex. J. Neural.
     Transm., Suppl. 14: 23-30.

26. Mrsulja, B.B., Djuricic, B.M., Cvejic, V., Mrsulja,
     B.J., Abe, T., Spatz, M. and Klatzo, I. (1979):
     Biochemistry of experimental ischemic brain edema.
     Proc. of the First Intern. Ernst Reuter Symposium,
     Brain Edema, Berlin (In press), Raven Press, New
     York.

27. Mrsulja, B.J., Spatz, M. and Klatzo, I. (1979):
     Cytochemistry of Hippocampus following cerebral
     ischemia. In: Pathophysiology of Cerebral Energy
     Metabolism, Eds. B.B. Mrsulja. Lj., M. Rakic, I.
     Klatzo and M. Spatz, Plenum Press, New York, London,
     pp. 73-90.

28. Mrsulja, B.J., Spatz, M., Walker, J.T., Jr., and
     Klatzo, I. (1979): Histochemical investigation
     of the Mongolian Gerbil's brain during unilateral
     ischemia. Acta Neuropathol. (Berl.) 46: 123-131.

29. Nishimoto, K., Wolman, M., Spatz, M. and Klatzo,
     I. (1978): Pathophysiologic correlations in the
     blood brain barrier damage due to air embolism.
     In: Pathology of Cerebral Microcirculation-Advances
     in Neurology, Vol. 20, Eds. J. Cervos-Navarro, E.
     Betz, G. Ebhardt, R. Ferszt, R. Wullenweber, Raven
     Press, New York, pp. 237-244.

30. Ostenda, M., Szumanska, G., and Gadamski, R. (1978):
     Specific hydrolases activity in blood vessels of
     rabbit brain after circulatory hypoxia. In: Patho-
     physiological, Biochemical and Morphological Aspects
     of Cerebral Ischemia and Arterial Hypertension. Eds.
     M.J. Mossakowski, I.B. Zelman, and H. Kroh, Warsaw,

pp. 60-66.

31. Rigoulet, M., Guerin, B., Cohadon, F., and
    Vandendreissche, M. (1979): Unilateral brain
    injury in the rabbit; reversible and irreversible
    damage of the membranal ATPases. J. Neurochem.
    32: 535-541.

32. Scrimgeour, K.G. (1977): Chemistry and control
    of enzyme reactions. Academic Press, London.

33. Skou, J.C. (1965): Enzymatic basis for active
    transport of Na^+ and K^+ across cell membrane.
    Physiol. Rev. 45: 596-617.

34. Somjen, G.G., Rosenthal, M., Cordingley, G.,
    Lamanna, J., and Lothman, E. (1976): Potassium,
    neuroglia, and oxidative metaboliam in central gray
    matter. Fed. Proc. 35: 1266-1271.

35. Stahl, W.L., and Broderson, S.H. (1976): Locali-
    zation of Na^+ K^+-ATPase in brain. Fed. Proc. 35:
    1260-1265.

36. Stahl, W.L., Spence, A.M., Coates, P.W., and
    Broderson, S.H. (1978): Studies on cellular
    localization on Na^+, K^+-ATPase activity in nervous
    tissue. In: Dynamic Properties of Glia Cells.
    Fds. E. Schoffeniels, G. Franck, L. Hertz, D.B.
    Tower, Pergamon Press, Oxford, New York, pp. 371-
    381.

THE ISCHEMIC AND POSTISCHEMIC EFFECT ON THE ACTIVITIES OF CEREBRAL MONOAMINE OXIDASE, CYTOCHROME OXIDASE AND ACETYLCHOLINESTERASE IN MONGOLIAN GERBILS

D. Mićić, K. Abe, W.D. Rausch, T. Abe
and M. Spatz

Laboratory of Neuropathology and
Neuroanatomical Sciences
National Institute of Neurological and
Communicative Disorders and Stroke
National Institutes of Health
Bethesda, Maryland 20205, U.S.A.

INTRODUCTION

Cerebral ischemia modifies many metabolites and enzymatic systems in the brain. However, little corre-lation was found between the altered levels of cerebral metabolites and the changes of some enzymatic activities such as cyclic nucleotide-related enzymes and ATPase (22). These investigations suggested that the enzyme activities in the cellular membranes especially the bound ones were more susceptible than the others to ischemia and/or re-circulation. A particular sensitivity of the membrane's properties to the ischemic injury was also implied by the observed synaptosomal decreased uptake and increased release of norepinephrine. Moreover, ischemia was found to depress greatly the activity of tyrosine hydroxylase and the content of biogenic neurotransmitters in the brain. Thus, the depletion of cerebral monoamines occur-ring in ischemia was accountable by the decreased syn-thesis and the augmented release of the amines (5, 15-20, 26, 28, 29).

The continuation of this work has been focused on the investigation of the MAO activity in ischemia, since MAO is a membranous enzyme involved in oxidative deamination and plays an important role in the metabolism of the monoamines. The simultaneously assayed mitochondrial MAO and cytochrome oxidase (CyO), as well as the synaptic acetylcholinesterase (AChE) levels from the same brain permitted the assessment of a possible specific vulnerability of the tested enzymes to the deprivation and reestablishment of cerebral blood supply. Regional cerebral ischemia induced by unilateral common carotid occlusion in Mongolian gerbils served as a model for the evaluation of these enzymatic activities since the same system was used for the neurotransmitters and other studies.

MATERIAL AND METHODS

Mature Mongolian gerbils (50-75 g) were anesthetized by intraperitoneal injection of Na pentobarbital (20 mg/kg body weight) prior to the induction of cerebral ischemia or to the sham operation. Unanesthetized animals served as additional controls. Each experimental group described below (symptom-positive only), as well as sham and controls consisted of 12-18 gerbils. Arterial blood samples were obtained from each representative group (4-5) for pH, PO_2 and PCO_2 analysis.

I. Gerbils subjected to ischemia only. The left common carotid artery was occluded with a Heifetz clip in the lower neck region for 1, 3 and 5 hour periods. The symptom-positive animals were separated from the symptom-negative gerbils according to the neurological signs described by Kahn (10). At the end of each experimental period the clipped animals were killed by decapitation.

II. Gerbils subjected to ischemia and various periods of recovery. The left common carotid artery was clipped the same way as in the first group but for 1 hour only. The symptom-positive gerbils were allowed to recover for various periods of time (Table 1 - 3 and Figs. 1 and 2). After decapitation the brains from each group of the animals were quickly removed and homogenized in 0.32m sucrose at pH 7.0. The mitochondrial and synaptosomal fractions were prepared according to the method of Whittaker and Barker (27).

Table 1. Blood gas analysis

	Ischemia				1 hour ischemia with recovery		
Duration in hours	pH	pCO$_2$(mm Hg) means $\pm$ S.E.	pO$_2$(mm Hg)	Duration in hours	pH	pCO$_2$(mm Hg) means $\pm$ S.E.	pO$_2$(mm Hg)
Control	7.3 $\pm$.02	26.1$\pm$2.4	98.1$\pm$4.3	72	7.38$\pm$.05	27.8$\pm$3.7	105.2$\pm$4.9
1	7.4 $\pm$.03	24.6$\pm$2.7	103.0$\pm$3.2	1	7.32$\pm$.05	20.6$\pm$1.6	97.3$\pm$3.4
3	7.37$\pm$.03	20. $\pm$3.3	104.8$\pm$3.5	5	7.35$\pm$.04	20.6$\pm$1.6	100.8$\pm$4.0
5	7.4 $\pm$.05	32.1$\pm$3.3	106.4$\pm$9.5	20	7.35$\pm$.02	20.6$\pm$.70	99.6$\pm$1.3

Table 2. Mitochondrial enzyme activities in continuous cerebral ischemia

Experimental Time	MAO				CyO			
	Ischemia		Sham		Ischemia		Sham	
	Left	Right	Left	Right	Left	Right	Left	Right
1 hour	20.26 ±.64	19.98 ±.69	19.53 ±1.09	19.79 ±.85	20.49 ±1.34	19.47 ±1.46	19.92 ±.60	20.08 ±.68
3 hours	16.39** ±.79	20.35 ±.77	19.83 ±.36	19.93 ±.41	17.97 ±1.40	18.20 ±1.22	18.51 ±.77	18.88 ±.66
5 hours	16.12** ±1.15	20.9 ±.55	21.1 ±.44	20.02 ±.18	16.02** ±.82	21.26 ±.55	21.54 ±.15	22.03 ±.36

Control values MAO 22.19+.69 (18) CyO 22.08+.75 (12) gerbils

Total MAO values are means ± S.E.M. of nmoles HOQ/mg protein/hour

CyO values are means ± S.E.M. of µg CyO/γ protein/hour } 6-9 animals

* p < .02 } above: significance between the affected and the unaffected side

** p < .001 } on the side: significance between the affected and sham operated cerebral hemisphere

Table 3. MAO activity in cerebral hemispheres of gerbils exposed to 1 hour ischemia and various periods of recovery

Recirculation Time	Ischemia		Sham	
	Left	Right	Left	Right
0	20.88+.64	19.98+.69 (9)	18.09+.84	19.79+1.15 (9)
1 hour	19.48+1.05	20.15+.51 (8)	17.71+.51	17.75+.64 (8)
3 hours	19.38+1.0	21.29+.98 (9)	20.74+.89	19.66+.67 (11)
5 hours	17.56+.49***	19.23+.66** (8)	21.90+.42	20.86+.48 (12)
10 hours	16.80+.96*** *	19.56+.72* (4)	21.41+.68	21.91+.63 (8)
20 hours	16.05+1.44** *	19.91+.78* (7)	21.75+.52	21.10+.60 (4)
72 hours	12.99+1.53*** ***	21.30+.72 (7)	20.13+.29	22.10+.60 (4)
1 week	13.23+.76*** ***	20.36+.82 (5)	21.83+.47	22.73+.85 (4)
4 weeks	20.96+.72	21.41+.67 (6)	—	—

Control values 22.19+.69 (18)

Mean + S.E.M. of 4 HOQ/mg protein/hr
Number of animals in parenthesis

* p < .05   above: significance between the affected and the unaffected side
** p < .01
*** p < .001   on the side: significance between the affected (or unaffected) and sham operated cerebral hemisphere

The total MAO activity was assayed by microfluorometric technique using spectrofluorometer which measures
the conversion of kynuramine to 4-hydroxyquinoline (12).
The activities of MAO type (A) and (B) were determined
by radioenzymatic technique of Tipton and Youdim using
serotonin and phenylethylamine as respective substrates
(24). The oxidation products were extracted into organic
solvent and counted in Beckman's LS-250 liquid scintillation counter.

The CyO activity was assayed by ultramicrophoto-
metric method of Hess and Pope based on the oxidation
of reduced cytochrome C (7).

AChE activity was measured by spectrophotometric
technique of Ellman et al. using acetylthiocholine as
substrate (4). The protein concentration was measured
by Lowry and biorad methods (1, 13).

RESULTS

Blood gas analysis. The level of pH, PO_2 and PCO_2
in the arterial blood of representative animals from
each experimental group was not significantly different
from the one obtained from the sham-operated and unanesthetized controls (Table 1).

I. Gerbils subjected to ischemia only

(1) Monoamine oxidase. The total MAO activity was
found to be decreased in the ischemic as compared to the
nonischemic and sham-operated hemisphere after 3 and 5
hours of continuous deprivation of blood supply. Prior
to this time, the activity of MAO in both groups was the
same but differed significantly from the unanesthetized
controls (Table 2). A similar ischemic effect was seen
on the activity of MAO (A) to the one observed on the
total MAO. However, the MAO (B) was somewhat differently
affected than either the total or the (A) form of MAO.
After an hour of occlusion, the bilateral depression
of the MAO (B) activity was greater than that of the
others. The bilateral levels returned to normal at 2
hours and dropped again at 3 hours of occlusion in the
ischemic hemisphere only (Table 2 and Fig. 1).

(2) Cytochrome oxidase. The effect of ischemia in
CyO activity was not seen until the third hour of arterial occlusion. During the entire period of observation
the level of CyO in the brain of sham-operated animals
was lower than the one observed in control unanesthetized
gerbils (Table 2).

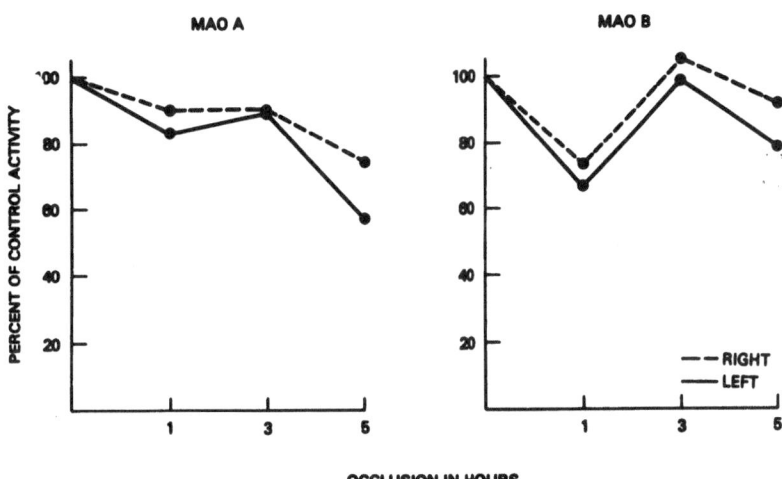

Fig. 1. The activity of MAO (A) and MAO (B) in unilateral
continuous cerebral ischemia expressed as percent
of controls [MAO (A) 174.8+6.6 nmoles/mg P/hr.
MAO (B) 18.3+.76 nmoles/mg P/hr. = 100 percent].
Each point represents the mean MAO values obtained from the animals. The levels of the
enzymes were significantly lower in both hemispheres at 1 hour and in the left hemisphere at
5 hours of ischemia (p < 0.01).

(3) Acetylcholinesterase. This enzyme activity
was not affected in the brain of gerbils subjected to
the unilateral deprivation of blood supply for 1-5 hours.
Ischemic values varied between 2.51+.06 and 2.57+.09 10^{-4}
moles substrate hydrol./min/mg P while sham and control
levels were found to be 2.54+.08 and 2.52+0.5 10^{-4} moles
hydrol./min/mg P, respectively.

II. Gerbils subjected to ischemia and various
periods of recovery

(1) Monoamine oxidase. In 1 hour of ischemia, the
level of either total or (A) or (B) form of MAO was not
significantly lower in the ischemic (ipsilateral) than

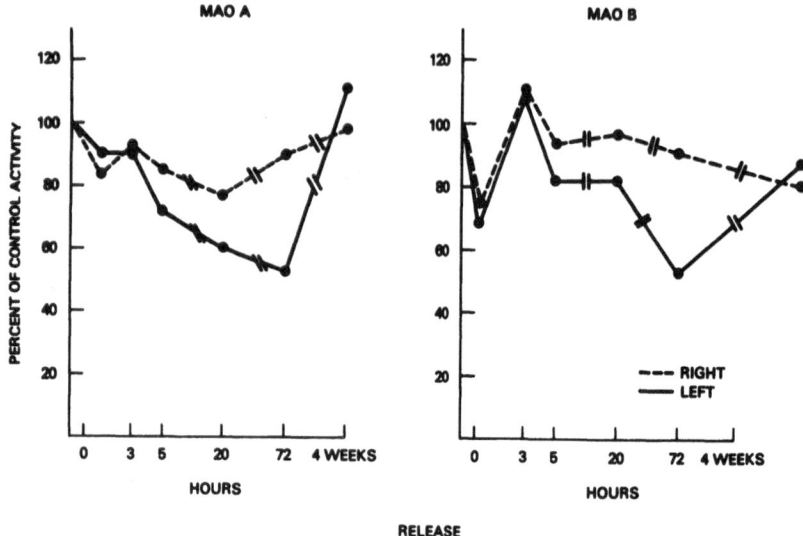

Fig. 2. The activity of MAO (A) and MAO (B) in various
periods of recovery after 1 hour of unilateral
cerebral ischemia. Each print represents the
mean MAO values of six animals. A significant
decrease in the activity of MAO (A) and (B) forms
occurred at 5 hours and thereafter the reduction
became more pronounced in the affected hemisphere
(left) (p < .01-0.001). The MAO (A) activity was
also significantly reduced in the hemisphere
contralateral to arterial occlusion after 5 and
20 hours of recirculation (p < .05 and p < .01,
respectively).

that in control (sham and unanesthetized gerbils) hemisphere until the clip was released for 5 hours. At this
time and at 20 hours of recovery, the activity of total
and (A) but not (B) form was also affected in the hemisphere contralateral to occlusion when compared to con-

trols. The greatest drop of the MAO level was found in 72 hours and 1 week after the reestablishment of circulation. One month later the activity of total (A) and (B) form returned either to normal or almost normal levels (Table 3 and Fig. 2).

(2) <u>Cytochrome oxidase</u>. The effect of the carotid artery occlusion of 1 hour's duration on the cerebral CyO was not different from the sham-operated animals until the cerebral blood supply was reestablished for several hours. However, the level of CyO activity in both groups was significantly lower than the one observed in the unanesthetized controls. The greatest reduction in the CyO activity was seen in the ischemic hemisphere 72 hours and 1 week after the arterial clip release. At no time was the CyO activity changed in the cerebral hemisphere contralateral to the occlusion (Table 4).

(3) <u>Acetylcholinesterase</u>. The activity of AChE was found to be significantly reduced in the ipsilateral as compared to the contralateral and sham-operated cerebral hemisphere at 72 hours after clip release. The concentration of AChE was $1.8 \pm .13(6)$ in the ischemic while the values of $2.55 \pm .13(6)$, $2.36 \pm .12(8)$ and $2.52 \pm .05(18)$ 10^{-4} moles/min/mg $\bar{P}$ were found in the opposite ($\bar{p} < .01$) and control hemispheres, respectively ($p < .001$).

DISCUSSION

The reduction in the activity of MAO [especially total and (A) form] prior to CyO and AChE observed in this study indicates a greater susceptibility of MAO than that of the other enzymes to cerebral ischemia. This concept is substantiated by the relative decreases in MAO and AChE activities reported in the bilateral cerebral ischemia at 5 and 10 minutes, respectively (15). Moreover, our results are in agreement with the earlier loss of MAO than that of CyO activity observed in the brain of rats recovering from an anoxic-ischemic insult (23). The different effect of ischemia on the MAO and CyO than that on the activity of AChE is not surprising since the activity of MAO and CyO but not that of AChE is oxygen-dependent. Furthermore, the diverse enzymatic reactivity to ischemic injury could be also based on the localization of these enzymes in the mitochondria (MAO and CyO) as opposed to the one in the synaptosomes (AChE) since mitochondria were described to be the first ones affected by cerebral ischemia (21). Obviously, this

Table 4. CyO activity in cerebral hemispheres of gerbils exposed to 1 hour ischemia and various periods of recovery

Recirculation Time	Ischemia		Sham	
	Left	Right	Left	Right
0	20.49+1.34	19.47+1.46 (9)	19.92+.60	20.08+.68 (9)
1	17.15+ .70	17.47+1.1 (6)	17.75+.77	18.26+.53 (6)
3	18.59+ .50 (4)	19.61+1.32 (9)	19.93+.42	19.96+.20 (4)
5	19.43+1.40	20.24+1.06 (8)	19.81+.43	20.56+.25 (8)
20	19.39+ .71*(4)	22.05+ .99 (7)	22.26+.81	22.52+.80 (6)
72	14.13+1.39***	21.56+ .44 (7)	22.20+.61	22.26+.32 (6)
1 week	13.07+ .06***	21.56+ .44 (7)	21.87+.74	22.57+.44 (6)
4 weeks	20.46+1.5	21.87+ .79 (6)	—	—

Mean + S.E.M. of μg CyO/γ protein/hour   Control values 22.08+.75 (12)
Number of animals in parenthesis

* p < .05   ⎫ above: significance between the affected and the unaffected side
*** p < .001 ⎬ on the side: significance between the affected and sham operated cerebral
             ⎭ hemisphere

possibility does not apply to the observation of greater MAO than CyO sensitivity to the ischemia unless these enzymes are differently arranged within the mitochondria. The different reactivity of total MAO and MAO (B) than that of MAO (A) to continuous ischemia may be based on their specific properties, irrespective of whether they represent two distinct proteins or variable sites in the same protein molecule (3).

Our results also indicate that the early stages of regional ischemia do not affect the activity of either MAO or CyO, since MAO [total and (B) form] and CyO drop was not seen prior to 3 and 5 hours of continuous arterial occlusion, respectively. Likewise, the observed reduction in MAO (irrespective of MAO form) and CyO levels seen in the brain of gerbils recovering from 1 hour of unilateral ischemia was not seen before 5 and 20 hours of recirculation, respectively. Similar observations were made in the studies of other cerebral enzymes and/or metabolites. Some of the postischemic changes were related to the "maturation phenomenon" (16). The discrepancies between the reported early morphological and late biochemical manifestation of ischemic changes in the mitochondria are most probably the results of the different technique used for these investigations (6). The advantage of the morphological examination is the detection of focal tissue changes which can be easily "masked" by the biochemical assays of the total mass; this is a well recognized entity.

These and previous studies have shown that the pentobarbital induced drop in the activity of CyO persists longer than that of MAO (14). Pentobarbital was also found to "protect" the brain against ischemia (8). It is therefore possible that the ischemic effect on the CyO could be either "delayed" and/or "spared" by the pentobarbital anesthesia. However, the behavior of these enzymes especially in the recovery period was unrelated to the pentobarbital activity, since the effect of pentobarbital was not seen in the sham-operated animals. Moreover, a similar time relationship in the appearance of the MAO and CyO was reported in the anoxic-ischemic brains of rats subjected to light anesthesia (23). The "delayed" reactivity of CyO as compared to the MAO could conceivably be related to their dissimilar localization. MAO was reported to be less active than CyO and succinic oxidase in most of the cellular areas (25). These are also the areas affected by ischemia (9) and therefore, the deficit in the activity of MAO might be manifested sooner than that of CyO. This could have been also the

reason for the transient drop of MAO but not of CyO in
the hemisphere contralateral to ischemia. The post-
ischemic manifestation of depressed MAO and CyO activity
in the presence of normal blood gases observed in these
animals could be due to an altered blood flow and/or to
an inability of the tissue to utilize the available
oxygen. It is of interest that the most critical period
for the function of MAO [total, (A) and (B)], CyO and
AChE appears to be 72 hours after clip release, since
all of them showed a marked decrease in their activity.
In other models of ischemia, the rate of protein synthesis was found to be 30-40% lower in postischemia than
in the controls (11). Whether the depression of the
observed enzymatic activities can be related to the rate
of protein synthesis at this particular time remains a
matter of conjecture.

Cerebral ischemia, in contrast to hypoxia, depletes
the biogenic amines' content, although both conditions
reduce the synthesis of the monoamines. This difference
was attributed to the increased release of the monoamines
in ischemia and the decreased degradation of the amine
in hypoxia (2, 19). The depression of the MAO activity
in the postischemic period observed in this study (especially 3-7 days after clip release) strongly suggests
that regional cerebral ischemia may also affect the degradation of the monoamines in the brain. The amines
have vasoactive properties and their liberation has been
implicated in the exacerbation of the brain injury (28).
On the other hand, the vasoactive function of these
amines may persist only as long as they are not inacti-
vated enzymatically. It is therefore conceivable that,
at a given time, the aggravated disease process attrib-
utable to the monoamines may be the result of both factors (the increased release and the depressed degradation).

Although the acetylcholinesterase was not affected
by the regional ischemia of 1-5 hours' duration, the
depression of this enzyme was seen in the late recovery
period after 1 hour of ischemia. Moreover, a reduction
of AChE was reported in bilateral ischemia of 5 and 10
minutes' duration (15). The drop in AChE activity could
probably influence the equilibrium of the acetylcholine
receptor complex and free the acetylcholine. Thus, the
additional change in the cholinergic metabolism in the
brain may also take part in modifying the recovery of
the animals from ischemia.

REFERENCES

1. Bradford, M. (1976): A rapid and sensitive method
    for the quantitation of microgram quantities of
    protein utilizing the principle of protein-dye
    binding. Anal. Biochem. 72: 248.

2. Davis, J.N. and Carlsson, A. (1973): The effect of
    hypoxia on monoamine synthesis, levels and metabo-
    lism in rat brain. J. Neurochem. 21: 783-790.

3. Diez, J.A. and Maderdrut, J.L. (1977): Development
    of multiple forms of mouse brain monoamine oxidase
    in vivo and in vitro. Brain Res. 128: 187-192.

4. Ellman, G.L., Courtney, K.D., Andres, V., Jr. and
    Featherstone, R.M. (1961): A new and rapid colori-
    metric determination of acetylcholinesterase activ-
    ity. Biochem. Pharmac. 7: 88-95.

5. Gaudet, R., Welch, K.M.A., Chabi E. and Wang, T.-P.
    (1978): Effect of transient ischemia on monoamine
    levels in the cerebral cortex of gerbils. J.
    Neurochem. 30: 751-757.

6. Ginsberg, M.D., Mela, L., Wrobel-Kuhl, K. and
    Reivich, M. (1977): Mitochondrial metabolism fol-
    lowing bilateral cerebral ischemia in the gerbil.
    Annals Neurol. 1: 519-527.

7. Hess, H.H. and Pope, A. (1953): Ultramicrospectro-
    photometric determination of cytochrome oxidase for
    quantitative histochemistry. J. Biol. Chem. 204:
    295-306.

8. Hoff, J.T., Smith, A.L., Hankinson, H.L. and Nielsen,
    S.L. (1975): Barbiturate protection from cerebral
    infarction in primates. Stroke 6: 28-33.

9. Ito, U., Spatz, M., Walker, J.T., Jr. and Klatzo, I.
    (1975): Experimental cerebral ischemia in Mongolian
    gerbils. I. Light microscopic observations. Acta
    Neuropathol. (Berl.) 32: 209-223.

10. Kahn, K. (1972): The natural course of experimental
    cerebral infarction in the gerbil. Neurology
    (Minneap.) 22: 510-515.

11. Kleihues, P. and Hossmann, K.-A. (1971): Protein
     synthesis in the cat brain after prolonged cerebral
     ischemia. Brain Res. 35: 409-418.

12. Krajl, M. (1965): A rapid microfluorimetric determi-
     nation of monoamine oxidase. Biochem. Pharmacol.
     14: 1684-1685.

13. Lowry, O.H., Rosebrough, N.J., Farr, A.L. and
     Randall, R.J. (1951): Protein measurement with the
     folin phenol reagent. J. Biol. Chem. 193: 265-275.

14. Micic, D., Klatzo, I. and Spatz, M. (1978): The
     effect of sodium pentobarbital on some mitochondrial
     enzymes. J. Neurochem. 30: 1627-1628.

15. Mrsulja, B.B., Mrsulja, B.J., Cvejić, V., Djuricić,
     B.M. and Rogac, Lj. (1978): Alterations of putative
     neurotransmitters and enzymes during ischemia in
     gerbil cerebral cortex. J. Neural Transm., Suppl.
     14: 23-30.

16. Mrsulja, B.B., Mrsulja, B.J., Spatz, M., Ito, U.,
     Walker, J.T., Jr. and Klatzo, I. (1976): Experi-
     mental cerebral ischemia in Mongolian gerbils. IV.
     Behaviour of biogenic amines. Acta Neuropathol.
     (Berl.) 36: 1-8.

17. Mrsulja, B.B., Mrsulja, B.J., Spatz, M. and Klatzo,
     I. (1975): Action of cerebral ischemia on decreased
     levels of 3-methoxy-4-hydroxyphenylethylglycol
     sulfate, homovanillic acid and 5-hydroxy-indoleacetic
     acid produced by pargyline. Brain Res. 98: 388-393.

18. Mrsulja, B.B., Mrsulja, B.J., Spatz, M. and Klatzo,
     I. (1976): Brain serotonin after experimental vascu-
     lar occlusion. Neurology 26: 785-787.

19. Mrsulja, B.B., Mrsulja, B.J., Spatz, M. and Klatzo,
     I. (1976): Catecholamines in brain ischemia - effects
     of α-methyl-p-tyrosine and pargyline. Brain Res.
     104: 373-378.

20. Mrsulja, B.B., Mrsulja, B.J., Spatz, M. and Klatzo,
     I. (1976): Monoamines in cerebral ischemia in rela-
     tion to brain edema. In: Dynamics of Brain Edema,
     H.M. Pappius and W. Feindel (eds.), pp.187-192,
     Springer Verlag, Berlin-Heidelberg-New York.

21. Rodriguez de Lores Arnaiz, G. and De Robertis, E.D.P.
     (1962): Cholinergic non-cholinergic nerve endings
     in the rat brain - II subcellular localization of
     monoamine oxidase and succinate dehydrogenase.   J.
     Neurochem. 9: 503-508.

22. Schwartz, J.P., Mrsulja, B.B., Mrsulja, B.J.,
     Passonneau, J.V. and Klatzo, I. (1976): Alterations
     of cyclic nucleotide-related enzymes and of ATPase
     during unilateral ischemia and recirculation in
     gerbil cerebral cortex. J. Neurochem. 27: 101-107.

23. Spector, R.G. (1963): Cerebral succinic dehydro-
     genase, cytochrome oxidase and mono-amine oxidase
     activity in experimental anoxic-ischaemic brain
     damage. Br. J. Exp. Pathol. 44: 251-254.

24. Tipton, K.F. and Youdim, M.B.H. (1976): Assay of
     monoamine oxidase. In: Monoamine Oxidase and its
     Inhibition, Ciba Foundation Symposium 39 (new
     series), Elsevier, Amsterdam.

25. Weiner, N. (1960): The distribution of monoamine
     oxidase and succinic oxidase in brain. J. Neurochem.
     6: 79-86.

26. Welch, K.M.A., Chabi, E., Buckingham, J., Bergin, B.,
     Achar, V.S. and Meyer, J.S. (1977): Catecholamine
     and 5-hydroxytryptamine levels in ischemic brain.
     Influence of p-chlorophenylalanine.   Stroke 8:
     341-346.

27. Whittaker, V.P. and Baker, I.A. (1972): The sub-
     cellular fractionation of brain tissue with special
     reference to the preparation of synaptosomes and
     their component organelles. In: Methods in Neuro-
     chemistry, Vol. II, pp. 2-52, F. Ranier (ed.),
     Dekker, Inc., New York.

28. Wurtman, R.J. and Zervas, N.T. (1974): Monoamine
     neurotransmitters and the pathophysiology of stroke
     and central nervous system trauma. J. Neurosurg.
     40: 34-36.

29. Zervas, N.T., Hori, H., Negora, M., Wurtman, R.J.,
     Larin, F. and Lavyne, M.H. (1974): Reduction in
     brain dopamine following experimental cerebral
     ischemia. Nature 247: 283-284.

OSCILLATORY PATTERN OF CATECHOLAMINE METABOLISM

FOLLOWING TRANSIENT CEREBRAL ISCHEMIA IN GERBILS

Vesna Cvejić, B. M. Djuričić and
B. B. Mršulja

Laboratory for Neurochemistry
Institute of Biochemistry
Faculty of Medicine
Belgrade, Yugoslavia

Cerebral ischemia in Monogolian gerbil (Meriones unguiculatus) results in an acute "energy crisis" followed by the marked changes in monoamines; norepinephrine is the first to be depleted in cerebral ischemia (6). In spite of a lot of data about the fate of monoamines in ischemia, only a few studies have evaluated the metabolism of amines in reflow (1, 7). Moreover, evidence is lacking about the regional changes of catecholamines and related degradative enzymes during long-term reflow after the transient ischemic attack. The data presented here reveal the alterations of the catecholamine and their degradative enzymes during long-term reflow; a pattern of oscillatory behavior of both amines and enzymes was evident in postischemia. Thus, the disorder of catecholamine metabolism in the ischemic brain and during the postischemic period perhaps contributes to the neurological dysfunction which follows an ischemic insult.

EXPERIMENTAL

Mongolian gerbils were used in the study. Both common carotid arteries in the neck region were occluded with Heifetz aneurysm clips for 15 min, and the gerbils were sacrificed 5, 30 and 60 min, and 2, 4, 6 and 12 hours, and 1, 3 and 7 days after the clips were removed. A complete group of sham-operated animals were run in order

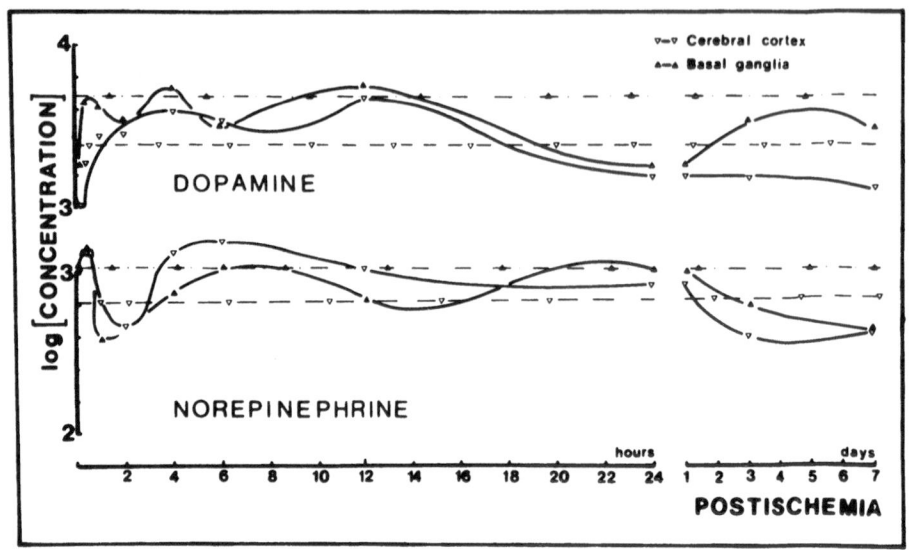

Fig. 1.   Oscillatory pattern of catecholamines in post-
           ischemia. Control levels were for dopamine
           2420 + 214 (n=7) (cerebral cortex) and 4938 +
           259 (n̄=6) (basal ganglia) and for norepinephrine
           667 + 62 (n=9) (cerebral cortex) and 1075 +
           127 (n=9) (basal ganglia) ng per g tissue (wet
           weight).

to avoid diurnal variations in catecholamine concentrations. Norepinephrine and dopamine, as well as monoamine oxidase (kynuramine as substrate) and catechol-O-
methyltransferase, were estimated by established fluori-
metric procedures (8). The results were compared with
the values of sham-operated animals for each experimental
period.

RESULTS

Norepinephrine and dopamine (Fig. 1).

    During 15 min of ischemia, the norepinephrine (NE)
content was reduced to 78% and 50% in the cerebral cortex
and the basal ganglia, respectively. With the onset of
reflow, an increase of NE levels was found. Thereafter
NE levels decreased further and were not recovered in
basal ganglia during the 7 days of reflow. In cerebral
cortex, an additional rebound was obtained at the 4th

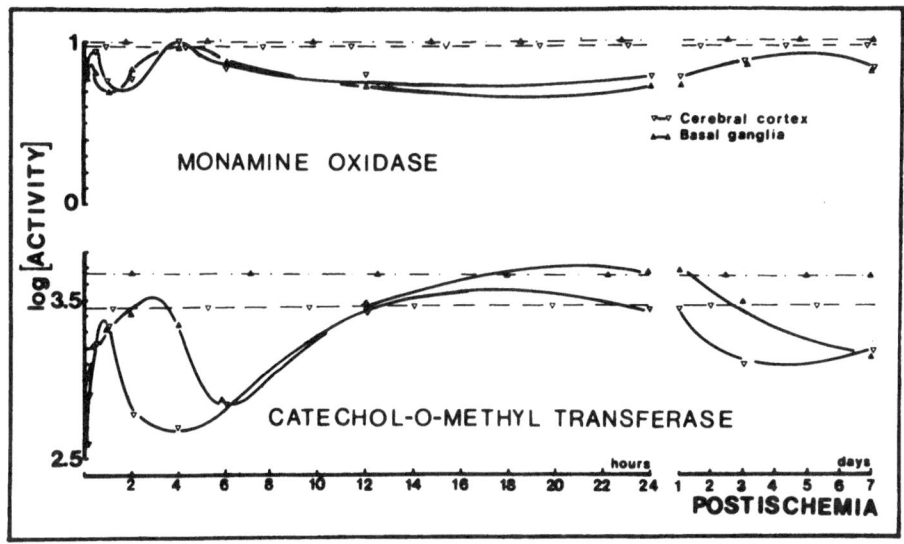

Fig. 2.   Oscillatory pattern of monoamine oxidase and
           catechol-0-methyl transferase in postischemia.
           Control activities were for monoamine oxidase
           9.5 + 0.8 (n=6) (cerebral cortex) and 10.2 +
           0.6 (n=6) (basal ganglia) nMoles per mg protein
           per hour (37°C); for COMT control activities
           were 2800 + 200 (n=5) (cerebral cortex) and
           4500 + 300 (n=5) (basal ganglia) pMoles per mg
           protein per hour (37°C).

and the 6th hour of postischemia, but thereafter the
level of NE decreased to about 50% of normal values.

    Dopamine (DA) was unchanged in the cerebral cortex
but decreased in basal ganglia during ischemia. With
the onset of reflow, DA content decreased in both the
cerebral cortex and the basal ganglia. The rebound of
DA was seen in cerebral cortex during the reflow period
(1-12 hours), but not in basal ganglia. However, an
oscillatory behavior in DA concentration was evident
during the postischemia, but the levels of this monoamine in the basal ganglia were lower than in the sham-
operated animals.

Monoamine oxidase and catechol-0-methyl transferase (Fig. 2).

The activity of monoamine oxidase (MAO) was reduced by 27% and 46% in the cerebral cortex and the basal ganglia, respectively. During 7 days of reflow, MAO activity in basal ganglia never recovered to the control values; however, oscillations in the enzyme activity were observed. In cerebral cortex an increase over the sham-operated values was found at the 4th hour of post-ischemia.

The activity of catechol-0-methyl transferase (COMT) was reduced to 62% of normal values (100%) in the basal ganglia but not in the cortex. In both brain structures, an oscillatory behavior of COMT activity was observed during the 7 days of reflow.

DISCUSSION

For a long time, the period following transient cerebral ischemia has been biochemically considered to be a reversal of the events occurring during ischemia. These conclusions are based on the finding that the metabolic-related energy-production metabolites are rapidly restored after the reestablishment of cir-culation (2, 5). Recently it was demonstrated that brain catecholamine levels, unchanged during ischemia, undergo significant alterations during reperfusion, con-sistent with the "maturation phenomenon" (3). Since neurological deficit may persist despite improved cerebral energy metabolism when reflow is restored after brief periods of ischemia, the deficit in neurological function may in part be due to the altered monoamine metabolism. The present data clearly show that alteration in monoamine metabolism persists for a long period following ischemia, both in cerebral cortex and basal ganglia.

It is evident that metabolism of the catecholamines is more affected in the basal ganglia than in the cerebral cortex during the postischemic period; the levels of DA and NE in basal ganglia were found to be reduced during the entire 7 days of reflow. Both MAO and COMT activities are depressed in the same postischemic period.

In the present study, a pattern of an "oscillatory behavior" was exhibited in the levels of catecholamine and the activity of MAO and COMT during the postischemic period (Figs. 1 and 2). The "up-and-down" changes ap-peared regardless of whether the amine levels or enzyme activity are enhanced or reduced; the oscillatory changes did not exhibit any similarities to the diurnal variations in catecholamine content. This pattern is more frequent and higher in amplitude in earlier periods of postischemia. These patterns coincide with the periods of hyperactivity and hypoactivity in postischemia (4).

CONCLUSION

Dysfunction of catecholamine metabolism, unlike that for the energy metabolism, persists during cerebral recirculation after 15 minutes of ischemia. Thus, the neurological deficit in postischemia is partially attributed to the altered function of neurotransmitters. Moreover, these findings suggest that the full recovery of the animals subjected to 15 min ischemia remains questionable.

ACKNOWLEDGMENT

This study was supported by the grant from the Union of Sciences of Republic Serbia (No. 40404-14).

REFERENCES

1. Gaudet, R., Welch, K.M.A., Chabi, E. and Wand, T.-P. (1978): Effect of transient ischemia on monoamine levels in the cerebral cortex of gerbils. J. Neurochem. 30: 751-757.

2. Kobayashi, M., Lust, W.D. and Passonneau, J.V. (1977): Concentrations of energy metabolites and cyclic nucleotides during and after bilateral ischemia in the gerbil cortex. J. Neurochem. 29: 53-59.

3. Mršulja, B.B. (1979): Some new aspects in the pathochemistry of the postischemic period. In: Pathophysiology of Cerebral Energy Metabolism. B.B. Mrsulja, Lj.M. Rakić, I. Klatzo and M. Spatz (eds.), pp. 47-59, Plenum Press, New York.

4. Mršulja, B.B., Lust, W.D., Mršulja, B.J. and Passoneau, J.V. (1977): Effect of repeated cerebral ischemia on metabolites and metabolic rate in gerbil cortex. Brain Res. 119: 480-486.

5. Mršulja, B.B., Lust, W.D., Mršulja, B.J., Passonneau, J.V. and Klatzo, I. (1976): Postischemic changes in certain metabolites following prolonged ischemia in gerbil cerebral cortex. J. Neurochem. 26: 1099-1103.

6. Mršulja, B.B., Mršulja, B.J., Cvejić, V., Djuričić, B.M. and Rogač, Lj. (1978): Alterations of putative neurotransmitters during ischemia in gerbil cerebral cortex. J. Neural. Transm., Suppl. 14: 23-30.

7. Mršulja, B.B., Mršulja, B.J., Spatz, M., Ito, U., Walker, T.J. and Klatzo, I. (1976): Experimental cerebral ischemia in Mongolian gerbils. IV. Behavior of biogenic amines. Acta Neuropath. (Berl.) 36: 1-8.

8. Nagatsu, T. (1973): Biochemistry of Catecholamines - The Biochemical Method. University Park Press, Baltimore-London-Tokyo.

INFLUENCE OF BARBITURATES, HYPOTHERMIA AND HEMODILUTION

ON POST-ISCHEMIC METABOLISM AND FUNCTIONAL RECOVERY

FOLLOWING CEREBRO-CIRCULATORY ARREST IN CATS

W. van den Kerckhoff, Y. Matsuoka, W. Paschen
and K.-A. Hossmann

Max-Planck-Institut für Hirnforschung
Forschungsstelle für Hirnkreislaufforschung
Köln (Merheim), West Germany

Increasing evidence provided by several laboratories over the past years has drawn attention to the fact
that recovery of the brain following a period of cerebro-
circulatory arrest depends not only on the duration of
ischemia but equally on post-ischemic events, above all
the quality of post-ischemic recirculation (1, 15, 26).
Two phases are particularly critical: the period immediately after ischemia, when blood flow is restored, and
a later period when spontaneous electro-cortical activity
begins to recover.

Disturbances of the restoration of flow immediately
after ischemia have been referred to as the "no-reflow
phenomenon" (1) and have been related to increased viscosity of stagnant blood (8), low perfusion pressure (3),
increased vascular resistance (5), and increased intracranial pressure (33). Under experimental conditions,
the no-reflow phenomenon can be fully prevented by re-
circulating the brain with high blood pressure (post-
ischemic hypertensive flush), combined with mild hemodilution and osmotherapy in order to decrease blood
viscosity and reduce critical brain swelling, respectively (15). With this approach major recirculation
defects can be avoided after complete ischemia up to
one hour, and instead reactive hyperemia ensues.

The second phase of delayed circulatory impairment which has been termed "post-ischemic hypoperfusion (15, 33), is more difficult to control. It appears after the termination of post-ischemic reactive hyperemia, and it apparently is the consequence of a disturbance of flow regulation. This disturbance is characterized by the complete suppression of CO_2 reactivity in the presence of autoregulation, and it results in an uncoupling of blood flow and metabolism (15, 33). During the recovery period, oxygen demands of the tissue are increased (post-ischemic hypermetabolism) (16, 19), but since this is not paralleled by an appropriate increase in blood flow, the available blood oxygen is not sufficient to cover the oxygen demands of the tissue, and relative hypoxia with stimulation of anaerobic glucose metabolism occurs (13). It has been suggested that this state of relative hypoxia might be responsible for the delayed functional disturbances which frequently occur after a period of beginning recovery and which may result in secondary irreversible brain damage (13).

From a therapeutical standpoint the misrelationship between oxygen requirements and oxygen supply to the brain could be ameliorated by either increasing blood flow or reducing the metabolic activity of the brain. Attempts to increase blood flow by systemic application of vaso-active substances have been unsuccessful because the cerebral vasculature seems to be less sensitive to such agents than the extracerebral vessels. Systemic blood pressure, therefore, tends to decrease before a substantial change in cerebro-vascular resistance occurs, resulting in further deterioration of cerebral blood flow (15, 36). Blockage of post-ischemic hypermetabolism with adrenergic blocking agents proved also to be unsuccessful (25). This was surprising because earlier investigations suggested that post-ischemic hypermetabolism was related to an increased concentration of cyclic AMP which in turn was supposed to be the consequence of a catecholamine-induced activation of adenyl cyclase (24).

In the present investigation, two other approaches were tested. In one series of experiments blood viscosity was lowered by hemodilution in an attempt to improve cerebral blood flow by improving the rheological properties of the blood. In another series an attempt was made to block post-ischemic metabolism with barbiturates or hypothermia. Since in earlier investigations highly controversial results had been obtained using this approach (13, 22, 31), we hoped to obtain new information by applying to this problem a broad methodological

battery consisting of physiological, biochemical and
electrophysiological techniques. Finally, a combination
of hemodilution, barbiturate loading and hypothermia was
tested in order to investigate whether a potentiation of
the positive effects of each of these approaches could be
obtained.

MATERIAL AND METHODS

The experiments were carried out in adult cats.
The animals were anesthetized with halothane/nitrous
oxide (halothane 0.2 %, N_2O 70 %, rest oxygen), immobil-
ized with gallamine triethiodide (Flaxedil[R]) and mechanically ventilated using an animal respirator. Arterial
PCO_2 was adjusted to 28-32 mm Hg, arterial oxygen content
was maintained above 100 mm Hg, and the negative base
excess was kept close to 5 meq/l. Ventilation was continuously monitored using a carbon dioxide analyser. In
the normothermic groups (see below) body temperature was
kept constant at 37° C, using a temperature controller
connected to a heating pad.

Catheters were placed into the following vessels:
a) into the innominate artery via the right brachial
artery. This catheter was used for bolus injection of
133-xenon during blood flow measurements.
b) into the sagittal sinus. The tip of the catheter was
directed towards the torcular; blood flow was continuously
withdrawn from this catheter using a roller pump and returned into a femoral vein. The shunt was used for
measuring 133-xenon radioactivity during blood flow measurements (see below) and for withdrawal of cerebral
venous blood samples.
c) into both femoral arteries for continuous measurement
of blood pressure, for sampling of arterial blood and for
bleeding the animal during induced hypotension (see below).
d) into both femoral veins for returning the blood withdrawn from the sagittal sinus, and for infusion of drugs.

Cerebral ischemia of 15 min duration was produced
by inflating a pneumatic cuff which was placed around the
animal's neck. The pressure in the cuff was 2000 mm Hg.
During inflation systolic blood pressure was lowered to
below 80 mm Hg in order to prevent a collateral blood
supply of the brain via the ascending spinal arteries.
Completeness of ischemia was controlled by injecting a
bolus of 133-xenon immediately before cuff inflation,
and recording the radioactivity of the skull with an
external scintillation detector. Ischemia was considered

to be incomplete when the decrease of radioactivity
during the period of ischemia was more than 2 %. At the
end of the ischemic period, the blood pressure was
abruptly raised to more than 150 mm Hg by infusion of
norfenefrin (Novadral[R]), the cuff was deflated, and blood
gases were rapidly re-adjusted to the normal by increasing
the speed of the ventilation pump and equilibrating post-
ischemic lactacidosis by titrated injection of sodium
bicarbonate.

Cerebral blood flow was measured by a modification
of the intra-arterial 133-xenon injection technique, the
xenon clearance being recorded in the cerebral venous
blood (16). This was achieved by passing the shunt which
connected the sagittal sinus with the femoral vein through
a glass coil brought into contact with the scintillation
detector. The clearance curves were evaluated by standard
bicompartmental analysis.

Cerebral metabolic rate of oxygen and glucose was
assessed during each flow measurement. Blood samples
were withdrawn from the femoral artery and the sagittal
sinus for determination of arterio-venous differences
of oxygen and glucose. Oxygen content was measured using
an oxygen analyser (Lex-O_2-Con, Lexington) and glucose
was determined using a glucose analyser (Glucotest, Beckman). An oxygen uptake quotient was calculated by expres-
sing oxygen uptake in percent of oxygen availability.

The electroencephalogram was recorded from both
hemispheres with bipolar silver ball electrodes brought
into contact with the exposed calvarium. EEG was recorded
together with the physiological parameters on a polygraph,
and stored on magnetic tape for subsequent evaluation.

At the end of the experiment the brain was frozen in
situ with liquid nitrogen, sawed into slices, and processed for various substrates of the energy-producing
metabolism, using standard enzymatic techniques.

Six groups of animals were compared:

1. Without thereapy (6 animals): only routine
stabilization of blood pressure, blood gases and of the
acid base balance of the blood were performed after
ischemia.

2. Post-ischemic hypothermia (30°C, 6 animals):
a drain was inserted intraperitoneally, and about 200 ml
of a cold dialysis solution were instilled into the

abdomen 5 min after the onset of post-ischemic recirculation. As soon as body temperature decreased to 30° C,
the solution was withdrawn and body temperature adjusted
at this level, using a heating bulb connected to a temperature controller.

     3. Post-ischemic hypothermia (26° C; 5 animals):
the same procedure was used as in group 2, but temperature was lowered to 26° C.

     4. Post-ischemic barbiturate loading (5 animals):
70 mg/kg thiopental were injected intravenously 5 minutes after the beginning of post-ischemic recirculation.
During barbiturate loading, the blood pressure was kept
as close as possible to normotensive levels by the additional infusion of norfenefrin (Novadral[R]).

     5. Post-ischemic hemodilution (6 animals): the
hematocrit of the blood was lowered to 20 vol. % by an
exchange infusion of about 40 ml 10 % dextran. During
dextran infusion, blood was withdrawn at the same rate
in order to avoid changes in blood volume.

     6. Post-ischemic combination therapy (6 animals):
hemodilution (lowering of hematocrit to about 20 vol. %),
mild barbiturate anesthesia (35 mg/kg) and mild hypothermia (30° C) were combined, using the above described
procedures.

     All animals were kept alive for 2-3 hours after
beginning of recirculation, and then processed as
described above.

RESULTS

     In all animals of the present series, ischemia was
performed under identical experimental conditions, in
particular under the same anaesthesia and at normothermic
body temperature. Within 15 sec after interruption of
cerebral blood flow the EEG flattened and remained iso-
electric throughout the ischemic period. Restoration of
blood flow was also performed under identical conditions.
Blood pressure was raised to at least 150 mm Hg immediately before deflating the pneumatic cuff, and deviations
in the acid-base balance of the blood and blood gases
were corrected as soon as possible.

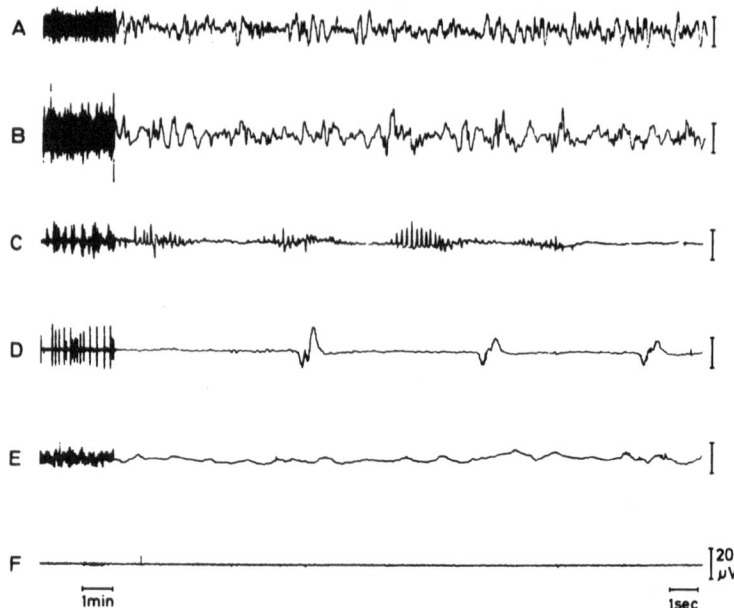

Fig. 1.   Recording of the EEG before (A) and after
           2 hours' recirculation following 15 min
           total ischemia of the cat brain (B-F).
           Animals with different degrees of recovery
           are compared. B: major continuous EEG
           activity, C: spindle wave activity,
           D: burst-suppression activity, E: minor
           low voltage activity, F: isoelectric EEG.

        Starting 5 min after the beginning of recirculation,
the animals were treated in different ways. In the hemodilution group, blood viscosity was lowered by decreasing
hematocrit to about 20 vol. %, in the hypothermia groups
body temperature was lowered to 30 and 26°C, respectively,
in the barbiturate group 70 mg/kg thiopental was given
intravenously, and in the combination therapy group hemodilution was combined with 30°C hypothermia and 35 mg/kg
thiopental. The results were compared with an untreated
group, in which the animals did not receive any special
therapy after ischemia except blood pressure stabilization and adjustment of respiration and arterial acid base
status. The physiological data recorded before and after
2 hours of recirculation are summarized in Table 1.

Table 1. Physiological variables before and 2 hours after ischemia

	Temp. °C	pH$_a$	P$_a$CO$_2$ mm Hg	P$_a$O$_2$ mm Hg	Hct Vol. %	mBP mm Hg	Gluc.a μmol/ml	O$_{2a}$ μmol/ml
Control before ischemia	36.7 ± 0.1	7.39 ± 0.01	29.8 ± 0.8	168 ± 9	34 ± 1	127 ± 3	13.4 ± 0.9	6.3 ± 0.2
Without therapy	36.8 ± 0.1	7.41 ± 0.03	32.0 ± 1.3	122 ± 21	32 ± 4	119 ± 10	14.8 ± 1.5	5.2 ± 0.4
Thiopental 70 mg/kg	36.7 ± 0.1	7.42 ± 0.04	32.7 ± 3.9	104 ± 18	32 ± 2	69 ± 12	20.0 ± 3.4	6.4 ± 0.6
Hypothermia 26°C	26.7 ± 0.7	7.39 ± 0.04	27.6 ± 1.6	123 ± 23	35 ± 4	70 ± 10	22.3 ± 3.4	5.1 ± 0.7
Hypothermia 30°C	29.8 ± 0.4	7.36 ± 0.03	29.1 ± 1.3	225 ± 35	39 ± 5	90 ± 23	18.9 ± 7.1	6.4 ± 1.2
Hemodilution	36.1 ± 0.4	7.45 ± 0.03	31.1 ± 1.9	182 ± 26	23 ± 2	119 ± 10	16.0 ± 2.2	2.95 ± 0.2
Combination therapy	30.9 ± 0.1	7.42 ± 0.03	28.8 ± 1.4	130 ± 24	20 ± 4	89 ± 9	16.1 ± 3.0	3.5 ± 0.5

The quality of resuscitation was estimated by determining the concentration of various substrates of
energy-producing metabolism in the brain tissue and by
recording the electroencephalogram (Table 2). A classification of the electrophysiological recovery was carried
out by scoring the EEG in the following way (Fig. 1):
normal EEG (5 points), major continuous EEG activity
(4 points), spindle-wave-activity (3 points), minor low
voltage activity (2 points), burst suppression activity
(1 point) and iso-electric EEG (0 points).

The results obtained in the various groups are summarized in Table 2. The lowest EEG score and the lowest
adenylate-energy charge were observed in the thiopental
group. This group had also the lowest concentration in
energy-rich phosphates, indicating that the energy-
producing metabolism was severely compromised. The
highest EEG score, the highest energy charge and the
highest concentration of energy-rich phosphates were seen
in the group with hemodilution. The other groups, including the animals without special therapy, were between
these two extremes, mild hypothermia and combination
therapy being slightly more efficient than no therapy at
all.

Recovery correlated with the degree of post-ischemic
hyperemia (Fig. 2). Thirty minutes after ischemia, the
hemodilution group exhibited an average flow rate of
about 180 % of control, whereas in the barbiturate group
flow was less than 40 %. At longer survival times the
relationship between blood flow and recovery was less
evident, with the exception of the thiopental and deep
hypothermia group which had distinctly lower flow values
than all the other animals.

The changes in the metabolic rate of oxygen ($CMRO_2$)
are summarized in Fig. 3. In the untreated animals,
hypermetabolism was present after 2 hours. The oxygen
consumption in these animals rose to more than 150 % of
the control value. In contrast, under barbiturate and
26° C hypothermia $CMRO_2$ decreased to 44 and 70 %, respectively. Post-ischemic hypermetabolism was also
absent during hemodilution and combination therapy, but
it was not influenced by 30° C hypothermia.

In the animals treated by 26° C hypothermia and
barbiturates, the low oxygen consumption apparently was
due to a lowering of the metabolic demands of the tissue.
Although blood flow in these animals had considerably
decreased, the oxygen uptake quotient was relatively

Table 2. Functional and biochemical recovery 2 hours after ischemia

	Functional EEG score	Creatine phosphate μmol/g	Adenosine triphosphate μmol/g	adenine nucleotides μmol/g	Energy charge	Lactate μmol/g
Control before ischemia	5.0 ± 0.0	4.3 ± 0.2	3.4 ± 0.4	4.2 ± 0.4	0.93 ± 0.01	1.3 ± 0.0
Without therapy	2.3 ± 0.6	3.3 ± 0.4	2.4 ± 0.6	2.8 ± 0.6	0.83 ± 0.06	10.8 ± 3.9
Thiopental 70 mg/kg	0.8 ± 0.6	2.4 ± 0.4	1.7 ± 0.3	2.5 ± 0.3	0.75 ± 0.10	5.3 ± 1.7
Hypothermia 26°C	2.4 ± 0.4	3.1 ± 0.6	2.2 ± 0.4	2.8 ± 0.3	0.83 ± 0.03	12.5 ± 2.5
Hypothermia 30°C	2.5 ± 0.3	3.7 ± 0.7	2.7 ± 0.2	3.3 ± 0.2	0.86 ± 0.05	8.5 ± 2.6
Hemodilution	3.0 ± 0.3	3.7 ± 0.2	3.0 ± 0.2	3.5 ± 0.1	0.90 ± 0.02	7.2 ± 0.8
Combination therapy	1.7 ± 0.5	3.5 ± 0.2	2.9 ± 0.2	3.5 ± 0.2	0.88 ± 0.03	10.5 ± 3.7

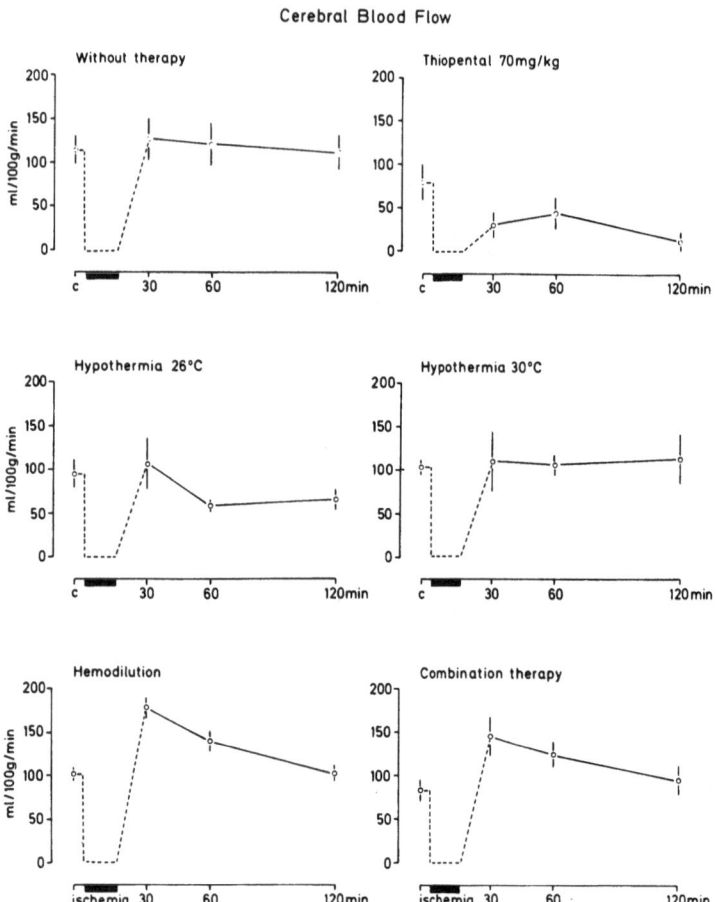

Cerebral Blood Flow

Fig. 2. Changes in blood flow during recirculation
         after 15 min complete ischemia. Comparison
         of different modes of treatment.

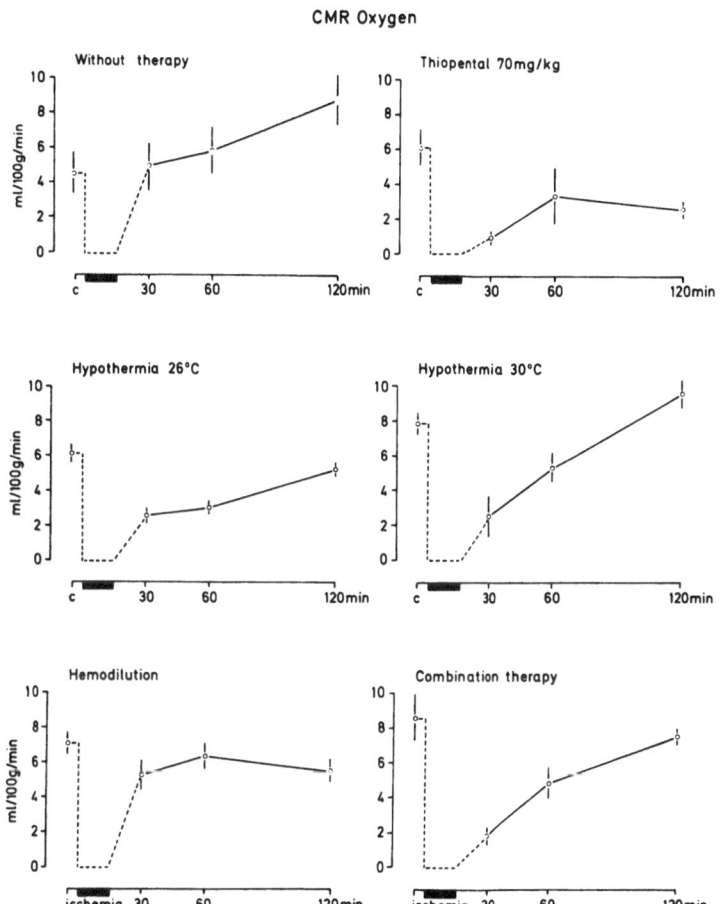

Fig. 3. Changes in metabolic rate of oxygen during recirculation after 15 min complete ischemia. Same treatment as in Fig. 2.

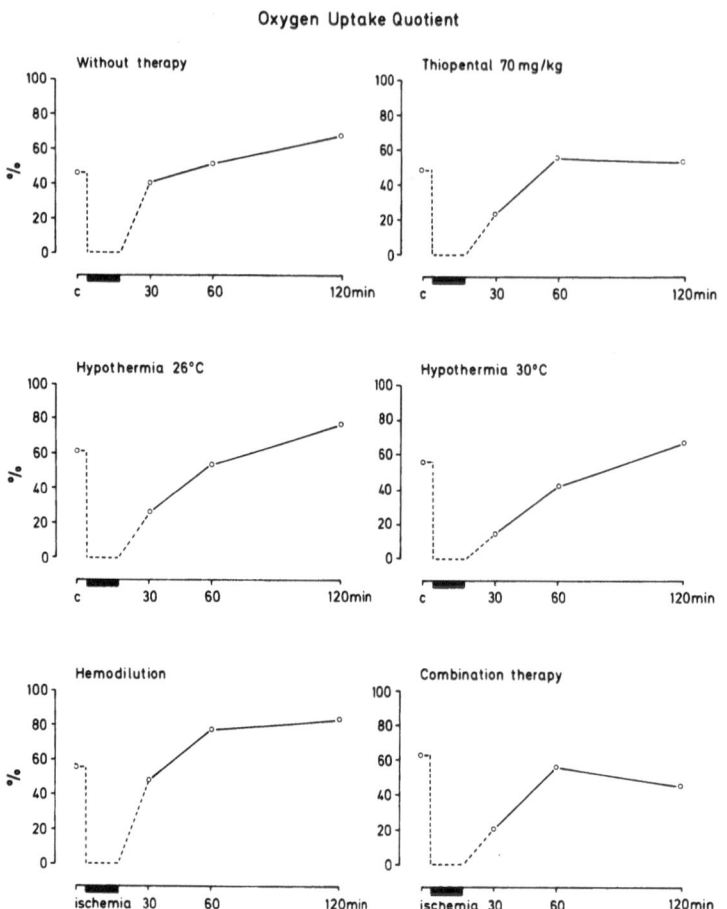

Fig. 4.   Changes of the oxygen uptake quotient during
           recirculation after 15 min complete ischemia.
           Same treatment as in Figs. 2 and 3. Oxygen
           uptake quotient is metabolic rate of oxygen
           in percent of oxygen availability.

low, indicating that the oxygen supply was in excess of
the oxygen demands (Fig. 4). This was in contrast to
the hemodilution group, in which despite a high flow
rate, the oxygen uptake quotient was 83 %.

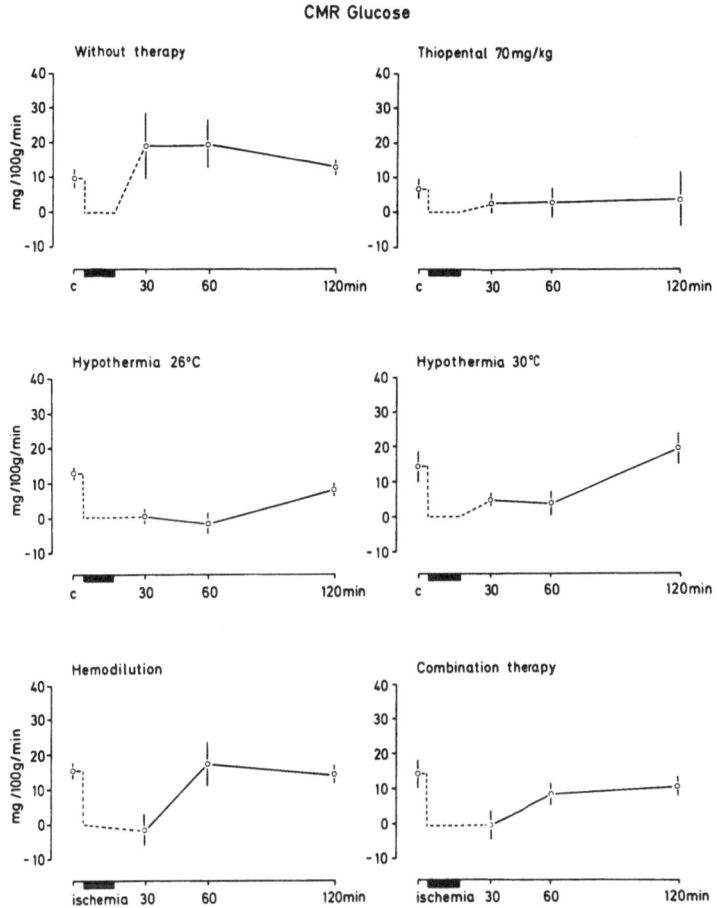

Fig. 5.   Changes in metabolic rate of glucose during
recirculation after 15 min complete ischemia.
Same treatment as in Figs. 2-4.

The changes in the uptake of glucose (Fig. 5)
must be interpreted with caution because rapid alterations of cerebral glucose content occurred following
ischemia which may introduce an error in the calculation
of the metabolic rate of this substrate. It should be
noted, however, that during hemodilution metabolic rate
of glucose did not increase above normal, and that the

glucose/oxygen uptake ratio also remained stable.   This
is of importance because it indicates that there was no
stimulation of anaerobic glucose metabolism, although
oxygen content in the (hemodiluted) arterial blood was
distinctly decreased (Table 1).

DISCUSSION

        The results of the present investigation demonstrate
that contrary to our assumptions, the inhibition of metabolic activity of the brain after ischemia did not ameli-
orate the post-ischemic recovery process.   The main denominator of functional recovery was blood flow, and it
was of little importance whether this was accompanied
by reduced oxygen content of the arterial blood or not.
Improving blood flow, in consequence, was the only therapeutic procedure which in the present experimental situation improved post-ischemic recovery.

        This conclusion is in contrast to several recent
communications about the beneficial effect of post-
ischemic hypothermia and post-ischemic barbiturate loading (2, 6, 10, 12, 24, 30) and therefore requires careful
interpretation.   An important consideration is the type
of experimental ischemia which has been studied.   Several
authors have reported different degrees of barbiturate
protection in complete and incomplete (30), total or
regional ischemia (2, 6, 12, 17, 24, 34), and there are
considerable differences between pre-ischemic and post-
ischemic barbiturate loading (10, 17, 24, 35).   Presumably similar differences exist for the application of
hypothermia (23, 37).   Our conclusion of a deteriorating
effect of post-ischemic metabolic inhibition on function
and metabolic recovery, therefore, may not be valid for
other types of ischemia.

        When we try to analyze our data, we have to differ-
entiate between at least four factors: the influence
of therapy on metabolic activity, on blood flow, on the
energy state of the brain and on functional recovery.

        In the untreated animals, metabolic rate of oxygen
increased above normal after ischemia, which confirms
earlier observations of post-ischemic hypermetabolism
(16, 19).   This increase was reversed by 26°C hypothermia and 70 mg/kg thiopental, but it was unaffected by
30°C hypothermia, 35 mg/kg thiopental (unpublished
observation), or a combination of both.

This is surprising because in non-ischemic animals even relatively light barbiturate anesthesia considerably reduces metabolic activity (7, 18, 27, 28). However, it has been suggested by Michenfelder (21) that this is mainly due to functional inhibition of electrocortical activity. Since during the early post-ischemic period recovery of electrocortical activity is poor, suppression of this activity may result in less pronounced saving of energy than in normal animals.

In the animals treated with deep hypothermia (26°C) and high barbiturate (70 mg/kg), metabolic rate of oxygen was substantially reduced, indicating that it is in fact possible to completely block post-ischemic hypermetabolism. However, the oxygen uptake quotient did not improve accordingly, because cardio-vascular insufficiency resulted in a decrease of blood flow and, in consequence, oxygen availability. It therefore remains doubtful whether this procedure is a useful approach for improving the mismatch between oxygen supply and oxygen demands of the tissue.

Another negative result of metabolic inhibition was the low energy state of the brain which was distinctly more disturbed than in the untreated animals. This was again in contrast to intact animals in which barbiturate anesthesia increases rather than lowers the energy stores of the brain (4, 20, 27). Post-ischemic metabolic inhibition, in consequence, seems to affect energy production more than energy consumption, and also for this reason is not suited for ameliorating post-ischemic recovery. This does not exclude that other metabolic effects such as free radical scavenging (9) might be of benefit, but if so, this did not show up during the observation time of our experiments.

A further unexpected result of our investigation was the beneficial effect of hemodilution on both functional recovery and the biochemical state of the brain. Hemodilution lowered the oxygen content of the arterial blood, and therefore did not improve oxygen availability although blood flow was increased. We therefore expected that the low oxygen availability was insufficient to cover the increased oxygen demands of the tissue during post-ischemic hypermetabolism, and that this would result in post-ischemic hypermetabolism, and that this would result in post-ischemic stimulation of anaerobic metabolism. However, this was not the case. Glucose uptake was not increased, and lactate content of the brain was lower than in untreated animals. The relatively low oxygen consumption,

consequently, appeared to be in equilibrium with the
actual oxygen needs of the brain. Since, moreover, both
biochemical and electrophysiological recovery in this
group were the most satisfactory of all animals, it has
to be concluded that post-ischemic hypermetabolism is not
a prerequisite of recovery. Instead, it might be an indicator of metabolic disturbance, e.g., uncoupling of
oxidative metabolism (32). If this is true, the positive
effect of hemodilution must be attributed to the prevention of such a disturbance. Since it is unlikely that
hemodilution has a direct effect on metabolism, it is
suggested that it acts indirectly by rendering microcirculation more homogeneous, and thus contributing to
both the better post-ischemic supply of substrates and
the removal of metabolic waste products from the brain.

This interpretation is also supported by the results
obtained with combination therapy. Recovery was slightly
worse than hemodilution alone but better than all the
other procedures. We suppose that, in this group, the
beneficial effect was due to hemodilution alone, and that
the additional treatment with hypothermia and barbiturate
--in contrast to other reports (11, 29)--was rather advers
to the recovery process.

In summary, we conclude from the present investigation that post-ischemic hypermetabolism is an indicator
of metabolic damage, and its treatment by metabolic inhibition contraindicated because this would affect the
symptoms rather than the cause of the disorder. Instead,
ways must be found to identify the factors contributing
to the development of hypermetabolism in order to design
a rational therapeutic procedure for preventing this
complication.

REFERENCES

1. Ames III, A., Wright, R.L., Kowada, M., Thurston, J.M.
    and Majno, G. (1968): Cerebral ischemia. II. The no-
    reflow phenomenon. Am. J. Pathol. 52: 437-453.

2. Black, K.L., Weidler, D.J., Jallad, N.S., Sodeman,
    T.M. and Abrams, G.D. (1978): Delayed pentobarbital
    therapy of acute focal cerebral ischemia. Stroke 9:
    245-249.

3. Cantu, R.C. (1969): Factors influencing postischemic
    cerebral vascular obstruction. Surg. Forum 20: 426-42

4. Carlsson, D., Harp, J.R., and Siesjö, B.K.   (1975):
    Metabolic changes in the cerebral cortex of the rat
    induced by intravenous pentothalsodium. Acta Anaes-
    thesiol. Scand. Suppl. 57: 7-17.

5. Chiang, J., Kowada, J., Ames III, A., Wright, R.L.,
    and Majno, G.   (1968): Cerebral ischemia, III. Vas-
    cular changes. Am. J. Pathol. 52: 455-475.

6. Corkill, G., Sivalingam, S., Reitan, J.A., Gilroy,
    B.A., and Helphrey, M.G.   (1978): Dose dependency of
    the post-insult protective effect of pentobarbital
    in the canine experimental stroke model. Stroke 9:
    10-12.

7. Crane, P.D., Braun, L.D., Cornford, E.M., Cremer, J.E.,
    Glass, J.M. and Oldendorf, W.H.   (1978): Dose depen-
    dent reduction of glucose utilization by pentobarbital
    in rat brain. Stroke 9: 12-18.

8. Fischer, E.G. and Ames III, A.   (1972): Studies on
    mechanisms of impairment of cerebral circulation
    following ischemia: effect of hemodilution and
    perfusion pressure. Stroke 3: 538-542.

9. Flamm, E.S., Demopoulous, H.B., Seligman, M.L. and
    Ransohoff, J.   (1977): Possible molecular mechanisms
    of barbiturate-mediated protection in regional
    cerebral ischemia. Acta Neurol. Scand. 56, Suppl. 64:
    150-151.

10. Goldstein, A. Jr., Wells, B.A., and Keats, A.S.
    (1966): Increased tolerance to cerebral anoxia by
    pentobarbital. Arch. Int. Pharmacodyn. Ther. 161:
    138-143.

11. Hägerdal, M., Welsh, F.A., Keykhah, M.M., Perez, E.
    and Harp, J.R.   (1978): Protective effects of combi-
    nations of hypothermia and barbiturates in cerebral
    hypoxia in the rat. Anesthesiology 49: 165-169.

12. Hoff, J.T., Smith, A.L., Hankinson, H.L. and Nielsen,
    S.L.   (1975): Barbiturate protection from cerebral
    infarction in primates. Stroke 5: 28-33.

13. Hossmann, K.-A.   (1978): Barbiturate protection of
    cerebral ischemia. In: International Conference on
    Atherosclerosis, L.A. Carlson et al. (eds.), pp. 251-
    256, Raven Press, New York.

14. Hossmann, K.-A. (1979): Cerebral dysfunction related to local and global ischemia of the brain. In: Brain Function in Old Age, F. Hoffmeister, C. Müller (eds.), pp. 385-393, Springer-Verlag, Berlin-Heidelberg-New York.

15. Hossmann, K.-A, Lechtape-Grüter, H. and Hossmann, V. (1973): The role of cerebral blood flow for the recovery of the brain after prolonged ischemia. Z. Neurol. 204: 281-299.

16. Hossmann, K.-A., Sakaki, S. and Kimoto, K. (1976): Cerebral uptake of glucose and oxygen in the cat brain after prolonged ischemia. Stroke 7: 301-305.

17. Hossmann, K.-A., Takagi, S. and Sakaki, S. (1977): Barbiturate loading following prolonged ischemia of the cat brain. Acta Neurol. Scand. 56, Suppl. 64: 376-377.

18. Lafferty, J.J., Keykhah, M.M., Shapiro, H.M., Van Horn, K. and Behar, M.J. (1978): Cerebral hypometabolism obtained with deep pentobarbital anesthesia and hypothermia. Anesthesiology 49: 159-164.

19. Levy, D.E.and Duffy, T.E. (1977): Cerebral energy metabolism during transient ischemia and recovery in the gerbil. J. Neurochem. 28: 63-70.

20. Lowry, O.H., Passonneau, J.V., Hasselberger, F.X. and Schulz, D.W. (1964): Effect of ischemia on known substrates and cofactors of the glycolytic pathway in brain. J. Biol. Chem. 239: 18-30.

21. Michenfelder, J. (1974): The interdependency of cerebral functional and metabolic effects following massive doses of thiopental in the dog. Anesthesiology 41: 231-236.

22. Michenfelder, J.D. and Milde, J.H. (1975): Cerebral protection by anaesthetics during ischaemia. Resuscitation 4: 219-233.

23. Michenfelder, J.D. and Milde, J.H. (1978): Failure of prolonged hypocapnia, hypothermia, or hypertension to favorably alter acute stroke in primates. Stroke 8: 87-91.

24.   Nemoto, E.M., Bleyaert, A.L., Bandaranayake, N.,
       Moossy, J., Rao, R.G. and Safar, P.   (1977):
       Amelioration of postischemic-anoxic brain damage by
       thiopental. In: Advances in Cardiopulmonary Resusci-
       tation, P. Safar (ed.), pp. 187-194, Springer-Verlag,
       New York-Heidelberg-Berlin.

25.   Nemoto, E.M., Hossmann, K.-A. and Cooper, H.K. :
       Postischemic hypermetabolism in cat brain (submitted
       for publication in Stroke).

26.   Nemoto, E.M., Snyder, J.V., Carroll, R.G. and Morita,
       H.   (1975): Global ischemia in dogs: Cerebrovascular
       CO_2 reactivity and autoregulation. Stroke 6: 425-431.

27.   Nilsson, L.   (1971): The influence of barbiturate
       anaesthesia upon the energy state and upon acid-base
       parameters of the brain in arterial hypotension and
       in asphyxia. Acta Neurol. Scand. 47: 233-253.

28.   Nilsson, L. and Siesjö, B.K.   (1975): The effect of
       phenobarbitone anaesthesia on blood flow and oxygen
       consumption in the rat brain. Acta Anaesthesiol.
       Scand. (Suppl): 18-24.

29.   Nordström, C.H. and Rehncrona, S.   (1978): Reduction
       of cerebral blood flow and oxygen consumption with a
       combination of barbiturate anaesthesia and induced
       hypothermia in the rat. Acta Anaesthesiol. Scand.
       22: 7-12.

30.   Nördstrom, C.H., Rehncrona, S. and Siesjö, B.K.
       (1976): Restitution of cerebral energy state after
       complete and incomplete ischemia of 30 min duration.
       Acta Physiol. Scand. 97: 270-272.

31.   Ping, F.C. and Jenkins, L.C.   (1978): Protection of
       the brain from hypoxia: a review. Can. Anae. Soc. J.
       25: 468-473.

32.   Schutz, H., Silverstone, P.R., Vapalahti, M., Bruce,
       D.A., Mela, L. and Langfitt, T.W.   (1973): Brain
       mitochondrial function after ischemia and hypoxia.
       I. Ischemia induced by increased intracranial pres-
       sure. Arch. Neurol. 29: 408-416.

33.   Snyder, J., Nemoto, E., Carroll, R., Morita, H.,
       Safar, P. and Kirimli, B.   (1973): Intracranial
       pressure, brain flow regulation and glucose and ox-
       ygen metabolism after 15 minutes of circulatory
       arrest in dogs. Stroke 4: 342.

34.   Steen, P.A., Milde, J.H. and Michenfelder, J.D.
       (1978): Cerebral metabolic and vascular effects of
       barbiturate therapy following complete global isch-
       emia. J. Neurochem. 31: 1317-1324.

35.   Steen, P.A., Milde, J.H. and Michenfelder, J.D.
       (1979): No barbiturate protection in a dog model of
       complete cerebral ischemia. Ann. Neurol. 5: 343-349.

36.   Takagi, S., Cocito, L. and Hossmann, K.-A.   (1977):
       Blood recirculation and pharmacological responsive-
       ness of the cerebral vasculature following prolonged
       ischemia of cat brain. Stroke 8: 707-712.

37.   White, R.J.   (1978): Cerebral hypothermia and circu-
       latory arrest. Review and commentary. Mayo Clin. Proc.
       Proc. 53: 450-458.

BEHAVIOR OF THE HABENULO-PINEAL COMPLEX IN DEEP HYPOTHERMIA (COLD, HYPOXIA, HYPERCAPNIA)

R. Miline, J. Milin, M. Matavulj, M. Vukovic
and D. Vujaskovic

Institute of Histology and Embryology
Faculty of Medicine
Novi Sad, Yugoslavia

Judging by our previous results, the habenula and the pineal gland represent a morphophysiological unity in the form of habenulo-pineal complex, a biological antagonist to the hypothalamo-hypophysial complex (9, 12, 13). According to our data, the pineal gland is a senso-neuroendocrine organ (8) which is highly activated in the defense processes of the organism (10, 11).

Our aim in this paper is to present the histophysiological changes of deep hypothermia in the habenulo-pineal complex.

MATERIAL AND METHODS

The experiments were conducted on 60 mature male Wistar rats. One group consisted of rats exposed to deep hypothermia, while the second group served as controls. Deep hypothermia was produced by the method of Giaja (1-3). Each experimental rat was daily immersed in a glass vessel containing water with ice for 3 hours. Once the rectal temperature of the animal reached 15°C, the rat was removed from the vessel and recovered at room temperature. The animals were sacrificed by decapitation at the end of one week.

For the light microscopic examination, the organs were fixed in Bouin, and in Bouin-Hollande sublimate liquids. The staining methods applied were: light

green-erythrosin-iron haematoxylin, Gomori-Bargmann, PAS-
McConail, Alcian blue-PAS-orange G, aldehyde thionin-
PAS-orange G.

The following histochemical and histoenzymological
investigations were carried out: Oil red; Sudan III;
nucleic acids (6); glucose-6-phosphate dehydrogenase (15);
lactate dehydrogenase, iso-citrate dehydrogenase (4);
succinic dehydrogenase (14); monoamine oxidase; and non-
specific esterases (17).

For electron microscopy, the tissues were fixed by
immersion in (6%) glutaraldehyde-(0.2 M) cacodylate
buffer (pH 7.4), then postfixed in (1%) osmic acid dehydrated with acetate, embedded in araldite and counter-
stained with uranyl acetate and lead citrate. The
examination was performed with an ISKRA LEM 4C electron
microscope.

RESULTS

A. Habenula

The medial habenular nucleus, which sends numerous
fibers to the pineal gland, was more clearly demarcated
in hypothermia than in the control animals. It displayed
irregularly shaped hypertrophic neuroganglionic cells
(Fig. 1) with eccentric, spherical nuclei and prominent
often hyperphloxinophilic peripheral nucleoli. Binuclear
cells were seen frequently. Most of the cells were pale
showing a distinct perinuclear zone of either homogeneous
hyperbasophilic or granular PAS and a Gomori-positive
reaction. This granular reactivity was found also between
the neuroganglionic and neuroglial cells and in the
pericapillary space as well as in the vicinity of the
ependyma. Other cells were characterized by dark homogeneous cytoplasm with distorted nuclei. Moreover there
were also some isolated cells which showed intracyto-
plasmic aldehyde-thionin-PAS-positive granules. The
neuroglial cells were found in close contact with the
neuroganglionic cells. They were also irregularly shaped
with densely packed nuclear chromatin. Throughout the
habenular nucleus, the enzymes belonging to the oxido-
reductase group showed increased activity: glucose-6-
phosphate dehydrogenase, lactate dehydrogenase, isocitrate dehydrogenase, succinic dehydrogenase and monoamine
oxidase. The most conspicuous was the elevation of the
nonspecific esterase activity (Fig. 2).

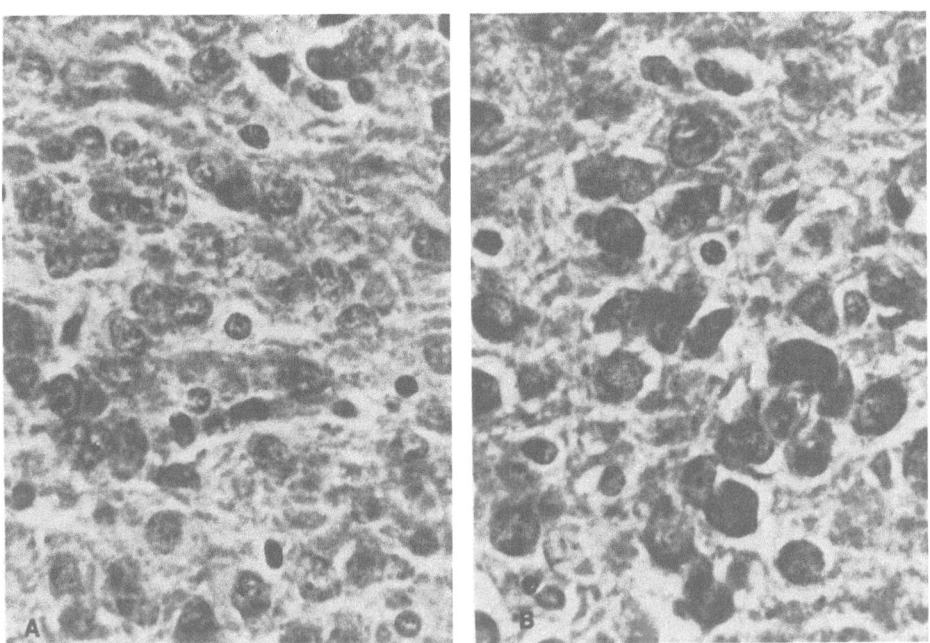

Fig. 1.   Habenula. Nucleus medialis. A. Control rat;
           B. Rat in deep hypothermia: hypertrophy of
           neuroganglial cells (Bouin, Florentin; oc.
           12.5; obj. 63).

B.   Pineal Gland

    Hypothermia led also to an obvious hypertrophy and
hyperplasia of light pinealocytes. Numerous hypertrophic
light pinealocytes were especially noticeable in the
peripheral zone of the distal segment: the cells were
arranged in the form of nodules or were palisading around
dilated capillaries (Fig. 3). Their nuclei were also
hypertrophic, irregular in shape, showing frequent in-
vaginations. The hypertrophic nucleoli varied in shape,
and the nuclei with an eccentrical nucleolus prevailed.
The light pinealocytes of the central zone displayed the
same features. Some of them contained small Gomori-
positive and others aldehyde-thionin-PAS-positive granules.
There were also cells with Alcian blue-PAS-orange G

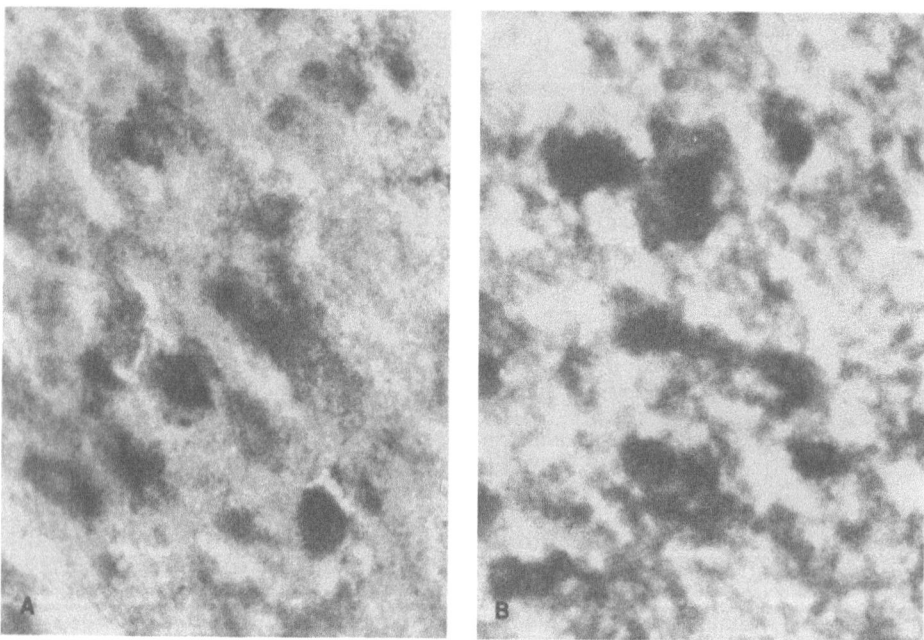

Fig. 2.   Habenula.   Nucleus medialis.   A. Control rat;
           B. Rat under influence of deep hypothermia:
           hyperactivity of nonspecific esterases (oc.
           12.5; obj. 40).

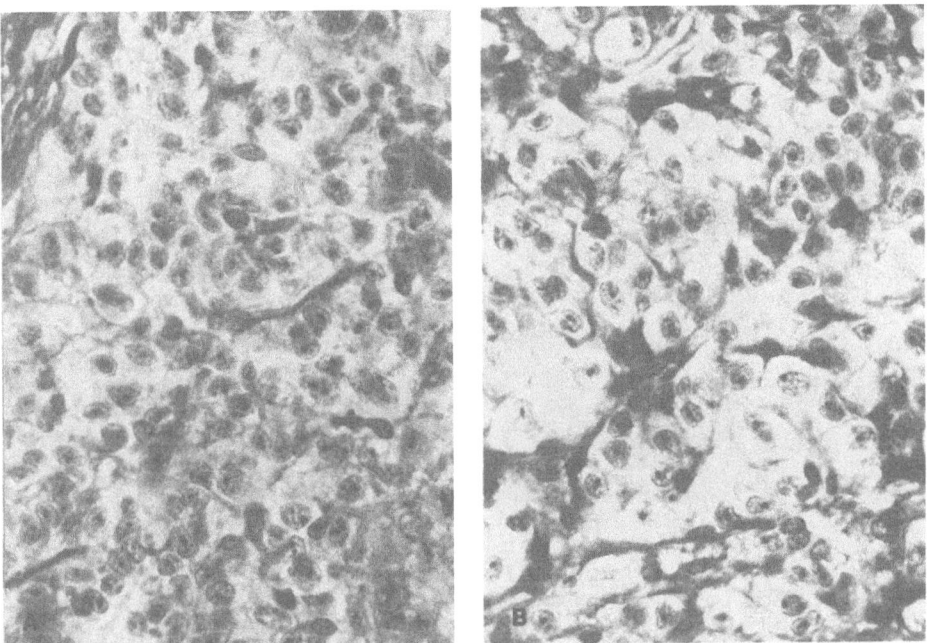

Fig. 3.   Pineal gland.   A.   Control rat; B. Rat in deep
           hypothermia: hypertrophy of light cells (Bouin,
           Florentin; oc. 12.5; obj. 40).

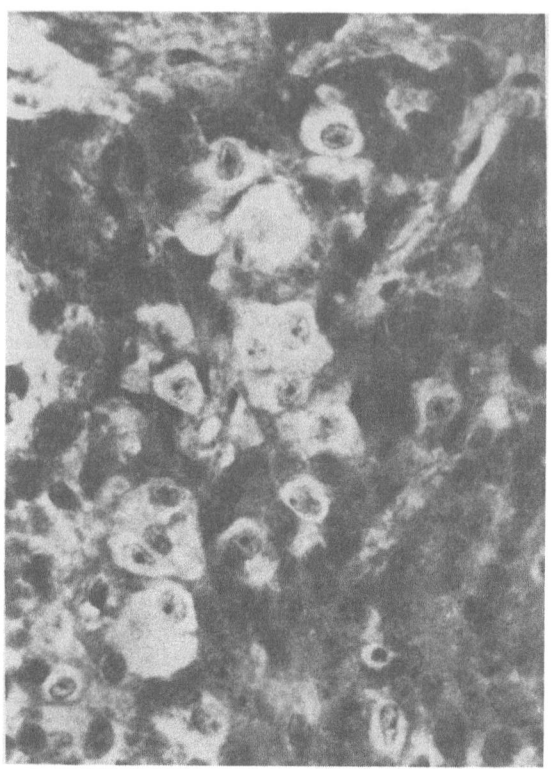

Fig. 4. Pineal gland. Rat in deep hypothermia; light
         cells; dark pinealocytes with unclear borders
         (Bouin, Gomori-Bargmann; oc. 12.5; obj. 40).

partly stained cytoplasm. Dark pinealocytes, those from
the central zone as well as those from the proximal, i.e.,
deep segment, possessed unclear borders (Fig. 4). Their
nuclei were hyperchromatic and were smaller than in cells
from the control animals. The entire gland contained
lipid droplets, but they were more numerous in the peripheral and central zone of the distal part of the gland.
A marked staining for ribonucleic acids (pyronine-m.
green) was also manifested in this part of the gland.
The activity of the enzymes belonging to the oxido-
reductase group was significantly increased, especially
the activity of isocitrate dehydrogenase as indicated
by the abundant and densely compacted formazan grains
(Fig. 5).

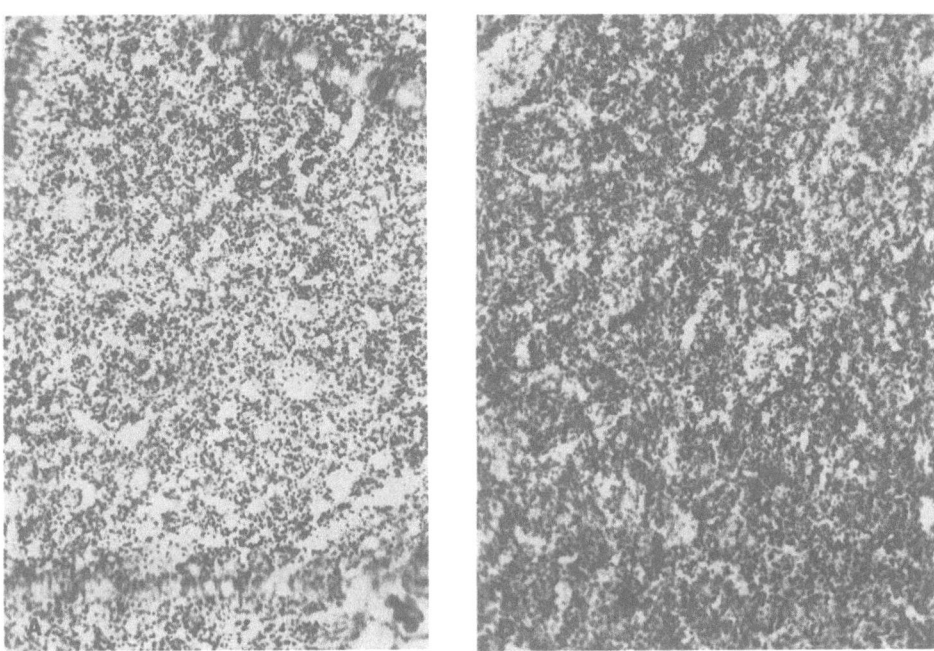

Fig. 5. Pineal gland. A. Control rat; B. Rat in deep
         hypothermia: hyperactivity of isocitrate
         dehydrogenase (oc. 12.5; obj. 40).

     The morphodynamics of the nuclei of the light
pinealocytes were markedly altered, showing deformed,
multilobular, and slightly segmented features. Frequently, parts of the indented nuclei were intercon-
nected by thin isthmuses of the nucleoplasm. In other
cases, the narrowed nuclei formed an internuclear bridge
deprived of the nucleoplasm (Fig. 6). Mitochondria were
numerous and polymorphous, and some were larger than
those of the control group. The granulated endoplasmic
reticulum was abundant and formed parallel oriented
tubuli in some of the light cells. The Golgi zone was
hypertrophic, and adjacent osmiophilic secretory granules
were well demarcated in many of the pinealocytes. Their
content of lipid droplets was either grouped in grape-like
structures or was individually distributed, with a variable size and osmiophilia.

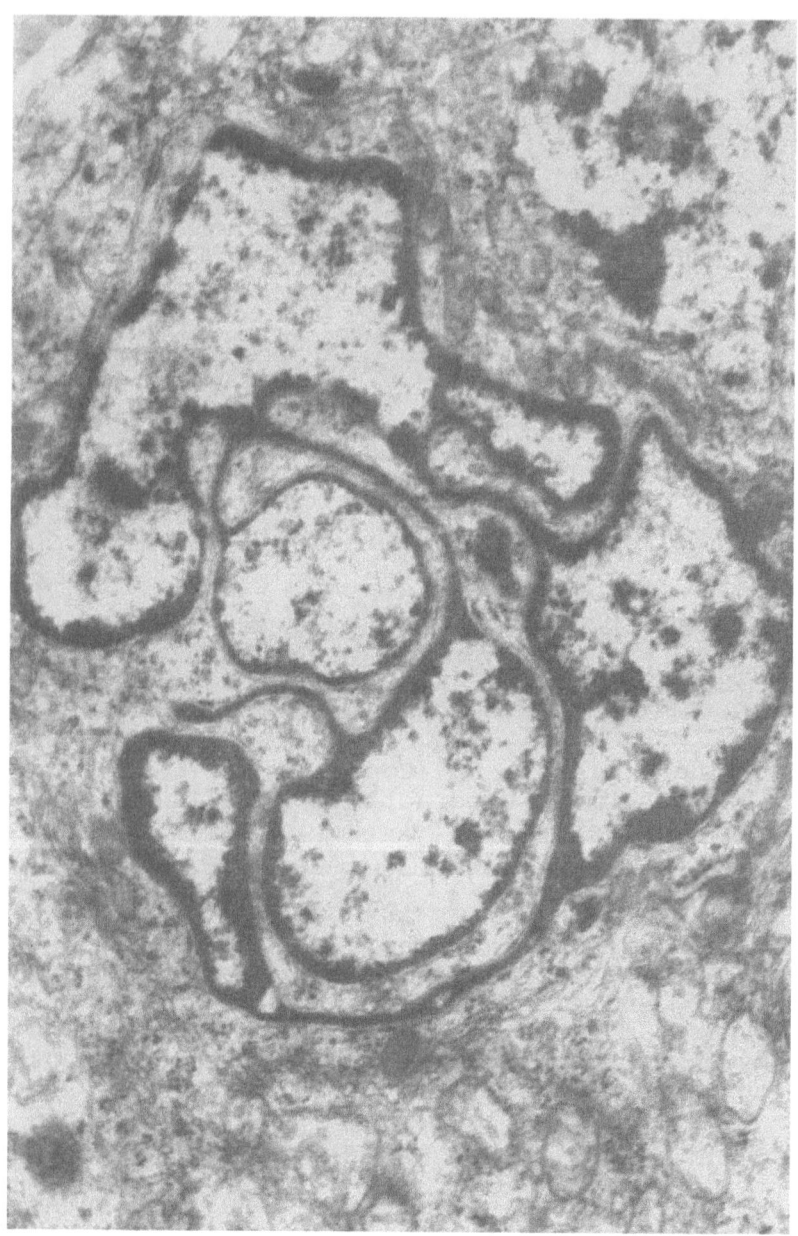

Fig. 6.   Pineal gland.   Rat in deep hypothermia: light
          cells, nucleus with nuclear bridges. 35,000 X.
          (Reduced 10% for reproduction)

Numerous dark pinealocytes were characterized by hyperplastic mitochondria densely packed in the cellular body and processes. The granular endoplasmic reticular, lipid droplets and lysosomes in the form of broad agglo- merates were frequently associated with densely aggre- gated mitochondria. There were also cells showing numerous lysosomal formations and mitochondria charac- terized by a discontinuous membrane and partially con- fluent crests.

DISCUSSION

Deep hypothermia, cold, hypoxia and hypercapnia have a common effect as stress-inducing factors that activate the defense mechanism of the neuroendocrine system. The induced hypothermic cellular hypertrophy and increased enzymatic reactivity found in the medial habenular nucleus indicate that neuroganglionic cells changed during hypothermia. The presence of granules in the cytoplasm of some cells, in intercellular spaces in the vicinity of capillaries, suggests a neuroglandular function of this part of the habenula.

Moreover, the cytological features of the light and dark pinealocytes, especially their ultrastructural characteristics described above, reflect their markedly stimulated activity. The presence of disintegrated dark pinealocytes is a reflection of an involutionary action taking place during deep hypothermia. Thus, the pineal gland undergoes functional dissociation of the gland parenchyma and displays some cells with a progressive and others with an involutionary reaction. The diverse cytological reactivity of the pinealocytes in hypothermia is indicative of the existence of more than one population of functionally different pinealocytes. These observations are in full accord with current ideas of polypeptide secretion by the pineal gland (7).

Since the pineal extract has many diverse properties it would be of interest to know the specific relationship between the cellular changes and antigonadotropic, anti- corticotropic and antithyrotropic activities. According to our preliminary studies, deep hypothermia led to a depression of the thyroid gland manifested by enlargement of the follicles with accumulation of colloids (unpub- lished observations). Therefore, one of the pinealocyte populations displaying the above described hyperactivity could be a site of the antithyrotropic secretion. There is also a functional correlation between the pineal gland and the hypothalamus, since the antihypophysial activity

of the pineal gland may be exerted through the hypothalamic
relay (13). In experimental animals under chronic stress
of deep hypothermia, a stimulated activity of the supraopti
nucleus takes place, hypertrophy of the neuroglandular
neurons, hypertrophy of nuclei and nucleoli and mobiliz-
ation of neuro-secretion (unpublished observations).

The cellular response of both the medial habenular
nucleus and the pineal gland to hypothermia suggests that
these two components of the habenulo-pineal complex are
functionally linked. The notion of this functional unit
of the neuroendocrine system as a morphophysiological unity
is supported by the studies of the pineal gland innervation, which originates not only from the superior
cervical ganglia (5) but from the habenula as well (16).

REFERENCES

1. Giaja, I. (1949): Hipotermija. Acta Medica
    Yugoslavica 3: 9-33.

2. Giaja, I. (1953): Hypotermie, hibernation et
    poikilothermie expérimentale. Biologie Médicale
    42: 1-36.

3. Giaja, I. (1955): Aclimation et hibernation. La
    Revue de Pathologie générale et comparée. 664:
    122-127.

4. Hess, R., Scarpelli, D.C. and Pearse, A.G.E.:
    (1958): Cytochemical localization of pyridine
    nucleotide-linked dehydrogenase. Nature 181:
    1531-1532.

5. Kappers, J.A. (1960): The development, topographical
    relations and innervation of the epiphysis cerebri
    in the albino rat. Z. Zellforsch. 52: 163-215.

6. Kurnick, N.B. (1952): Histological staining with
    methyl-green pyronin. Stain Technol. 27:
    233-242.

7. Lukaszyk, A. and Reiter, R. (1975): Histophysio-
    logical evidence for the secretion of polypeptides
    by the pineal gland. Am. J. Anat. 143: 451-464.

8. Milin, R., Devečerski, V. and Krstić, R. (1969):
    Corpus pineale - glande de nature sensoneuroendo-
    crine. Radovi 37: 69-84.

9. Milin, R., Devečerski, V. and Milin, J. (1972):
    Uticaj buke na semenik u epifizektomisanih pacova.
    In: Aktuelni Problemi Iz Endokrinologije, ed.
    Srpsko lekarsko društvo, pp. 5-13, Galenika,
    Beograd.

10. Miline, R. (1957): La part de l'épiphyse dans le
    syndrome d'adaptation. In: Congrès National des
    Sciences Médicales. Académie de la Republique
    populaire Roumaine (ed.), Bucarest.

11. Miline, R., Krstić, R. and Devečerski, V. (1968):
    Sur le comportement de la glande pinéale dans des
    conditions de stress. Acta Anat. (Basel) 71:
    352-402.

12. Miline, R. and Šćepović, M. (1959): La part du
    complexe habénulo-épiphysaire dans l'histo-
    physiologie de la glande thyroide. Annales
    d'Endocrinologie 20: 512-518.

13. Miline, R., Werner, R., Šćepović, M., Devečerski,
    V. and Milin, J. (1969): Influence du froid sur
    le comportement du noyau supraoptique chez les
    rats épiphysectomisés. Bulletin de l'Association
    des Anatomistes 145: 289-293.

14. Nachlas, M.M., Tsou, K., de Sousa, E., Cheng, C.
    and Seligman, A.M. (1957): Cytochemical demonstration
    of succinic dehydrogenase by the use of a new
    p-nitrophenol substituted ditetrazole. J. Histochem.
    Cytochem. 5: 420-423.

15. Nachlas, M.N., Walker, D.G. and Seligman, A.M.
    (1958): The histochemical localization of triphos-
    phopyridine diaphorase. J. Biophys. Biochem. Cytol.
    4: 467-470.

16. Nielsen, J.T. and Møller, M. (1975): Nervous con-
    nections between the brain and the pineal gland in
    the cat (Felis catus) and the monkey (Cercopithecus
    aethiops). Cell Tiss. Res. 161: 293-301.

17. Spannhof, L. (1967): Einführung in die Praxis der
    Histochemie. pp. 122-123, G. Verlag Fischer, Jena.

MECHANISMS OF ISCHEMIC BRAIN DAMAGE

IN ACUTE ARTERIAL HYPERTENSION

I. V. Gannushkina and M. V. Baranchikova

Institute of Neurology
Academy of Medical Science
Moscow, USSR

During the last 8-10 years, it has been shown that the breakdown of the cerebral blood flow autoregulation is the main reason for the damage of the cerebral vessels and brain tissue in acute arterial hypertension (2, 7, 9, 13, 17, 19). Consequently, the passive dilatation of the superficial arteries of brain increases the permeability of the blood-brain barrier (BBB) to all the constituents of blood including blood proteins (i.e., filtration edema develops). The spotty distribution of the damages among the normal unchanged brain structures is characteristic for all the processes studied: brain blood supply and filling changes in the pial arteriolar caliber, BBB permeability to the Evans blue-bound proteins, accumulation of proteins in astrocytes and neurons as well as the parenchymatous spread of the tracer (1-3, 7, 10, 14).

In our previous works, we reported that the arteries of the white matter are less vulnerable than those of the cortex, due to their better developed muscle. We also explained that the greatest vulnerability of the boundary zones as compared to the base of the main cerebral arteries was a result of a maximal increase of intraluminal pressure in the "dead points" during the acute increase of the arterial pressure (4, 6). According to our data (6), the arteries of the occipital region of the arterial boundary zones have more often a straight course, branch at acute angles, and show "end to end" anastomoses between the middle and posterior cerebral arteries. Therefore the above-mentioned

anatomical properties allow us to conclude that these
arteries have lower hydraulic resistance.

It is also well known from literature that there is
a severe damage of arterioles, capillaries and venules
in the areas of the increased permeability of BBB to
Evans blue-bound albumin or horseradish peroxidase; at
the same time pathological changes of nervous tissue can
also be observed (5, 8, 15, 16).

This brief survey of the modern literature indicates
that quite a lot is known about the pathogenesis of the
cerebral tissue and vessel damage in acute hypertension.
However, there are still numerous questions to be clari-
fied concerning the heterogenous filling of intracerebral
vessels found in the areas of the increased BBB permeability and the mechanism of the damages in the white
matter, where such an increase of BBB permeability to
proteins is extremely rare.

To examine further the mechanisms of increased BBB
permeability with regard to various compounds of plasma,
measurements of brain water content in the regions of
increased BBB permeability to Evans blue bound-albumin
as well as in the unchanged areas were carried out.

MATERIAL AND METHODS

Thirty normal rabbits weighing 2.5-3 kg were used.
Acute hypertension was induced by the intravenous infusion of 2 mg noradrenalin in 9 ml saline at a speed of
0.33 mg/min. Mean arterial blood pressure (MABP) was
continuously recorded from the femoral artery. The BBB
dysfunction was studied with Evans blue (4 ml of a
solution 2%/kg) and sodium fluorescein.

In 10 normal and 15 hypertensive rabbits, total
tissue water content was measured separately in the
gray and the white matter of various parts of the brain
by drying the samples at 100°C to constant weight for
4-6 days. All the wet samples, weighing not more than
10 - 30 mg, were dissected immediately after decapitation.

The cortical brain tissue samples of the hypertensive animals were taken from the main areas showing increased BBB permeability, from the boundary zones and
from the symmetric regions. The samples of the white
matter from the hypertensive animals were also taken
from the regions below the opening of the BBB to serum
proteins, i.e., from under the "blue spots", from the

boundary zones and from the symmetric regions too.   The
results were statistically analyzed by a computer.

The brains of 5 rabbits with a standardized increase of MABP were sliced for histological examination,
using cryostat sections for the Pickworth Lepehne benzi-
dine staining, and paraffin embedded samples for the
routine histological stains.

RESULTS

Our measurements showed that the water content in
control animals is 81.157 ± 0.264% and 70.286 ± 0.454%-
in the gray and white matter, respectively.   The differences found in the water content between boundary
zones and the base of main cerebral arteries were insignificant in both the gray and the white matter.

Five-six minutes after the MABP was elevated, the
water content was found to be markedly increased in the
cortical "blue spots" and in the unstained gray matter
samples obtained from different parts of the brain.

The water content values of 73-74% were frequently
observed in these areas.   However, normal values of
water content were also found in the cortical "blue spots"
of the gray matter.   Similar changes in the water content
were also observed in the regions of unstained tissues
(free of "blue spots") obtained from various parts of
the cortex.

As far as the white matter is concerned, the values
of the water content decreased, increased or remained
unchanged, irrespective of the tissue origin i.e.,
whether the samples were taken from the normal or the
"blue spot" or from below the blue stained region.

Studying the interrelations of the brain areas
permeable to Evans blue-bound albumin and to sodium-
fluorescein, we found no correlation between them. Thus,
the "blue spots" were mostly found in the cortical
boundary zones and macroscopically did not spread to the
underlying white matter.   The extravasation of sodium-
fluorescein was observed in other regions of the brain,
in the cortex and underlying white matter.   This leakage
was not limited to the base of one or several cerebral
arteries and appeared like an oily paper spot.

During the acute arterial hypertension, besides the
reported heterogenous filling of the intracerebral

vascular network, we also observed an increase and then
a decrease of vascular blood filling in the "blue spots"
as a result of edema formation. Our present study also
showed that these events do not always develop in the
same sequence: the decrease in the vascular blood content in the "blue spots" was also seen in the early
stages when the majority of the vessels and capillaries
were still dilated. At the same time the well filled
vessels and the small areas of brain with dilated vessels
(veins were more often seen) showed typical leakage of
serum proteins and a pronounced decrease of vascular
filling in the area of the "blue spots". All these
changes were characteristic also for the stain-free
regions of the brain. There was no obvious correlation
between the localization of the "blue spots" and the
regions showing a decreased filling of the brain vessels.

DISCUSSION

It can be concluded from our findings that the acute
increase of MABP leads to the local breakdown of cerebral
blood flow autoregulation followed by diverse changes
in tissue water content. The acute hypertension causing
the breakdown of cerebral blood flow autoregulation,
filtration edema and the increase of intracranial pressure induced an uneven increase in local tissue pressure.
The observed reduction in the tissue water content in
some cerebral areas could be the reflection of this
process.

We assume that it is due to the redistribution of
the free tissue water, which is a constituent of the
total tissue water. The process is more frequent and
more pronounced in the white than in the gray matter,
this assumption being based on indirect evidence that
the large extracellular space of the white matter
allows the migration of water.

The marked difference in the molecular weight of
Evans blue-bound albumin and that of sodium-fluorescein
is most likely responsible for the observed dissimilar
extravasation of these compounds. The molecular weight-
dependent mobility of BBB tracers in brain tissue has
been thoroughly studied by Klatzo et al. (11, 12, 18)
in experimental models of vasogenic brain edema. However, the data concerning the shifts of the tissue
water content, with the subsequent development of foci
with decreased water content, were not reported in the
previous studies.

We believe that the existing viewpoints on the pathogenesis of the cerebral parenchymatous and vascular injury after acute hypertension are incomplete. By now we know that the breakdown of cerebral blood flow auto-regulation due to acute hypertension occurs in the hydraulically most vulnerable parts of the vascular system of the brain surface. It occurs mainly in the occipital parts of the arterial boundary zones (6). It is followed by an increased blood flow, pathological vasodilatation, elevation of intracapillary pressure and intense transudation of fluid into the brain tissue. The developing local brain edema compresses the intra-cerebral vessels, thus leading to a secondary decrease of vascular filling and brain blood supply.

Hence, the well known cerebral ischemia typical for hypertensive encephalopathy is to be considered only as a consequence of the above-mentioned processes. Con-sidering the pathogenesis of brain ischemia after acute hypertension, we should not forget that the vessels can be compressed by fluid accumulation in the area. The accumulated fluid could draw off more fluid from the extracellular space of adjacent regions, causing a compression of the vessels in the areas with decreased tissue water content, with subsequent damage of the brain tissue. Assuming that it is possible for tissue edema to develop in the regions with initial decreased water content, it would primarily be of cytotoxic rather than of vasogenic origin. It is also possible that the increased tissue pressure could influence local venous return only. In support of this our studies concerned with the filling of the intracerebral vessels in the experimental hypertensive model have shown some dilated vessels, mainly veins, in ischemic areas among the damaged neurons.

These studies vividly show that the cause for the development of ischemic foci resulting from the break-down of the cerebral blood flow autoregulation during the acute arterial hypertension is rather a complicated one. Therefore, the pathogenetic factors of the hyper-tensive encephalopathy required re-evaluation. We con-sider that the breakdown of cerebral blood flow auto-regulation plays a leading role in this process, but its subsequent manifestations are diverse in various parts of the brain.

REFERENCES

1. Dinsdale, H.B., Robertson, D.M. and Haas, R.A. (1974): Cerebral blood flow in acute hypertension. Arch. Neurol. 31: 80-87.

2. Ekström-Jodal, B., Häggendal, E., Linder, L.E. and Nilsson, N.J. (1971): Cerebral blood flow autoregulation at high arterial pressures and different levels of carbon dioxide tension. Europ. Neurol. 6: 6-10.

3. Farrar, J.K., Jones, I.V., Graham, D.I., Strandgaard, S. and MacKenzie, E.T. (1976): Evidence against cerebral vasospasm during acutely induced hypertension. Brain Res. 104: 176-180.

4. Gannushkina, I.V., Galayda, T.V., Shafranova, V.P and Rjassina, T.V. (1977): Further studies of the role of geometry of arterial system of the brain in "breakthrough" of cerebral blood flow autoregulation. In: Cerebral Function, Metabolism and Circulation, D.H. Lassen, N.A. Lassen, (eds), pp. 21.12-21.13, Munksgaard, Copenhagen.

5. Gannushkina, I.V. and Shafranova, V.P. (1975): The differences of arterial autoregulation in gray and white matter in acute hypertension. In: Blood Flow and Metabolism in the Brain. A. M. Harper, W.B. Jennet, J.D. Miller and J.O. Rowan (eds.), pp. 5.31-5.35, Churchill Livingstone, Edinburg, London, New York.

6. Gannushkina, I.V. and Shafranova, V.P. (1976): Morphological evidence and pathogenesis concerning the spotty nature of brain tissue damage in experimental hypertensive encephalopathy. In: Cerebral Vascular Disease. J.S. Meyer, H. Zechner, M. Reivich, (eds.), pp. 61-65, Georg Thieme Publ., Stuttgart.

7. Gannushkina, I.V., Shafranova, V.P., Dadiany, L.N., and Galayda, T.V. (1973): Mechanisms related to decrease of CBF during acute increase of arterial pressure in hypertensive animals. In: Cerebral Vascular Disease. 6th International Conference Salzburg 1972, J.S. Meyer, H. Zechner, M. Reivich, O. Eichhorn. Georg Thieme Publ., Stuttgart.

8. Hansson, H.A., Johansson, B. and Blomstrand, C.
    (1975): Ultrastructural studies on cerebrovas-
    cular permeability in acute hypertension. Acta
    Neuropath. (Berl.) 32: 187-198.

9. Johansson, B. (1974): Blood-barrier barrier
    dysfunction in acute arterial hypertension.
    Thesis, Göteborg.

10. Johansson, B., Li Ch.-L., Olsson, Y. and Klatzo, I.
    (1970): The effect of acute arterial hypertension
    on the blood-brain barrier to protein tracers.
    Acta Neuropath. (Berl.) 16: 117-124.

11. Katzman, R., Clasen, C. H., Klatzo, I., Meyer,
    J.S., Pappius, H.M. and Waltz, A.G. (1977): Brain
    edema in stroke. Stroke 8:, 4: 509-540.

12. Klatzo, I., Miquel, J. and Otenasek R. (1962):
    The application of fluorescein labeled serum
    proteins (FLSP) to the study of vascular perme-
    ability in the brain. Acta Neuropath. (Berl.)
    2: 144-160.

13. Lassen, N.A. and Agnoli, A. (1972): The upper limit
    of hypertensive encephalopathy. Scand. J. clin. Lab.
    Invest. 30: 113-116.

14. MacKenzie, E.T., Strandgaard, S., Graham, D.I.,
    Jones, J.V., Harper, A.M. and Farrar, J.K. (1976):
    Effects of acutely induced hypertension in cats on
    pial arteriolar caliber, local cerebral blood flow,
    and the blood-brain barrier. Circulation Res. 39:
    33-41.

15. Nag, S., Robertson, D.M. and Dinsdale, H.B. (1977):
    Cerebral cortical changes in acute experimental
    hypertension. An ultrastructural study.
    Laboratory Investigation 36: 2, 150-161.

16. Olsson, Y. and Hossmann, K. (1970): Fine structural
    localization of exudated protein tracers in the brain.
    Acta Neuropath. (Berl.) 16: 103-116.

17. Skinhøj, E. and Strandgaard, S. (1973): Pathogenesis
    of hypertensive encephalopathy. Lancet 1: 461-462.

18.   Steinwall, O. and Klatzo, I. (1966): Selective
       vulnerability of the blood-brain barrier in
       chemically induced lesions. J. Neuropath. Exp.
       Neurol. 25: 542-559.

19.   Strandgaard, S. (1978): Autoregulation of cerebral
       circulation₀in hypertension. FADL's forlag
       København. Arhus. Odense.

THE EFFECT OF CHEMICALLY AND ELECTRICALLY INDUCED ACUTE

HYPERTENSION ON THE PERMEABILITY ACROSS CEREBRAL VESSELS

Erik Westergaard

Anatomy Department C
University of Copenhagen
Copenhagen, Denmark

The term, blood-brain barrier (BBB), was introduced by Ehrlich (7) who observed that intravenously injected dyes stained all of the animals' organs except their brains. Likewise, it was observed that trypan blue did not pass the BBB (6) and the intravenously injected fluorescein behaved similarly (16, 24). These chemical compounds bind to albumin in the blood (28). The cerebral microvasculature therefore constitutes a barrier to dye-albumin complexes. However, it cannot be excluded that very small amounts in fact could have passed the barrier in these experiments since the sensitivity of the methods employed is limited.

In contrast, the tracers horseradish peroxidase (HRP) and microperoxidase (MP) can be detected in very small amounts. Furthermore, they allow studies of the BBB at the ultrastructural level. The BBB to HRP was first described by Reese and Karnovsky (26) and Brightman and Reese (3). They found that HRP did not pass the BBB, due to the effective hindrance of the tight junctions between adjacent endothelial cells. Furthermore, since the number of endothelial vesicles was small, it was assumed that they were not sufficient to transfer HRP from blood to brain. Reexamination of the BBB to HRP demonstrated that insignificant amounts of HRP cross the small cerebral arterioles in very few, short segments (29, 31), and therefore it was concluded that vesicular transfer could have taken place. Recently, transfer of microperoxidase has been demonstrated, similar to that observed for HRP, indicative of vesicular transport existence (Westergaard,

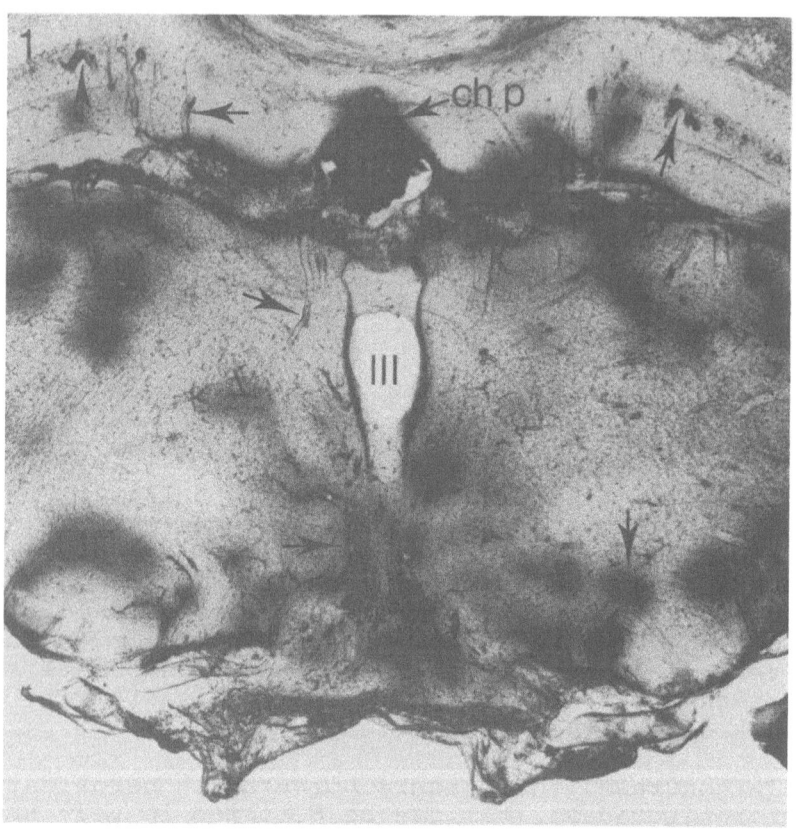

Fig. 1. Light microscopical picture of a one mm thick,
         cleared, frontal section from a rat with Ara-
         mine-induced acute, systemic arterial hyperten-
         sion. Reaction product occurs in the choroid
         plexus (ch p) as under normal conditions. The
         number of segments of vessels containing HRP
         in their walls (arrows) has been significantly
         increased. Furthermore, there is a pronounced
         spread of reaction product into the neuropil.
         The third ventricle (III) is in the center of
         the picture. X 20
         (Reduced 10% for reproduction.)

in press). The vesicular transfer of HRP is increased
under various experimental conditions, as previously
reported (30).

The increased permeability of the BBB to dye-albumin
complexes after chemically induced acute hypertension has
been described in several papers (1, 10, 12-14, 23). The
same observations were made by Olsson and Hossmann (23)
using HRP at the ultrastructural level but without elucidation of the route(s) taken by HRP across the cerebral
endothelium. The increased permeability to HRP observed
by Hansson, et al. (9) after intravenous injection of
metaraminol bitartrate (Aramine) was explained by the
possible existence of three pathways; namely, channels
through the endothelial cells, vesicular transfer, and
interendothelial movement. The enhanced permeability to
HRP after chemical induction of acute hypertension was
also found by Nag, et al. (21, 22), Westergaard and
Brondsted (32), Brondsted and Westergaard (4, 5) and by
Westergaard, et al. (34). However, they never found
injured endothelial cells which could have been responsible
for the passage of HRP across the endothelium. The observed reaction product did not form a continuous column
between two adjacent endothelial cells, from the first
luminal to the first abluminal tight junctions. Therefore,
the opening of tight junctions by the high blood pressure
and the subsequent interendothelial movement could be
excluded. Furthermore, opened tight junctions were never
seen. However, the reaction product was present in endothelial vesicles, the number of which was markedly increased. The vesicles were open to the vessel lumen or
to the subendothelial basement membrane or were freely
situated in the cytoplasm of the endothelial cells(figs.1-4)
Chains of vesicles, simultaneously open to the vessel
lumen and the subendothelial basement membrane, were never
observed. In addition, channels through the endothelial
cells were not seen. Based on these observations it was
assumed that acute hypertension increased the normally
occurring vesicular transfer of HRP. Since the tracer
used is a plant enzyme with a molecular weight of 40,000,
it is likely that the plasma proteins could behave in the
same way. At this point, it should be mentioned that
experiments in our laboratories have shown that HRP which
circulates freely in the blood stream does not break down
into smaller components by the peroxidase activity
(Bundgaard and Moleer, in preparation). Moreover, Westergaard and coworkers carried out control experiments in
order to elucidate the influence of the high systemic
blood pressure per se. When Aramine was given silultan-
eously with Regitin (an α-blocker) to avoid the elevation

Fig. 2-4.   Electron micrographs of parts of vessel walls
             containing HRP. The lumen (lu) is upwards.
             This is followed by the endothelial cell(s)
             (en). The subendothelial basement membrane
             (bm) is filled with reaction product. The
             pictures are from rats with acute, systemic
             chemically induced hypertension.

Fig. 2.   The luminal surface of the endothelia cells
             is covered by a thick layer of reaction
             product. There is no HRP from the first
             luminal to the first abluminal tight junctions
             (arrows). X 118,800

Fig. 3.   The luminal, endothelial plasma membrane
             forms deep invaginations (arrows). Maybe
             vesicles have been under formation. The
             extracellular spaces of the neuropil (np)
             are filled with HRP.   X 82,200

Fig. 4.   The luminal, endothelial plasma membrane
             forms deep invaginations (arrows 1, 2). The
             endothelial cell contains vesicles filled
             with reaction product. One of these (arrow 3)
             is open to the basement membrane. Chains of
             vesicles (arrows 4, 5) appear to be open to
             the basement membrane.   X 105,900
             (Reduced 10% for reproduction)

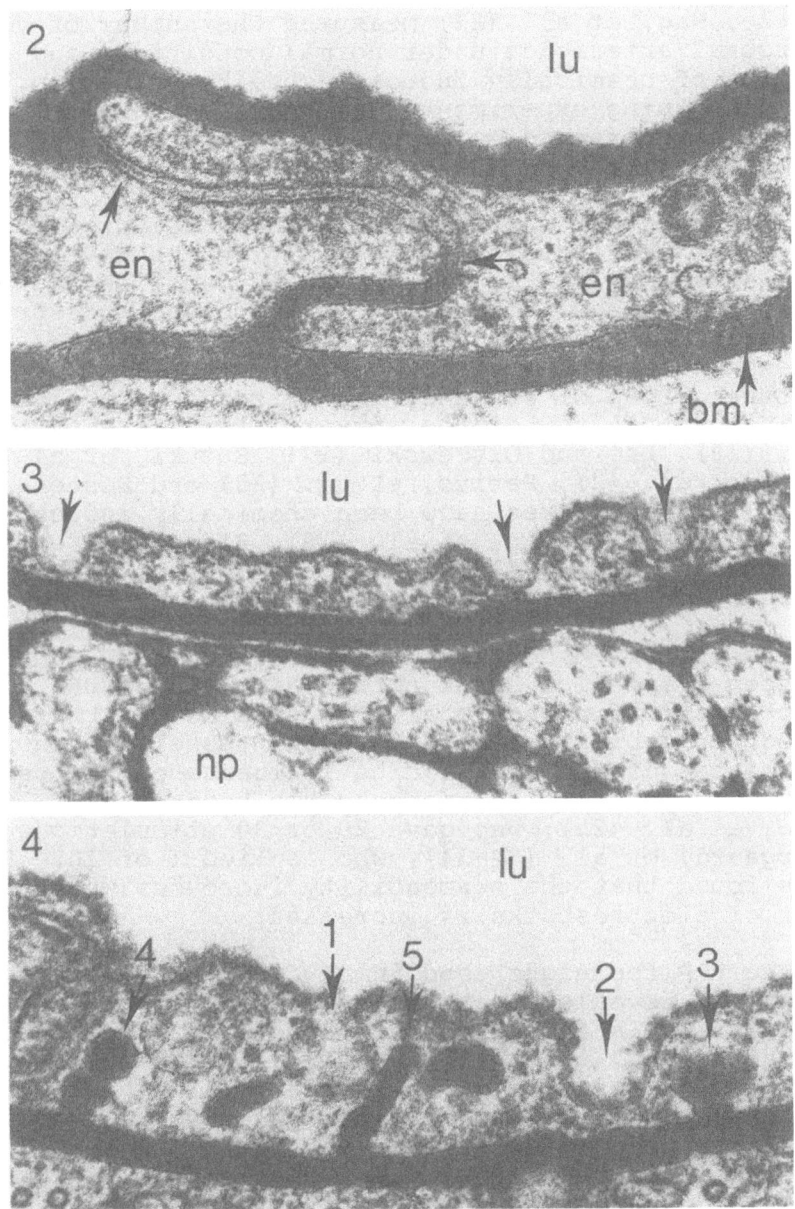

of blood pressure, the increase in the permeability to
HRP was absent. Therefore, we strongly suggest that the
systemic blood pressure increases the local blood pressure in the brain, resulting in enhanced permeability.
Recently, Nag, et al. (22) measured the number of vesicles
in cerebral arterioles under normal conditions and after
induction of chemically induced acute hypertension.
Their convincing experiments demonstrated that the number
of vesicles increases by a factor of 8 when acute hypertension is evoked.

The effect of electrically induced seizures (electroshocks) has been studied extensively during the last
decade after application of the newly developed methods
(tracer-techniques, monitoring of the blood pressure in
small animals, measuring arterial pH, pCO_2 and pH, and
the xenon-technique for controlling cerebral blood flow).
In another group of studies, the seizures were electrically induced. A few of the investigations are by Lending,
et al. (18), Lee and Olszewski (17), Suzuki, et al. (27),
Bolwig, et al. (2), Petito, et al. (25) and Westergaard,
et al. (33). Seizures have been chemically induced in
experiments by Lending, et al. (18), Eisenberg, et al.
(8), Lorenzo, et al. (19, 20), Hedley-Whyte, et al. (11)
and Johansson and Nilsson (15).

The tracers used for the evaluation of BBB changes
have been Evans blue, radioactive albumin and HRP. A
characteristic finding in all the studies was the increased permeability across cerebral vessels in the various regions of the brain. The influence on the permeability of repeated electroshocks has been studied by
Petito, et al. (25), who gave 20 or 30 stimulations, and
Westergaard, et al. (32-34), who applied 1 or 10. Both
groups found that the permeability increases when the
number of electroshocks is increased.

The HRP technique used in studies where seizures
were evoked revealed that the reaction product was in the
arteriolar walls and in the surrounding neuropil's extracellular spaces. The pathways taken by HRP across the
BBB are important from physiological, pathophysiological
and therapeutical points of view. Electron microscopial
examination never demonstrated any damage to the endothelial cell in the HRP studies. Therefore, another
pathway or pathways for transport of HRP across the BBB
must exist. The possible routes could be movement through
the cleft between endothelial cells after disruption of
tight junctions, through channels simultaneously open to
the vessel lumen and to the subendothelial basement

membrane, or vesicular transfer. However, channels were not observed in any of the cited studies. The most significant amount was believed, by Lorenzo, et al. (19) and Hedley-Whyte, et al. (11), to pass interendothelially in contrast to Petito, et al. (25), who considered vesicular transfer to be the main mechanism. Westergaard, et al. (33) assume that vesicular transfer is the only pathway, since the reaction product never formed a continuous line from the first luminal to the first abluminal tight junctions. The reaction product occurred in vesicles open to the lumen, open to the subendothelial basement membrane, or freely situated in the cytoplasm of the endothelial cells. With the purpose of elucidating the factors that might influence the effect of electroshock, the arterial pH, pCO_2, pO_2 and blood pressure were measured and observed within normal ranges. The cerebral blood flow was significantly increased. This observation, in addition to the finding that the arterial, systemic blood pressure was markedly increased, allows the hypothesis that electrical stimulation raises the systemic blood pressure, which then elevates the local blood pressure in the brain and therefore increases the cerebral blood flow. That this may be the situation is suggested by the fact that electroshocks given to rats with transected spinal cord (cervical part) do not increase the systemic blood pressure and do not change the cerebral blood flow. In addition, when the spinal cord transection precedes the stimulation of the brain, the permeability to HRP is exactly the same as under control conditions.

In conclusion, it has been demonstrated that acute hypertension and chemically or electrically-induced seizures increase the local cerebral blood pressure, the cerebral blood flow, and the permeability to proteins across cerebral vessels. The mechanism by which blood pressure influences the plasma membranes of endothelial cells in smaller arterioles to increase the formation of endothelial vesicles remains to be explained.

REFERENCES

1. Blomstrand, C., Johansson, B. and Rosengren, B. (1975) Blood-brain barrier lesions in acute hypertension in rabbits after unilateral x-ray exposure of brain. Acta neuropath. (Berl.) 31: 97-102.

2. Bolwig, T.G., Hertz, M.M. and Westergaard, E. (1977): Acute hypertension causing blood-brain barrier break-

down during epileptic seizures. Acta neurol. scand.
56: 335-342.

3. Brightman, M.W. and Reese, T.S. (1969): Junctions
    between intimately apposed cell membranes in the
    vertebrate brain. J. Cell Biol. 40: 648-677.

4. Brøndsted, H.E. and Westergaard, E. (1975): Vesic-
    ular transport of proteins from blood to brain
    during acute hypertension. Acta physiol. scand.
    95:   67 A-68 A.

5. Brøndsted, H.E. and Westergaard, E. (1978):
    Increased vesicular transport of horseradish perox-
    idase across the blood-brain barrier after chemical
    induction of hypertension. In: Advances in
    Neurology, J. Cervós-Navarro, et al. (eds.), Vol. 20,
    pp. 347-348, Raven Press, New York.

6. Clasen, R.A., Pandolfi, S., Laing, I. and Casey, D.
    (1974): Experimental study of relation of fever to
    cerebral edema. J. Neurosurg. 41: 576-581.

7. Ehrlich, P. (1887): Zur therapeutischen Bedeutung
    der substituirenden Schefelsäuregruppe. Therapeut.
    Mh. 1: 88-90.

8. Eisenberg, H.M., Barlow, C.F. and Lorenzo, A.V.
    (1970): Effect of dexamethasone on altered brain
    vascular permeability. Arch. Neurol. 23: 18-22.

9. Hansson, H.-A, Johansson, B. and Blomstrand, C.
    (1975): Ultrastructural studies on cerebrovascular
    permeability in acute hypertension. Acta neuropath.
    (Berl.) 32: 187-198.

10. Häggendal, E. and Johansson, B. (1972): On the
    pathophysiology of the increased cerebrovascular
    permeability in acute arterial hypertension in cats.
    Acta neurol. scand. 48: 265-270.

11. Hedley-Whyte, E.T., Lorenzo, A.V. and Hsu, D.W.
    (1976): The role of arteries and endothelial
    vesicles in breakdown of the blood brain barrier
    with drug-induced seizures. J. Neuropath. exp.
    Neurol. 35: 331.

12. Johansson, B. (1974): Blood-brain barrier dysfunc-
    tion in acute arterial hypertension. Thesis. Printed
    in Sweden, Gotab, Kungälv 74.12707.

13.   Johansson, B. (1974): Blood-brain barrier dysfunc-
      tion in acute arterial hypertension after papaver-
      ine-induced vasodilation. Acta neurol. scandinav.
      50: 573-580.

14.   Johansson, B., Li, C.-L., Olsson, Y. and Klatzo, I.
      (1970): The effect of acute arterial hypertension
      on the blood-brain barrier to protein tracers.
      Acta neuropath. (Berl.) 16: 117-124.

15.   Johansson, B. and Nilsson, B. (1977): The patho-
      physiology of the blood-brain barrier dysfunction
      induced by severe hypercapnia and by epileptic
      brain activity. Acta neuropath. (Berl.) 38: 153-158.

16.   Klatzo, I., Miquel, J. and Otenasek, R. (1962): The
      application of fluorescein labeled serum proteins
      (FLSP) to the study of vascular permeability in the
      brain. Acta neuropath. (Berl.) 2: 144-160.

17.   Lee, J.C. and Olszewski, J. (1961): Increased cere-
      brovascular permeability after repeated electroshock.
      Neurology 11: 515-519.

18.   Lending, M., Slobody, L.B. and Mestern, J. (1959):
      Effect of prolonged convulsions on the blood-cerebro-
      spinal fluid barrier. Am. J. Physiol. 197: 465-468.

19.   Lorenzo, A.V., Hedley-Whyte, E.T., Eisenberg, H.M.
      and Hsu, D.W. (1975): Increased penetration of
      horseradish peroxidase across the blood-brain barrier
      induced by Metrazol seizures. Brain Res. 88: 136-140.

20.   Lorenzo, A.V., Shirahige, I., Liang, M. and Barlow,
      C.F. (1972): Temporary alteration of cerebrovascular
      permeability to plasma protein during drug-induced
      seizures. Am. J. Physiol. 223: 268-277.

21.   Nag, S., Robertson, D.M. and Dinsdale, H.B. (1977):
      Cerebral cortical changes in acute experimental
      hypertension. An ultrastructural study. Lab. Invest.
      36: 150-161.

22.   Nag, S., Robertson, D.M. and Dinsdale, H.B. (1979):
      Quantitative estimate of pinocytosis in experimental
      acute hypertension. Acta neuropath. (Berl.) 46:
      107-116.

23. Olsson, Y. and Hossmann, K.-A. (1970): Fine struc-
     tural localization of exudated protein tracers in
     the brain. Acta neuropath. (Berl.) 16: 103-116.

24. Olsson, Y., Klatzo, I., Sourander, P. and Steinwall,
     O. (1968): Blood-brain barrier to albumin in embry-
     onic new-born and adult rats. Acta neuropath. (Berl.)
     10: 117-122.

25. Petito, C.K., Schaefer, J.A. and Plum, F. (1977):
     Ultrastructural characteristics of the brain and
     blood-brain barrier in experimental seizures.
     Brain Res. 127: 251-267.

26. Reese, T.S. and Karnovski, M.J. (1967): Fine
     structural localization of a blood-brain barrier to
     exogenous peroxidase. J. Cell Biol. 34: 207-217.

27. Suzuki, O., Takanohashi, M. and Yagi, K. (1976):
     Protective effect of dexamethasone on enhancement
     of blood-brain barrier permeability caused by
     electroconvulsive shock. Arzneim.-Forsch. (Drug
     Res.) 26: 533-535.

28. Tschirgi, R.D. (1950): Protein complexes and the
     impermeability of the blood-brain barrier to dyes.
     Am. J. Physiol. 163: 756.

29. Westergaard, E. (1974): Transport of protein tracers
     across cerebral arterioles under normal conditions.
     In: Pathology of Cerebral Microcirculation,
     J. Cervós-Navarro, (ed.), pp. 218-227, Walter de
     Gruyter, Berlin-New York.

30. Westergaard, E. (1977): The blood-brain barrier to
     horseradish peroxidase under normal and experimental
     conditions. Acta neuropath. (Berl.) 39: 181-187.

31. Westergaard, E. and Brightman, M.W. (1973): Trans-
     port of proteins across normal cerebral arterioles.
     J. comp. Neurol. 152: 17-44.

32. Westergaard, E. and Brøndsted, H.E. (1975): The
     effect of acute hypertension on the vesicular trans-
     port of proteins in cerebral vessels. In: Proc.
     VII Int. Congress of Neuropathology, Budapest 1974,
     pp. 619-622, Excerpta Medica, Amsterdam.

33. Westergaard, E., Hertz, M.M. and Bolwig, T.G. (1978):
     Increased permeability to horseradish peroxidase
     across cerebral vessels, evoked by electrically
     induced seizures in the rat. Acta neuropath. (Berl.)
     41: 73-80.

34. Westergaard, E., van Deurs, B. and Brønsted, H.E.
     (1977): Increased vesicular transfer of horseradish
     peroxidase across cerebral endothelium, evoked by
     acute hypertension. Acta neuropath. (Berl.) 37:
     141-152.

THE PATHOLOGY OF INTRACEREBRAL ARTERIOLAR SPASM

W. Roggendorf and J. Cervós-Navarro

Institut für Neuropathologie
Klinikum Stegliz
Freie Universität Berlin
Federal Republic of Germany

INTRODUCTION

Spastic constriction of subarachnoid arteries is a frequent sequela of subarachnoid hemorrhage, especially after aneurysmal ruptures (13, 14, 2, 4). Since the early work of Dreszer & Scholz (11) and Echlin (12), there has been a controversy as to whether a spasm may also occur in the small intracerebral arteries or even in the arterioles. More recent ultrastructural studies by Matakas et al. (17) and Cervós-Navarro et al. (8) have revealed spastic constriction of the intracerebral arterioles under various experimental conditions. Therefore, it appeared of interest to determine the influence of vasospasm on the regional blood supply and to investigate the morphological changes of the vessel after metabolic and electric stimuli.

MATERIAL AND METHODS

In the first experimental group, six adult cats were anesthetized and artificially ventilated. Blood gases were kept normal, and the blood pressure was monitored. After a left side craniotomy, the indifferent electrode was attached to the right temporal muscle, and an agar-gel silver electrode was used to stimulate the left meningeal vessels through the intact arachnoidea. The animals were killed 20 minutes after the last stimulation by a rapid injection of 50 ml carbon suspension into the inferior vena cava.

Specimens of the parietal lobe and a part of the caudate nucleus were processed for electron microscopy. The brain was then removed, fixed in formalin and treated as previously described (8).

In the second experimental group, an electroshock was repeatedly applied in 9 animals by 2 metal electrodes attached to the temporal muscle on each side (220 V, 50 Hz for 500 msec) as previously described (17).

Two groups of cats were used as controls. The first group consisted of 6 cats, which were treated as above, but the cerebral vessels were not subjected to electric stimulation and the animals were not artificially ventilated. The second group consisted of 9 cats in which respiratory alkalosis was produced by hyperventilation (9), but the brains were not perfused with carbon suspension.

RESULTS

Macroscopic and Light Microscopic Findings of Spastic Constriction. In the first experimental group, there was no clear correlation between the intensity of the direct current (DC) and pial vessel reaction. Whenever the gel electrode was brought into direct contact with the adventitial surface of a pial artery, a DC of 5 V for 2 sec (which corresponded to about 10 mA) was sufficient to produce a powerful constriction of the vessel (Fig. 1). However, for the induction of a constant and spastic constriction of arterial vessels, a DC of 20 - 100 V for 30 sec (20 - 150 mA) was necessary. Stimulation was first performed with 20 V. When none or only moderate constrictions of pial arteries occurred, stimulation was repeated with 40 V. If again the vessel reaction seemed only moderate, 100 V were applied. There were 2 trials in 2 of the animals and 3 in 4 of the animals. When the spastic constriction occurred, its location was usually not influenced by movement of the electrode along the same vessel. Besides the segmental constrictions, a quick and marked reaction of the branching segments of arteries was observed. The smaller branch of the external segment, just distal to the branching point, showed a prompt constriction upon electric stimulation. Moderate constrictions varied in duration, while spastic constrictions usually persisted the entire 20 minutes of the experimental period.

In the second experimental group, all the brains showed a moderate swelling within a few minutes after

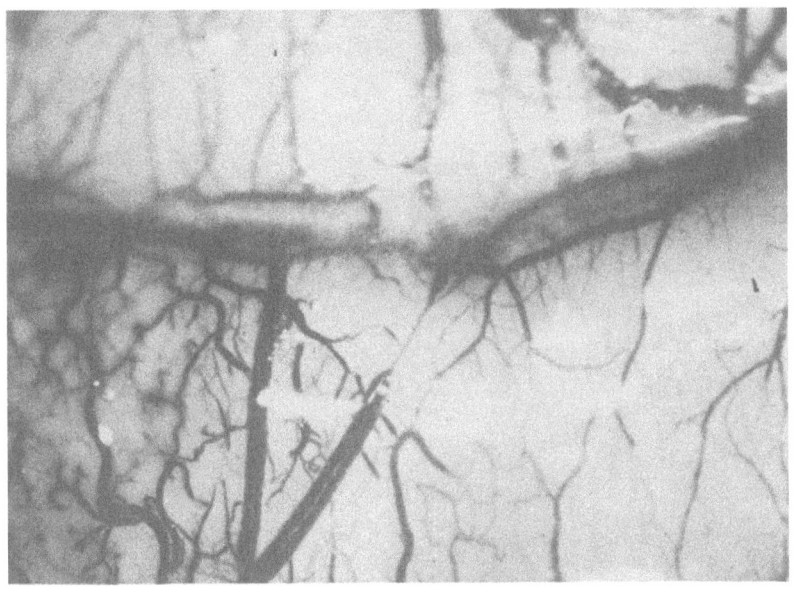

Fig. 1.   Segmental spasm of pial artery about 5 min after
          electric convulsive treatment, the cat still
          living.

current application.   The brain volume increased visibly.
In the area of stimulation, the brain surface protruded
out of the skull; however, the brain swelling did not
last longer than 5 to 10 min, and the brain volume decreased to normal before the animals were sacrificed.
In 5 cases, the large pial arterial vessels observed
through the skull window reacted strongly to the current.
In 3 cases, powerful constriction occurred, whereas in
the other two, the constriction was moderate.   The constrictions appeared within 1 to 3 min after current
applications and did not disappear until the animals were
sacrificed.   The spasm was segmental, varying from 0.5
to 5 mm long, and occurred only in 1, or at most, 2
arterial vessels.

     In two-thirds of the animals belonging to both
groups, small patches of the cerebral cortex free of the
carbon staining were observed mainly in the frontal or
temporal region of the brain on either side (Fig. 2a).

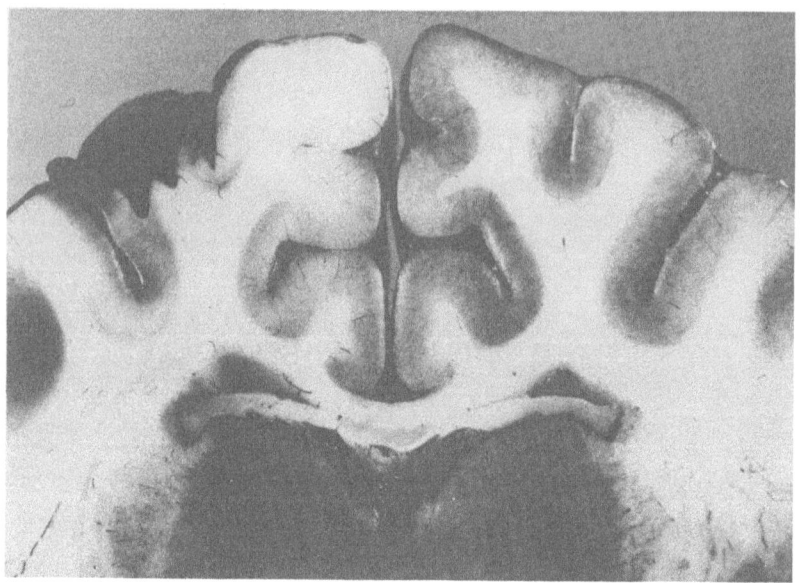

Fig. 2a.   20 min after electric convulsive treatment.
           There are areas with no blackening by the
           carbon suspension in both hemispheres.

        Frozen sections (200 μm thick) revealed variable
diameters of all large and medium-sized parenchymatous
vessels. Each animal had some narrowed segments which
varied in length and were distributed irregularly.
Capillaries showed no variations in diameter. The spastic
segments were exclusively observed in the gray matter of
the cortex or basal ganglia (Fig. 2b). There were no
significant differences between the two experimental
groups. Furthermore, the occurrence of spastic segments
was no higher in the frontal part of the brain at the
site of electric stimulation than in the rest of the
brain.

        In semi-thin sections, most of the arterial vessels
were found to be moderately constricted but not collapsed.
It was estimated that spastic constriction occurred in
not less than 1% and not more than 5% of all the cross-
sectioned vessels. There were no significant differences
among the individual animals or beween the two experimental groups. Venules were also constricted, but not
collapsed. The perivascular zone of many muscle containing

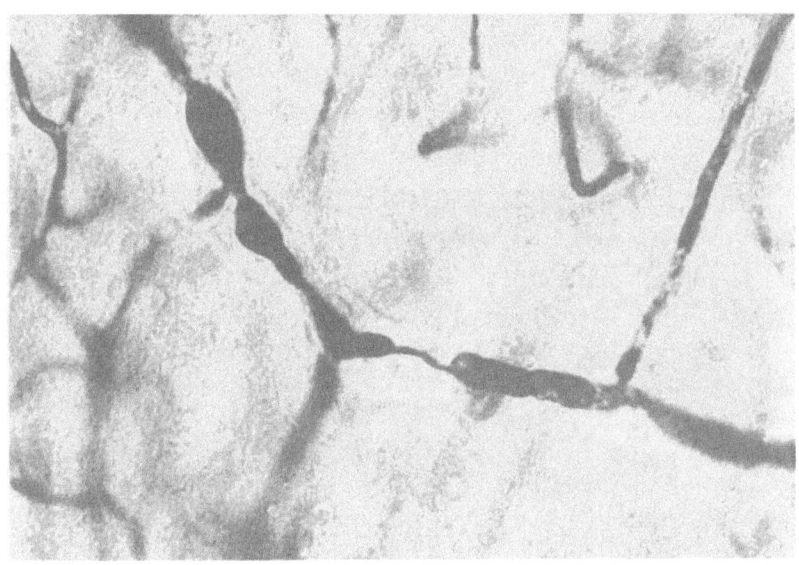

Fig. 2b. Frozen sections of brain cortex. The large
vessel shows some constriction and a segmental
spasm. Capillaries exhibit discontinuous
filling by the carbon suspension due to the
presence of erythrocytes, but no change in
diameter has occurred. Frontal lobe. x 200.
(Reduced 10% for reproduction)

vessels was destroyed, but this never occurred around
capillaries, nor was there any general edema of the
tissue.

Electron Microscopy of Physiological Constriction.
In the following description, we refer to spastic constriction or spasms when a strong deformation of the
entire vessel wall leads to a completely obstructed
vessel, in contrast to the physiological constriction,
in which the profile of the vessel is retained. In the
constricted vessels the endothelial cells overlapped at
the site of endothelial tight junctions. In contracted
arterioles (second control groups), these tight junctions
were set upright in such a position that they ran per-
pendicular to the basal lamina. The endothelial cell
nuclei showed a protrusion far into the vessel's lumen
during physiological constriction (Fig. 3). The subendothelial space was extremely undulated. The media

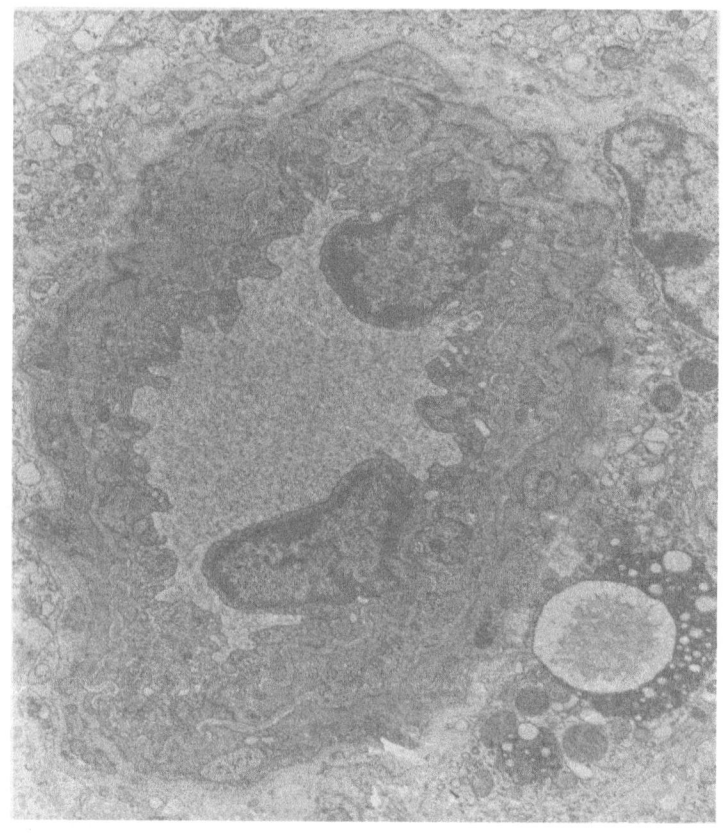

Fig. 3.   Normally constricted arteriole of cortex.
          Nuclei of endothelial cells protrude into the
          vessel lumen. The subendothelial space is
          undulated. Control group 2, fixation by
          immersion. x 5,200.
          (Reduced 10% for reproduction)

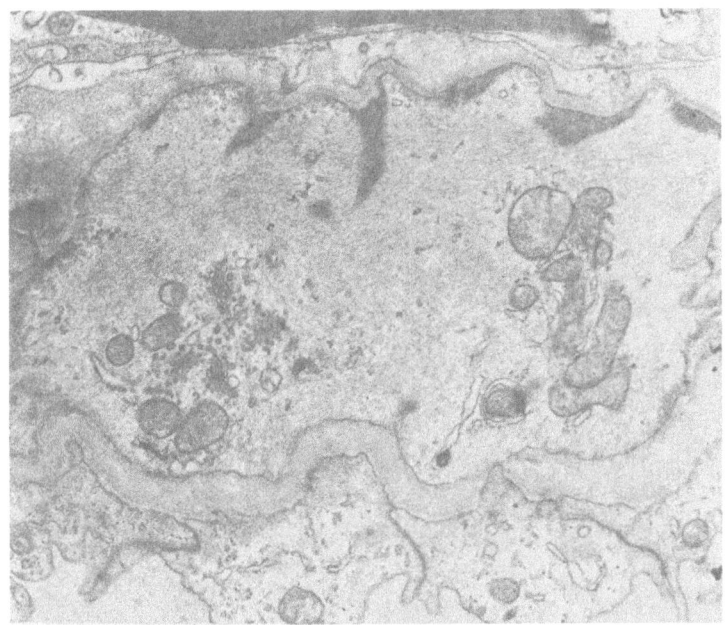

Fig. 4. Smooth muscle cell with physiological constriction. The abluminal cell membrane is
extremely scalloped and shows deformation by
attachment devices. Control group 2,
immersion fixation. x 13,500.
(Reduced 10% for reproduction)

showed moderate deformation of smooth muscle cells, compression and indentation of the cell nuclei. The muscle
cell membrane was extremely scalloped and showed characteristic deformation of the prominent attachments: their
broad bases were situated in the cell membrane and extended into the cell like a supporting belt. The cytoplasm with the pinocytotic vesicles lined up along the
cell membrane bulged out between the attachment devices
toward the abluminal side (Fig. 4).

The perivascular space was normal without perivascular glial sheath damage.

Electron Microscopy of Spastic Constriction. Areas
with spastic or moderately constricted vessels in semithin sections were selected for electron microscopy.
A total number of 50 arterial vessels of both experimental
groups, ten of which were spastic, were observed under

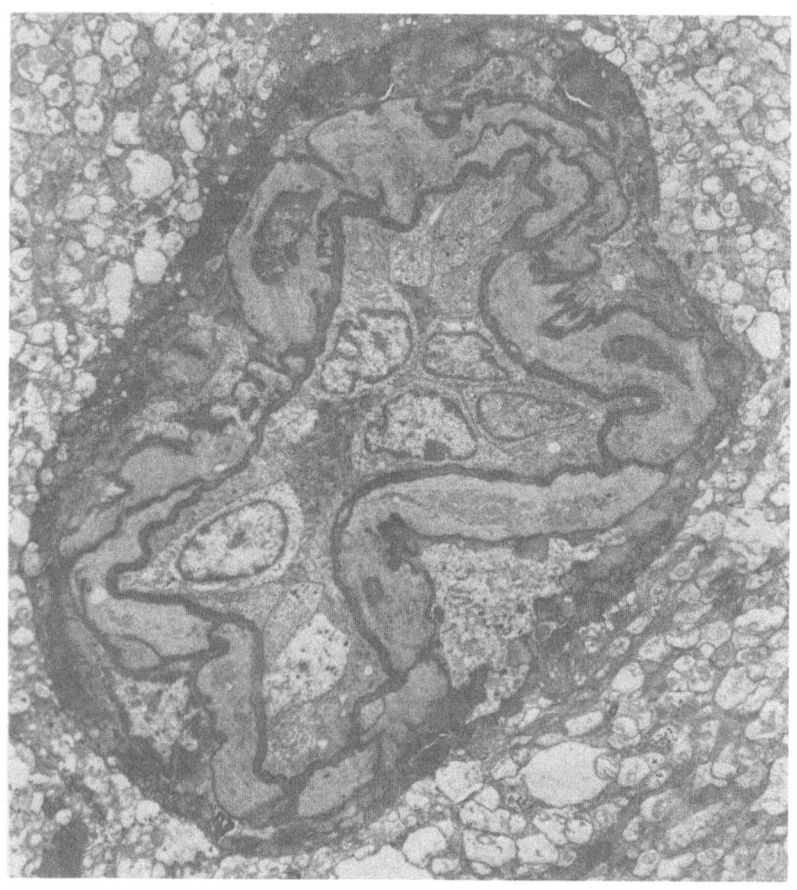

Fig. 5. Intercerebral arteriole in spasm. The lumen
         of the vessel is nearly completely collapsed.
         The endothelial cells touch each other.
         Experimental group 1. x 5,520.
         (Reduced 10% for reproduction)

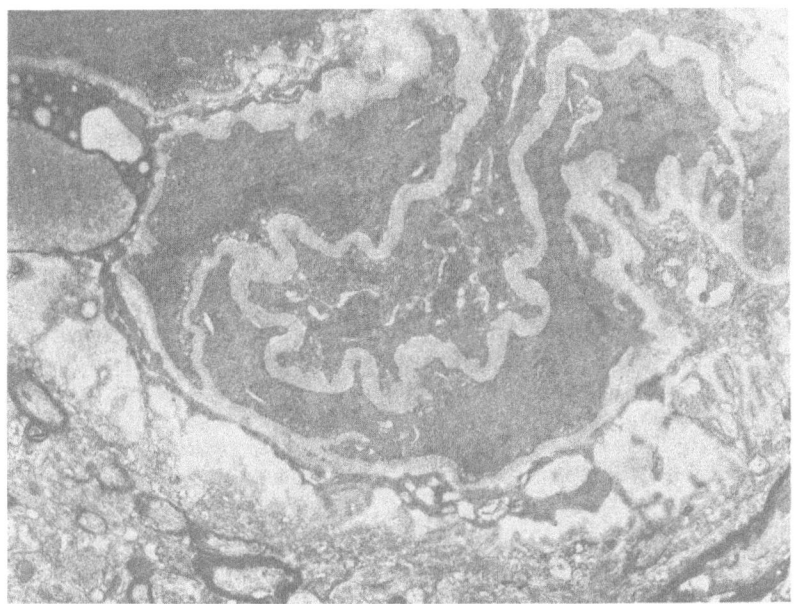

Fig. 6.   Arteriole from cortex after stimulation by
          direct current. The lumen is completely col-
          lapsed, extreme deformation of smooth muscle
          cells. Narrow adventitial branches as well
          as collagen can be found regularly. Some
          glial processes protrude into the peri-
          vascular space. x 8,500.
          (Reduced 10% for reproduction)

the electron microscope. The muscular sheath of all
vessels was one-, two- or three-layered. Spastic constriction occurred only in vessels with a muscular
sheath. In spastic vessels, the wall appeared convoluted.
The overall profile of these vessels was, in most cases,
star-shaped, the whole vessel wall having two, three or
four indentations. The lumen was reduced to a small
cleft unless white or red blood cells had been trapped
within.

     Intima: The round oval endothelial cell nuclei
in both groups had a narrow cytoplasmic border and protruded far into the vessel lumen. The endothelial cells
from the opposite sides of the intimal circumference
touched each other, often with an even surface. On
cross-sections, the inner basal lamina was convoluted

and appeared, in some cases, thickened and more homogeneous. Myoendothelial tight junctions were normal in
number and appearance.

In contrast to the endothelium, the media differed
conspicuously in both groups. The intracerebral arterioles of group one showed a completely occluded lumen,
while endothelial cells, subendothelial space and smooth
muscle cells were only slightly scalloped (Fig. 5). The
nucleus of smooth muscle cells was moderately scalloped.
The abluminal membrane of smooth muscle cells was flattened in large segments. The triangular attachment
devices were situated on the cell membrane without
leading to deformation characteristic of simple contrac-
tion. Occasionally, there was a narrow, electron-dense
marginal border, which was possibly formed by contiguous
attachments. Besides spastically contracted vessels,
one also found sections where the vessel wall was only
moderately folded in.

In the arterioles of the 2nd group that were
spastically contracted by DC, a bizarre and extreme
deformation of smooth muscle cells occurred (Fig. 6). It
is here that the normally regular width of 0.15 to
0.20 µm between the smooth muscle cells became irregular
and sometimes doubled. The muscle cells were very narrow
and formed spicular cell processes both luminally and
abluminally. There was also no deformation of the abluminal cell membrane by the attachments. Myomyal tight
junctions were rare and limited to narrow muscle processes.

The perivascular space in all groups appeared as a
gap. Adventitial vascular nerve endings were frequently
observed. In both groups, about 2/3 of all the cross-
sectioned arterial vessels displayed concentric zones
of perivascular damage of the neuropil. The basement
membrane and a small neighboring ring of the neuropil
appeared intact. The injured tissue was always outside
of these small rings but was not present in all cases in
which a spastic vessel was observed. The injury was even
observed around moderately constricted vessels. It was
never found around capillaries. In only a few cases was
there severe perivascular astrocytic edema around capillaries. In these cases, the capillary lumen was patent.

DISCUSSION

Dreszer & Scholz (11) produced cardiazol shocks in
cats and afterwards observed areas with decreased cerebral perfusion. They concluded that spastic occlusion

of small arterial vessels must be responsible for this
effect. Later, Echlin (12) observed the reaction of
pial vessels to electric shock stimulation through a
cranial window. He reported that an electric stimulus
produced segmental spasm of pial arteries. The drawings
of Echlin correspond exactly to our pictures. Authors
have referred to observations by Echlin in many studies
concerned with the effect of electric convulsive treatment on brain tissue (3, 15, 21, 22), but only Alexander
& Löwenbach (1) reported their own observations of marked
constrictions of arterial vessels which might have
corresponded to spasms.

It was most important in this study to verify that
the spastic constrictions of the vessel were not artificially produced by the methods used for the histological preparation of the tissue. The absence of spasm
in the control groups provides evidence that the reaction
was due to electric stimulation. Our experiments clearly
demonstrate that electric convulsive treatment and
direct current application cause spastic constriction
of pial, as well as parenchymatous, vessels.

Spastic constriction, as shown in our preparations,
differs from physiological constriction. The morphological criteria of physiologic vasoconstriction has
been established by Citters et al. (10) and Phelps and
Luft (18). The vascular lumen, after physiologic vaso-
constriction, is concentrically narrowed and the vessel
wall thickened. The smooth muscle cells show a corset-
like deformation of the abluminal side at the cell
membrane which never occurs in spastic vessels. Spastic
vessels, however, are not concentrically narrowed. In
contrast to the physiological constriction, the spasm
induced by electroshock or direct current is not a
characteristic deformation of the single smooth muscle
cell, but rather a folding and indentation of the whole
vessel wall that leads to an occlusion of the lumen.

A consistent and important observation was the
segmental nature of the spastic constriction in pial
vessels. This was already mentioned by Echlin (12).
Our histological findings in the brain tissue confirmed
this result for the parenchymatous vessels. It seems
probable that this segmental arrangement of spasm can be
attributed to the innervation and constriction mechanism
of the brain vessels. Nerve terminals or varicosities
in brain vessels are found only at rather long intervals
(7). Moreover, the innervation models of smooth muscle
cells in brain arterioles seem to be segmental. According

to Burnstock (6), only one smooth muscle cell is stimulated by the transmitter released out of the nerve
varicosty. A number of neighboring smooth muscle cells
are then stimulated by myomyal tight junctions so that
only a segment of the vessel wall reacts to neurogenic
stimuli. Support for this assumption seems to lie in
the fact that the segments of the small arterial vessel
wall, just distal to the point where they branch off a
large vessel, are particularly susceptible to vasoconstriction due to widespread electric stimulation. Rhodin
(20) observed that myomyal tight junctions are particularly frequent at these branching points.

Whenever a vessel was spastic, as observed electron-
microscopically, the lumen of these vessels certainly did
not allow any perfusion. However, it is difficult to esti
mate what kind of physiological consequences may have
been due to the spastic segments of vessels. Physiological studies, both in man and animals, revealed that
global cerebral blood flow increases during electrically
induced seizures (16, 19). Brodersen et al., (5) found
an increase of cerebral blood flow of more than 200% and
a simultaneous increase of cerebral oxygen metabolic rate
of about 100%. However, the increase of blood flow and
metabolic rate was of short duration. In the experiments
of Brodersen, the blood flow dropped to subnormal and the
metabolic rate to normal values within 3 min and 15 min,
respectively, after electric convulsive treatment. It
must also be considered that an increase of global cerebral blood flow does not exclude the possibility that
circumscribed areas of the brain are underprofused. The
final solution of these problems needs methods more subtle
and exact than those which have been used so far.

REFERENCES

1.   Alexander, L. and Löwenbach, H. (1944): Experimental
     studies on electro-shock treatment: the intracerebral
     vascular reaction as an indicator of the path of the
     current and the threshold of early changes within the
     brain tissue. J. Neuropath. Exp. Neurol. 3: 139-171.

2.   Allcock, J. M. (1966): Arterial spasm in subarachnoid
     hemorrhage. A clinical and experimental study. Acta
     Radiol. (diagnosis) 5: 73-83.

3.   Alpers, B. J. and Hughes, J. (1942): Changes in the
     brain after electrically induced convulsions in cats.
     Arch. Neurol. Psychiatr. 47: 385-398.

4. Arutiunow, A. I., Baron, M. A. and Majorova, N. A. (1970): Experimental and clinical study of the development of spasm of cerebral arteries related to subarachnoid hemorrhage. J. Neurosurg. 32: 617-625.

5. Brodersen, P., Paulson, O. B., Bolwig, T. G., Rogon, Z. E., Rafaelsen, O. and Lassen, N. A. (1973): Cerebral hyperemia in electrically induced epileptic seizures. Arch. Neurol. (Chic.) 28: 334-338.

6. Burnstock, G. (1970): Structure of smooth muscle and its innervation. In: Smooth Muscle, E. Bülbring, A. F. Brading, A. W. Jones and T. Tomita (eds.), pp. 1-69, Edward Arnold, London.

7. Cervós-Navarro, J. and Matakas, F. (1974): Electron microscopic evidence for innervation of intracerebral arterioles in the cat. Neurology 24: 282-286.

8. Cervós-Navarro, J., Matakas, F., Roggendorf, W. and Christmann, U. (1978): The morphology of spastic intracerebral arterioles. Neuropath. Appl. Neurol. 4: 369-379.

9. Cervós-Navarro, J., Seipert, M., Valencak, E., Matakas, F. and Sieber, V. (1971): La morphologie ultrastructurale du S.N. dans "la perfusion de Luxe" de l'acidose respiratoire. Neurochirurgie (Paris) 17, 1: 595-600.

10. Citters van, R. L., Wagner, B. M. and Rushmer, R. F.: (1962): Architecture of small arteries during vaso-constriction. Cir. Res. 10: 668-675.

11. Dreszer, R. and Scholz, W. (1939): Experimentelle untersuchungen zur frage der hirndurchblutungs-störungen beim generalisierten krampf. Z. ges. Neurol. Psychiatr. 164: 140.

12. Echlin, F. A. (1942): Vasospasm and focal cerebral ischemia. Arch. Neurol. Psychiatry 47: 77-96.

13. Echlin, F. A. (1965): Spasm of the basilar and vertebral arteries caused by experimental sub-arachnoid hemorrhage. J. Neurosurg. 23: 1-11.

14. Lende, R. A. (1960): Local spasm in cerebral arteries. J. Neurosurg. 17: 90-103.

15.   Lidbeck, W. L. (1944): Pathological changes in the
      brain after electric shock. J. Neuropath. Exp.
      Neurol. 3: 81-86.

16.   Lovett Doust, J. W., Barchia, R., Lee, R. S. Y.
      Little, M. H. and Watkinson, J. S. (1974): Acute
      effects of ECT on the cerebral circulation in man.
      Europ. Neurol. 12: 47-62.

17.   Matakas, F., Cervós-Navarro, J., Roggendorf, W.,
      Christmann, U. and Sasaki, S. (1977): Spastic
      constriction of cerebral vessels after electric
      convulsive treatment. Fortschr. Psychiatr.
      Nervenhlk. 224: 1-9.

18.   Phelps, P. C. and Luft, J. H. (1969): Electron
      microscopical study of relaxation and constriction
      in frog arterioles. Am. J. Anat. 125: 399-428.

19.   Plum, F., Posner, J. B. and Troy, B. (1968): Cerebral
      metabolic and circulatory responses to induced con-
      vulsions in animals. Arch. Neurol. (Chic.) 18:
      1-13.

20.   Rhodin, J. A. G. (1967): The ultrastructure of
      mammalian arterioles and precapillary sphincters.
      J. Ultrastruct. Res. 18: 181-223.

21.   Scholz, W. and Jötten, J. (1951): Durchblutungs-
      störungen im katzenhirn nach kurzen serien von
      elektrokrämpfen. Arch. Psychiatr. Z. Neurol. 186:
      264-279.

INTRAVASCULAR AGGREGATION OF PLATELETS

AND CEREBROVASCULAR INSUFFICIENCY

Tsukasa Fujimoto, Yutaka Inaba;
Takeshi Motomiya and Hiroh Yamazaki

Department of Neurosurgery
Tokyo Medical and Dental University
Tokyo, Japan

Division of Cardiovascular Research
Tokyo Metropolitan Institute of Medical Science
Tokyo, Japan

Platelet aggregation undoubtedly plays a major role in the pathogenesis of many types of ischemic cerebrovascular disease. When platelet aggregates occlude a vessel's lumen, the vasoactive substances released from platelets, such as 5-hydroxytryptamine (serotonin), histamine and prostaglandin endoperoxide, may play a role in the genesis of the ischemic disease. Among these substances, prostaglandin endoperoxide and thromboxane A_2 have recently drawn attention, because of their ability to contract the vessels and aggregate platelets. These prostaglandins (PGs) are produced from the arachidonic acid that probably originates from phospholipids in the platelet's membrane.

Furlow and Bass (4) injected arachidonic acid into the carotid artery of rats and observed an appearance of intravascular platelet thrombi in the brain and a dramatic suppression of electroencephalographic activity, although all neural elements appeared normal except for an occasional hyperchromatic neuron. Since they did not observe any vascular lesions, we have examined the vascular changes induced by substances released from platelet aggregates using the same model.

MATERIALS AND METHODS

Thirty rabbits of Japanese white strain weighing 3.2 to 4.9 kg were used in this study. Animals were divided into two groups, consisting of 22 and 8 animals, respectively. Rabbits of the latter group were injected with anti-rabbit-platelet sheep serum repeatedly to reduce the platelet count to around $5 \times 10^4/\mu l$ (12).

All of the animals were anesthetized with 40-50 mg/kg of pentobarbital intraperitoneally, and additional doses given as needed. A tracheostomy was performed in order to support respiration, with the help of a Harvard respirator when necessary. The right carotid artery was exposed, a segment of P-10 polyethylene tubing was inserted within 3-4 minutes into the right lingual artery and its tip positioned at the bifurcation of the carotid artery for the injection of a bolus of sodium arachidonate solution (0.7 mg/kg). Internal carotid artery and main branches of the external carotid artery were left completely intact. After injection of 100 unit/kg of heparin, polyethylene tubing was inserted into the right femoral artery for continuous monitoring of the arterial blood pressure and periodic collection of blood. Another intubation of the internal jugular vein was also used for periodic blood sampling. The samples were collected immediately, 10, 30 and 60 minutes after the injection of the arachidonate.

Following the blood collection at 10 minutes, 2% Evans blue (0.7-1.0 ml/kg) was injected into the femoral artery. In 5 of the rabbits skull windows were made in the mid-parietal region bilaterally, and the brain surface was observed through this window.

At the end of the 60 minute experimental period the animals were perfused with 2.5% glutaraldehyde or 10% neutral paraformaldehyde solution through the left cardiac ventricle and ascending aorta under the pressure of 150 CmH$_2$O. Thereafter, a part of the middle, anterior and posterior arteries dissected from the subarachnoid space under the surgical microscope was prepared for scanning electron microscopical examination using Hitachi SEM. The paraformaldehyde-fixed brains were prepared for light microscopical examination using Hematoxylin and Eosin, MSB (Martius-Scarlet-Blue staining) and Tri-C (Masson-Goldner's Trichrome) staining methods.

The platelet count and volume were measured by Coulter counter model ZBI and Coulter channelyzer. EEG, ECG and respiration were monitored during the experiment.

RESULTS

Immediately after the injection of sodium arachidonate, the ipsilateral eye turned pale and became myotic, recovered slowly and was still visible at the end of the experiment. In 2 out of the 5 animals with skull windows, movement of small white substances (platelet aggregated mass) was seen and the narrowing or interruption of the blood stream was observed, especially near the arterial branching, in the first 10 to 30 minutes.

Moreover, immediately after the injection, the animals developed a Cheyne-Stokes type respiration which sometimes required the aid of the respirator. In most cases, the respiration recovered in 10 to 20 minutes and the respirator could be discontinued.

The pulse rate, which was reduced after the injection, returned to normal in a few minutes. The blood pressure initially was elevated but recovered in the later phase. However, 60% of the animals did not return to the pre-injection level until the end of the experiment.

A low voltage, slow wave EEG was exhibited several seconds after the injection on the ipsilateral hemisphere, and a slight but significant suppression of the EEG was observed on the opposite side. In 10 to 30 seconds after the injection, flat waves appeared on the right (ipsilateral) side, and low amplitude, slow waves were seen on the left side. Seventy percent of the animals displayed abnormal changes of EEG and, except in 5 cases, suppression of EEG continued until the end of the experiment.

The platelet counts were $51.7 \pm 8.4 \times 10^4/\mu l$ in the group without pre-treatment and $5.6 \pm 0.8 \times 10^4/\mu l$ in the thrombocytopenic group. After the injection of the arachidonic acid, platelet count decreased to $71 \pm 8\%$ but increased to the pre-injected value within 10 to 30 minutes. Similarly, the platelet's volume distribution curve shifted to the small size; then, again returned to the large size. These changes were absent in the thrombocytopenic group.

HISTOLOGICAL FINDINGS

I. Macroscopic findings

The penetration of Evans blue albumin complex was seen in 77% (17 out of 22) of the non-pretreated group

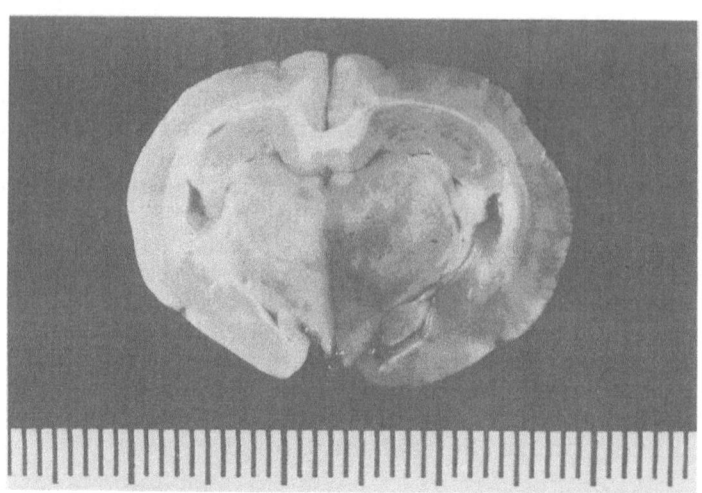

Fig. 1.   Marked extravasation of Evans blue albumin
           complex is seen only on the side with injected
           sodium arachidonate. Brain swelling is also
           clearly visible on the same side.

and in 50% (4 out of 8) of the thrombocytopenic group
(Fig. 1). The most marked extravasation was seen consistently in the region of the right middle cerebral
artery. However, a variable staining was also observed
in the area of the anterior and posterior cerebral artery.

    Generally, the extravasated Evans blue albumin stain
was more pronounced in the hippocampus and the basal
ganglia than in the cortex. Sometimes the extravasation
extended to the basal ganglia of the opposite side; however, it was not seen in the brain stem except occasionally in the upper part. The pons, medulla and spinal
cord were completely free of Evans blue staining.

II. Light microscopical findings

    On the injected side of the hemisphere, a marked
congestion was seen since many capillaries and small
vessels were filled with blood corpuscles. Extravasation
of an amorphous substance was observed in the middle or
large sized vessels (Fig. 2-1). Large platelet thrombi
were seldom seen, but adherence of the platelet to the
surface of the middle or large sized vessels was found
frequently.

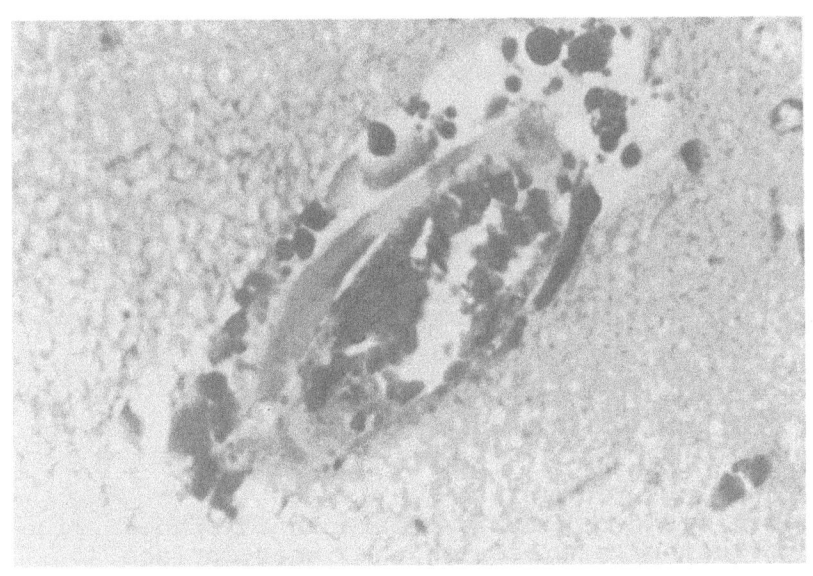

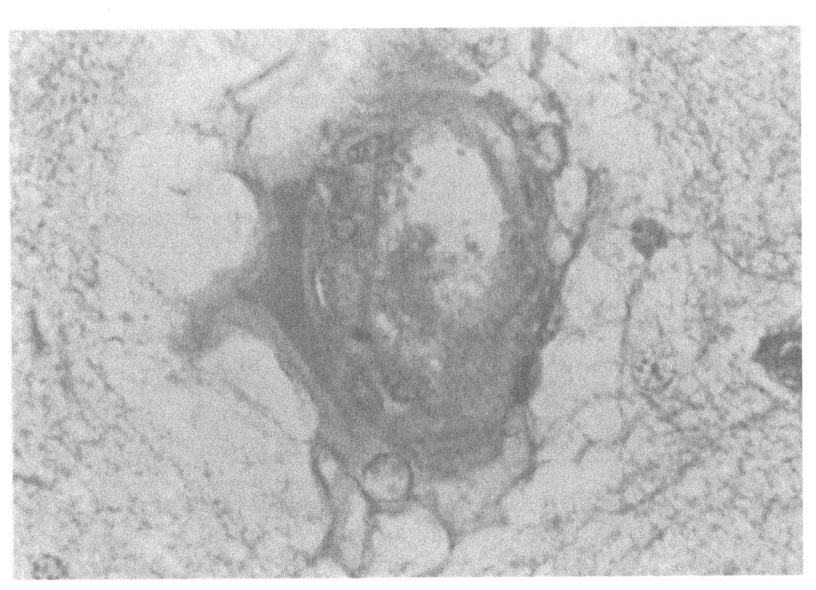

Fig. 2. Medium sized artery on the injected side of the brain. The inner surface of vessels adherent platelets. Extravasated amorphous substance, probably representing plasma, is also seen in the lumen of the vessel (Fig.2-1). X 300 Edematous changes with an irregularly thickened vessel wall (Fig. 2-2). X 300 (Reduced 10% for reproduction)

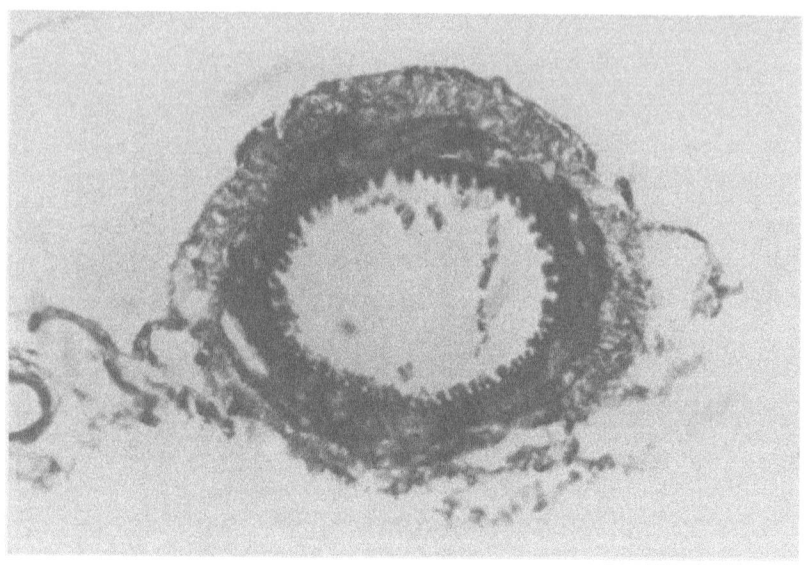

Fig. 3.   Large sized artery in subarachnoid space.   Fine
           protrusions are seen on the inner surface.
           X 200 (Reduced 10% for reproduction)

     Vascular wall damage, mainly of the media, was seen
frequently (Fig. 2-2), and the platelets adhered to the
inner surface of these vessels.   Thickened wall and undu-
lation of internal elastic layer were also found spora-
dically.   Some large arteries in subarachnoid space
showed regular and fine protrusions on the inner surface
(Fig. 3).   The surrounding parenchyma was vacuolated,
showing many dark cells with pycnotic nuclei which were
more marked in the blue stained area.   These changes were
rarely observed on the opposite side of the brain and in
the thrombocytopenic rabbits.

III.   Scanning electronmicroscopical findings

     On the inner surface of the artery on the injected
side, adherence of platelets was seen frequently.   In
some parts, adherent fibrin and red blood cells were also
seen.   The inner surface of the artery covered by platelet aggregates was indistinguishable from the surroundings
However, the area adjacent to the adherent platelets
showed a deranged endothelial lining (Fig. 4).   The surface revealed many protrusions of endothelial cells, and

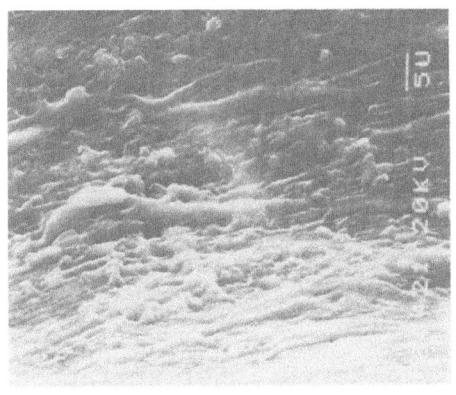

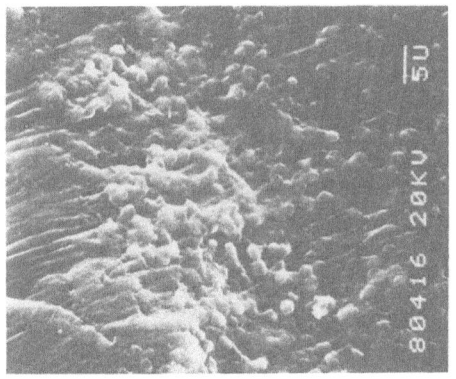

Fig. 4. Scanning electronmicroscopic findings of inner
         surface of the middle cerebral artery on the
         side of injection. The derangement of endo-
         thelial lining is seen below or near the
         adherent platelets. Platelets form round
         structures and aggregate in several places.

these cells appeared flat, showing a mosaic arrangement
in the markedly affected area (Fig. 5). These characteristic changes were not seen on the opposite side, except
for an occasional adherence of platelets (Fig. 6-1).
Moreover, they were seen infrequently in the thrombocytopenic group (Fig. 6-2).

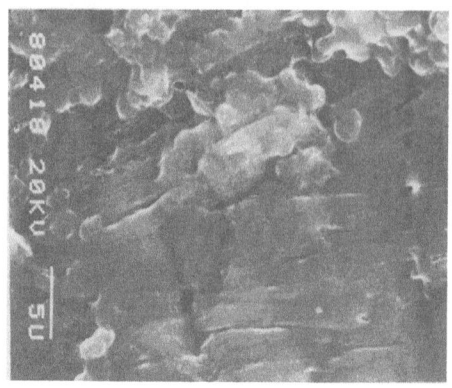

Fig. 5.   Markedly altered inner surface area of the
           middle cerebral artery on the injected side.
           Flattened endothelial cells with a mosaic
           arrangement are seen.

DISCUSSION

     Furlow and Bass produced cerebrovascular-occlusion
with intravascular platelet aggregates by injection of
arachidonic acid into the carotid artery of the rat
(4, 5, 6). Following these reports, we injected sodium
arachidonate into the bifurcation of the carotid artery
through the lingual artery without disturbing the blood
flow of the internal carotid artery.

     Similar changes in cardiorespiratory and electro-
encephalographic findings to those reported by Furlow
and Bass were observed in the present experiment. Also,
we have found injuries of cerebrovascular vessels associated with the intravascular platelet aggregates. The
characteristic findings of the vascular injuries were as
follows: (a) extravasation of the Evans blue albumin
complex on the injected side of the hemisphere indicating
a destruction of blood-brain barrier, which was observed
both macro- and microscopically; (b) derangement and
injuries of endothelial lining associated with adhesion
of platelet thrombi seen under the scanning electron
microscope; (c) injuries of medium and large sized arteries on the injected side of the brain, characterized by
an undulated internal elastic layer and a thickened wall.
These findings were only seen on the injected side and
especially in the region of the middle cerebral artery.

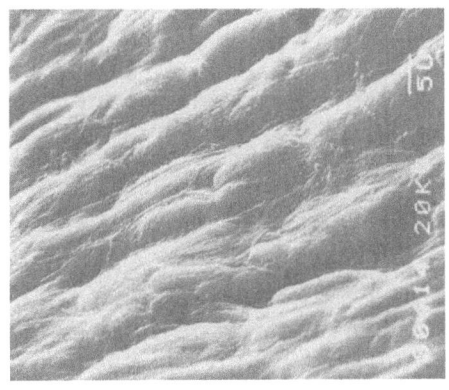

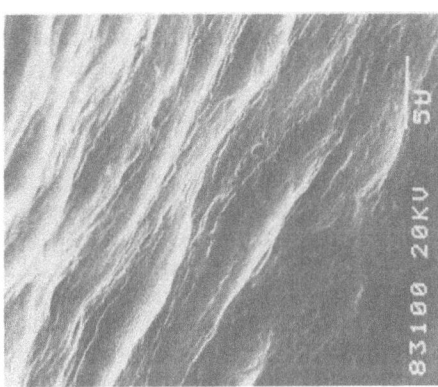

Fig. 6.   Inner surface of the middle cerebral artery of
          the uninjected (Fig. 6-1) and the injected side
          (Fig. 6-2) of thrombocytopenic animal.

Generally, these abnormal findings were fewer in the
thrombocytopenic than normal rabbits injected with sodium
arachidonate.

     Contrary to previous reports of extensive intravas-
cular platelet aggregates (4, 14), we have seen only
scattered platelet thrombi in the histological preparations, although white thrombi were observed in the cortical vessels immediately after the injection of arachidonic
acid.   Therefore, it is conceivable that the platelet
thrombi disaggregated or underwent lysis 60 minutes after
injection.   In contrast to these findings in the small
vessels and capillaries, injuries in the medium and large
sized arteries were observed as described above.   Previously,

we found that medium sized pulmonary arteries showed increased thickness of the wall, convoluted and separated
internal elastic lamina, vacuolated endothelial cells,
and vacuolated smooth muscle cells associated with platelet thrombi at 45 to 60 minutes after an intravenous injection with ADP (12). Platelet thrombi in capillaries
were scarcely seen in these preparations. As this phenomenon was not observed in the thrombocytopenic rabbits, we
have considered that the constriction and injuries of the
medium sized pulmonary arteries could have been induced
by a release of substances from platelets (12).

     In the present study, we have found injuries of
vessel walls in brain after the appearance of platelet
thrombi induced by an injection of arachidonic acid and
these changes were scarcely seen in the thrombocytopenic
rabbits. According to these results, it may be suggested
that cerebral vessels were injured by the released substances from platelets as described in the experiments
on the pulmonary arteries (12). Recently, activated
platelets in vivo were implicated to play an important
role in the genesis of vascular injuries (13), since they
released several vasoactive substances (9, 10, 13). In
addition to serotonin, histamine and prostaglandin endo-
peroxides, prostaglandins and thromboxane A_2 have also
been included in the group of released substances (7, 8,
15). Prostaglandins and thromboxane A_2 released from
platelets are produced from the arachidonic acid, the
source of which is probably the platelet membrane phos-
pholipid (1). These substances show potent vasoconstric-
tive and platelet-aggregating effects (3, 7, 8). Thromboxane A_2 showed a strong constriction of cerebral arteries (2). These substances may not only induce a vascular
constriction but may also increase vascular permeability.
The enlargement of endothelial junctions due to a constriction of endothelial cells (11) could be responsible
for the mechanisms of the induced vascular injuries.

     Although the activated platelets were implicated to
induce vascular injuries (10), direct evidence was not
obtained until now. The present results indicate a possible role of released substances from platelets to induce vascular injuries. However, the possibility that
the ischemic changes in brain induced by occlusion of
platelet thrombi are due to a toxic effect of arachidonic
acid as a non-specific detergent (16) cannot be excluded.

SUMMARY

   Sodium arachidonate (0.7 mg/kg) was injected into
the right cerebral artery of 30 rabbits without disturbing cerebral blood flow. Eight rabbits were pre-treated
with anti-platelet serum to produce thrombocytopenic
state. Sixty minutes after the injection, vascular injuries associated with platelet thrombi were observed on
the injected side of the brain. The characteristic histological findings consisted of a flattened and convoluted
endothelial lining of the inner cerebral arteries of surface, extravasation, undulated internal elastic layer and
an occasional arterial wall thickening of medium and large
sized arteries. These changes were infrequently seen on
the non-injected side of the brain and in the thrombocytopenic rabbits. These results favor the possibility
that the substances being released from platelet aggregates are responsible for the vascular injuries in the
brain.

REFERENCES

1. Caen, J.P., Cronberg, S. and Kubisz, P. (1977):
    Platelets Physiology and Pathology, Stratton, p. 16,
    Intercontinental Medical Book Corp., New York.

2. Ellis, E.F., Nies, A.S. and Oates, J.A. (1977):
    Cerebral arterial smooth muscle contraction by
    thromboxane A_2. Stroke 8: 480-483.

3. Ellis, E.F., Oels, O., Roberts, L.J., Payne, N.A.,
    Sweetman, B.J., Nies, A.S. and Oates, J.A. (1976):
    Coronary arterial smooth muscle contraction by a
    substance released from platelets, evidence that it
    is thromboxane A_2. Science 193: 1135-1137.

4. Furlow, T.W. and Bass, N.H. (1975): Fatty acids,
    platelets, and microcirculatory obstruction.
    Science 190: 491.

5. Furlow, T.W. and Bass, N.H. (1975): Stroke in rats
    produced by carotid injection of sodium arachidonate.
    Science 187: 658-660.

6. Furlow, T.W. and Bass, N.H. (1976): Arachidonate-
    induced cerebrovascular occlusion in the rat.
    Neurology 26: 297-304.

7. Hamberg, M., Svensson, J. and Samuelsson, B. (1975):
    Thromboxanes, a new group of biologically active
    compounds derived from prostaglandin endoperoxides.
    Proc. Natl. Acad. Sci. USA 72: 2994-2998.

8. Hamberg, M., Svensson, J., Wakabayashi, T. and
    Samuelsson, B. (1974): Isolation and structure of
    two prostaglandin endoperoxides that cause platelet
    aggregation. Proc. Natl. Acad. Sci. USA 71: 345-349.

9. Haslam, R.J. (1964): Role of adenosine diphosphate
    in the aggregation of human blood platelets by
    thrombin and by fatty acids. Nature 202: 765-768.

10. Hovig, T. (1963): Release of a platelet aggregation
    substance (adenosine diphosphate) from rabbit blood
    platelets induced by saline "extract" of tendon.
    Thrombi. Diath. Haemorrh. 9: 264-278.

11. Majno, G., Shea, S.M. and Leventhal, M. (1969):
    Endothelial constriction induced by histamine-type
    mediators. An electronmicroscopic study. J. Cell.
    Biol. 42: 647.

12. Motomiya, T., Matsubara, O. and Yamazaki, H. (in
    press): Constriction and damage of the pulmonary
    artery by suddenly induced intravascular aggregation
    of platelets. Cardiovascul. Surgery.

13. Mustard, J.F. and Packham, M.A. (1975): The role of
    blood and platelets in atherosclerosis and the com-
    plication of atherosclerosis. Thromb. Diath.
    Haemorrh. 33: 444-456.

14. Silver, M.J., Hoch, W. and Kocsis, J.J. (1974):
    Arachidonic acid causes sudden death in rabbits.
    Science 183: 1085-1087.

15. Smith, J.B. and Willis, A.L. (1970): Formation and
    release of prostaglandins by platelets in response
    to thrombin. Br. J. Pharmacol. 40: 545.

16. Spector, A.A. and Hoak, J.C. (1975): Fatty acids,
    platelets and microcircirculatory obstruction.
    Science 190: 490-491.

FURTHER STUDIES ON ISOLATED BRAIN CAPILLARIES: SOME CHARACTERISTICS OF THE ADENOSINE TRIPHOSPHATASE, ADENYLATE- AND GUANYLATE CYCLASE

F. Joó, I. Karnushina, I. Tóth and E. Dux

Laboratory of Molecular Neurobiology
Institute of Biophysics
Biological Research Center
Szeged, Hungary

In light of recent results emphasizing the primary importance of capillary endothelium in the maintenance of the blood-brain barrier (4, 24, 41, 43), it has become important to perform biochemical studies on a subcellular fraction enriched in brain capillaries. A procedure for the isolation of brain capillaries was originally developed in our laboratory (26) and has been modified by several authors (5, 15, 32, 47). The availability of the micromethod has promoted different studies aiming at the better elucidation of regulatory mechanisms in relation to the adaptations of cerebral blood flow and the blood-brain barrier.

Because of the well-known participation of adenosine triphosphatases (ATPases), adenylate- and guanylate cyclases in regulatory processes, it seemed of interest to study the properties of these enzymes in isolated brain capillaries. This paper shows that: 1) The ATPase activity confined to the brain capillaries corresponds mainly to Ca^{++} - Mg^{++} - ATPase. This enzyme activity could be inhibited by dibutyryl-cyclic AMP in a dose-dependent manner. 2) The capillary adenylate cyclase is coupled to histamine receptors, especially of H_2-type. 3) The brain capillaries contain a membrane-bound guanylate cyclase which can be activated by Triton X-100.

MATERIALS AND METHODS

Histochemistry

Adult rats weighing 150-200 g were anesthetized with diethyl ether and perfused with a paraformaldehyde solution in a sodium cacodylate buffer (pH 7.4). Brain samples from the cerebral and cerebellar cortex were sectioned by a Vibratom [R] and incubated at 37°C in the presence of ATP, 5'-adenylyl- (42) or 5'-guanylyl-imidodiphosphate (44), respectively. For light microscopy, the end product was developed in buffered sodium sulphide solution. For electron microscopy, the sections were postfixed in osmic acid solution buffered with s-collidine. To establish the specificity, incubations were carried out by omitting the substrates or using enzyme inhibitors. Durcupan-embedded sections were cut on a Reichert OMU-2 ultramicrotome, stained with lead citrate or viewed without staining in a JEOL 100 B electron microscope.

Fluorescent microscopy

The permeability of brain capillaries was studied by the sensitive fluorescent microscopic method of Hamberger and Hamberger (16).

Electron microscopy

Certain metal salts reacting strongly with the sulfhydryl groups have been extensively used for the inhibition of ATPase activity (34, 55). In our studies, mercury chloride and nickel chloride were given intravenously at a concentration of 10^{-1} and 10^{-4}M, respectively. The fine structural changes developing after the in vivo inhibition of ATPase activity were studied in samples fixed in Millonig's (36) buffered osmic acid solution.

Biochemistry

Isolation of capillaries from brain. The fraction of isolated capillary vessels was prepared by a micromethod described earlier (26). Prior to decapitation, the anesthetized animals -- five guinea pigs and ten rats in each experiments -- were perfused through the heart with 0.9 percent NaCl solution. All the following procedures were performed at 4°C. The cortex was homogenized

manually, passed through nylon sieves suspended in 10 vol
of 0.25 M sucrose, 50 mM Tris-HCl (pH 7.2), 5 mM EDTA,
and centrifuged at 1000 g for 10 minutes. The pellet was
suspended in 0.25 M sucrose buffer and centrifuged repeatedly. The final pellet was suspended in 3 vol of sucrose
and placed on a stepwise sucrose gradient 1.0 M. 1.3 M
and 1.5 M and centrifuged in a Beckman SW 25.1 rotor at
20,000 rpm for 30 minutes. The capillaries were recovered
from the bottom of the tube. The whole procedure took
2.5 hours and yielded from 6 g of cortex approximately
3 mg protein in the capillary fraction, as determined by
the Lowry method (33).

   Adenosine triphosphatase. The ATPase activity was
determined according to Sun and Samorajski (53) with
minor modifications. The incubation mixture contained the
monovalent or bivalent metal ions required by the different
ATPase activity. The samples were incubated for 30 minutes
at 37°C, the reaction mixture was stopped by addition of
trichloroacetic acid (TCA), was centrifuged and the inor-
ganic phosphate formed in the course of the incubation was
determined in the supernatant by the method of Rathbun and
Betlach (40).

   Adenylate cyclase. The capillary-rich fractions in
2 mM Tris maleate buffer (pH 7.8) with 2 mM EGTA were
manually homogenized using a teflon-glass homogenizer.
Adenylate cyclase was measured according to Hegstrand et
al. (18). Incubations containing 30-50 µg proteins were
performed in triplicate and the cyclic AMP content was
determined by the protein-kinase binding method (6).

   Guanylate cyclase. The assay was performed according
to the method of Arnold et al. (1). The capillary homogenate (200 µl) from rat brain was incubated at 37°C for
10 minutes in triplicate samples in the presence of 4 mM
Mn^2, GTP as substrate, and 1 mM isobutylmethylxanthine
was added for inhibition of phosphodiesterase. The reaction was stopped by addition of 50 mM acetate buffer
(pH 4.0) and by heating at 90°C for 5 minutes. The cyclic
GMP formation was determined by radioimmunoassay (50)
using Amersham cyclic GMP RIA kits.

Studies on histamine receptors

   Adenylate cyclase measurements were performed to
monitor the activation elicited by different concentrations of histamine. Properties of the histamine receptors
coupled to the capillary adenylate cyclase were character-

ized by using specific agonists and antagonists of histamine H_1 and H_2 receptors.

RESULTS

Histochemistry

   Localization of the adenosine triphosphatase. Figures 1-3 show the localization of ATPase in the cerebellar cortex of rat. Similar capillary staining was observed in the cerebral cortex. The reaction product was
confined mainly to the basal lamina (Fig. 2). In the
histochemical controls, the basal lamina did not contain
electron dense precipitate (Fig. 2). Ouabain pretreatment
(in $10^{-3}M$) did not inhibit this staining. Certain diva-
lent cations injected in vivo into animals strongly inhibited the capillary adenosine triphosphatase, increased
the permeability (Figs. 4-6) and resulted in significant
ultrastructural changes of the basal lamina (Fig. 7). It
is therefore most likely that the capillary ATPase activity
corresponds to the $Ca^{++} - Mg^{++} - ATPase$ (EC 3.6.1.3).

   Localization of the adenylate cyclase. Figures 8-10
demonstrate the localization of adenylate-cyclase positive
structures in the cerebral cortex of rat. Under the light
microscope, in addition to the strong staining of capillaries, the enzyme activity was apparent in astrocytes
(Fig. 8). Under the electron microscope, the reaction
product was observed in the luminal and basal membranes
of capillaries, as well as in the surface membranes of
astrocytes (Fig. 9). In the histochemical controls, no
electron dense precipitate was seen in capillaries or
astrocytes (Fig. 10).

   Localization of the guanylate cyclase. The reaction
end product was, as a rule, observed in capillaries
(Figs. 11-12). Under the light microscope, staining of
astrocytes was not observed (Fig. 11). Under the electron
microscope, dense reaction product was present in the
luminal and basal membranes of capillaries (Fig. 12).
Figure 13 shows the substrate-free control.

Biochemistry

   Adenosine triphosphatase. The light microscopic
appearance of the capillary-rich fraction is shown in
Figure 14. The results of ATPase determinations indicate

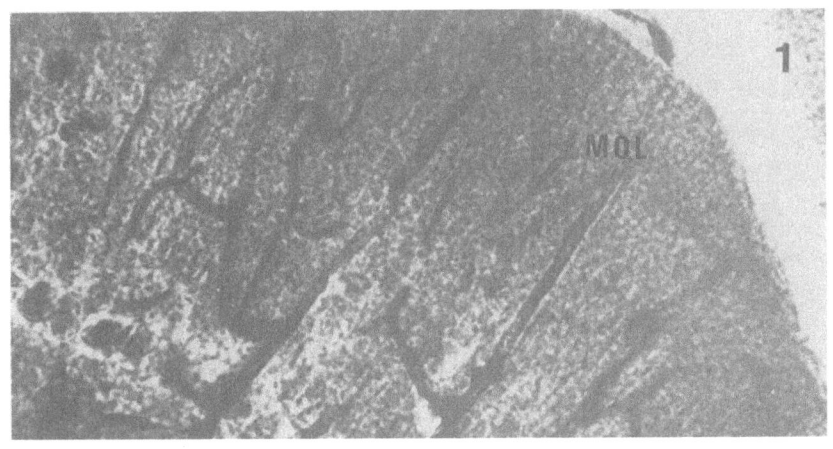

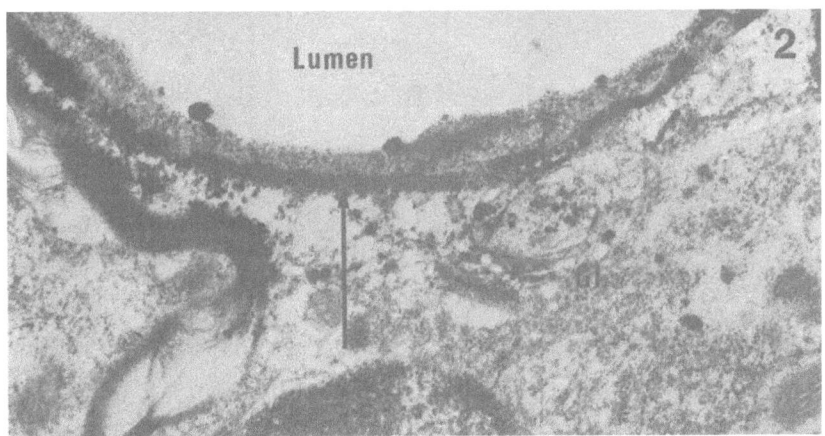

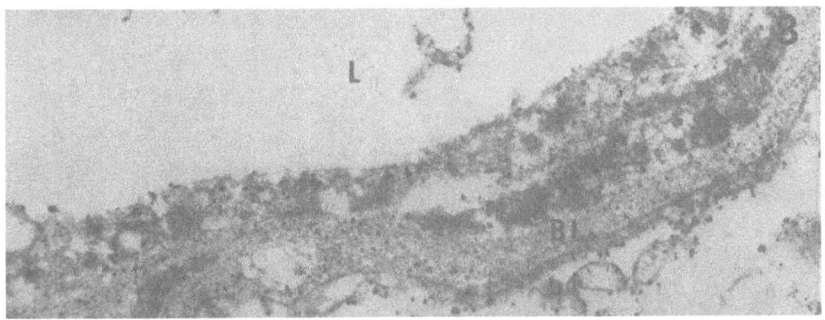

Figures 1-3.   Localization of adenosine triphosphatase.
               Fig. 1: Light microscopy, X 400.
               Fig. 2: Electron microscopy, X 20,000
               Fig. 3: Electron microscopy, substrate-
               free control, X 40,000
               (Reduced 10% for reproduction)

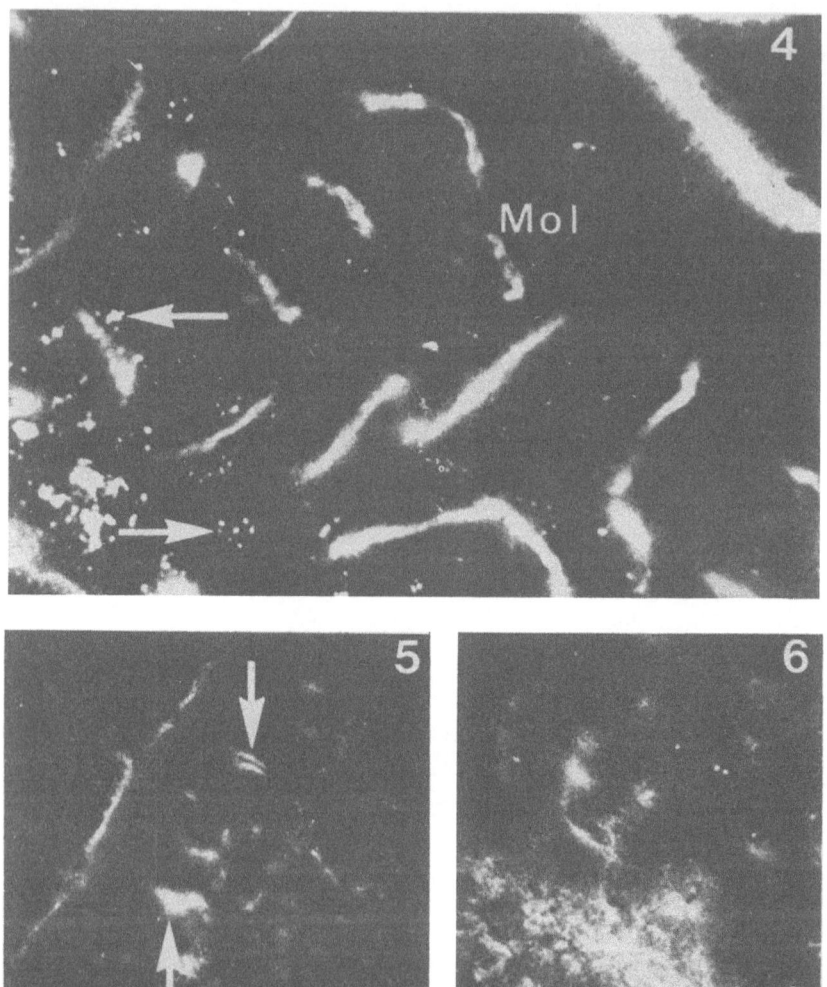

Figures 4-6.   Fluorescent microscopy for tracing changes
                of permeability after in vivo ATPase
                inhibition.
                Fig. 4: Control, X 400
                Fig. 5: 15 minutes after intravenous injec-
                tion of nickel chloride (0.01 g/kg), X 250
                Arrows indicate the extravasations.
                Fig. 6: 30 minutes after intravenous injec-
                tion of mercuric chloride (1.5 mg/kg),
                X 400 (Reduced 10% for reproduction)

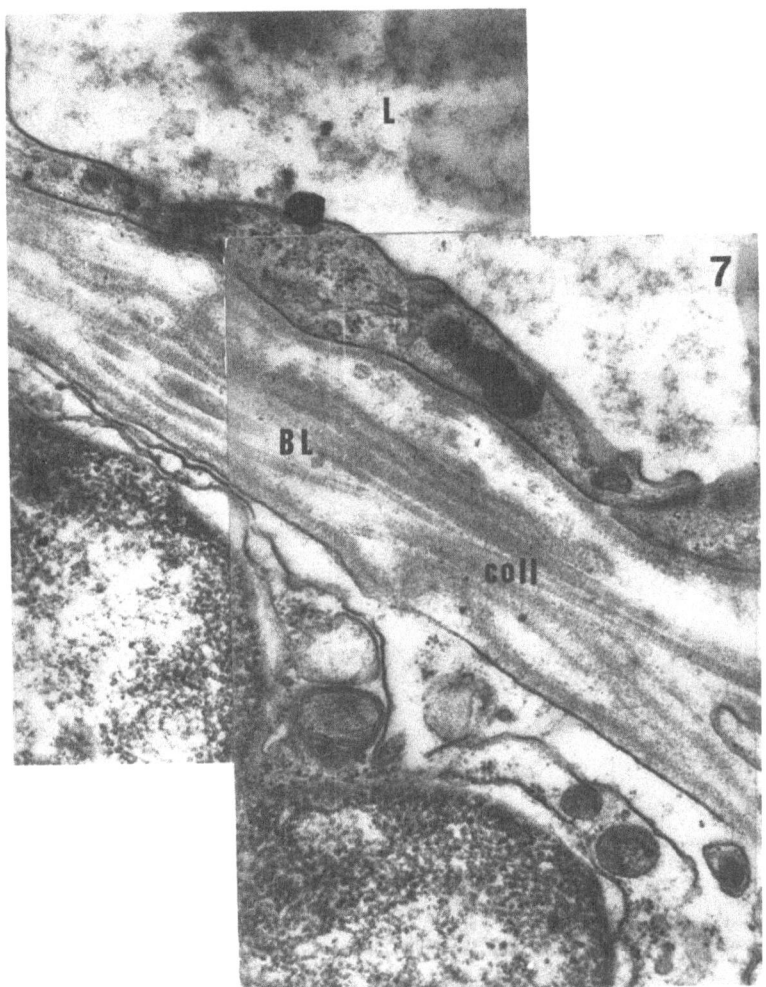

Figure 7.   Development of collagen-like (coll) in the
             swollen basal lamina (BL). L = lumen,
             X 50,000 (Reduced 10% for reproduction)

the presence of all three types of ATPases in the isolated
capillaries (Table 1).   In comparison with the initial
homogenate, a 26-fold enrichment was detected for the
Ca^{++} - Mg^{++} - ATPase in the capillaries. The effect of
different concentrations of dibutyryl cyclic AMP on the
Ca^{++} - Mg^{++} - ATPase activity was tested in vitro. As
shown in Figure 15, a concentration-dependent inhibition
was observed.

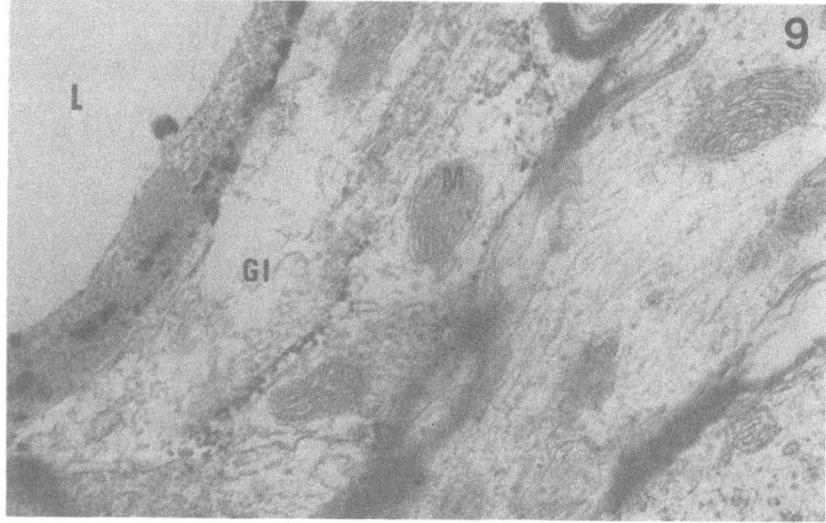

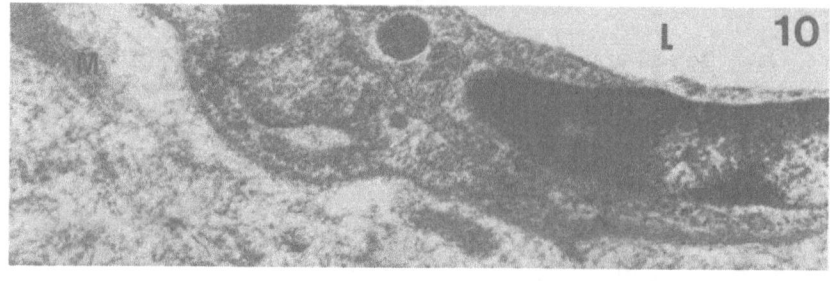

Figures 8-10.   Localization of adenylate cyclase.
                 Fig.   8: Light microscopy, X 400
                 Fig.   9: Electron microscopy, X 20,000
                 Fig. 10: Electron microscopy, substrate-
                 free control, X 30,000
                 (Reduced 10% for reproduction)

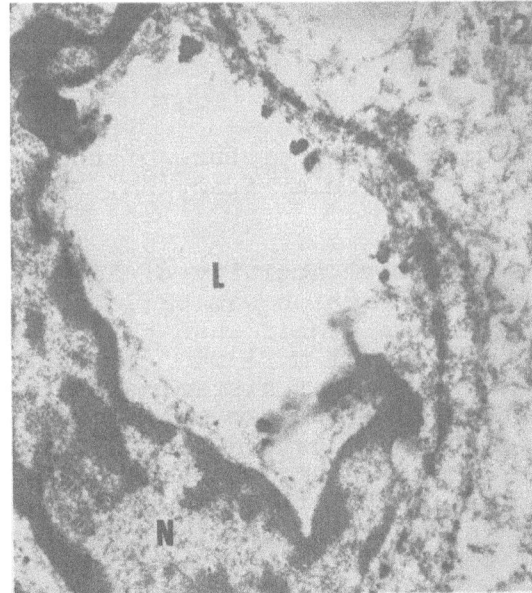
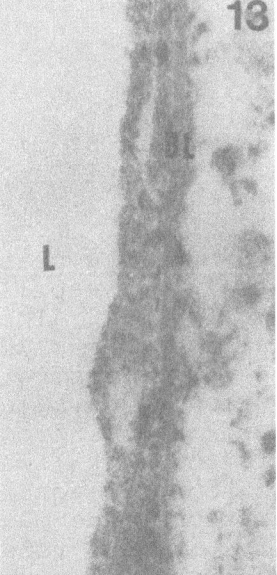

Figures 11-13.   Localization of guanylate cyclase.
                 Fig. 11: Light microscopy, X 400
                 Fig. 12: Electron microscopy, X 15,000
                 Fig. 13: Electron microscopy, substrate-
                 free control, X 30,000
                 (Reduced 10% for reproduction)

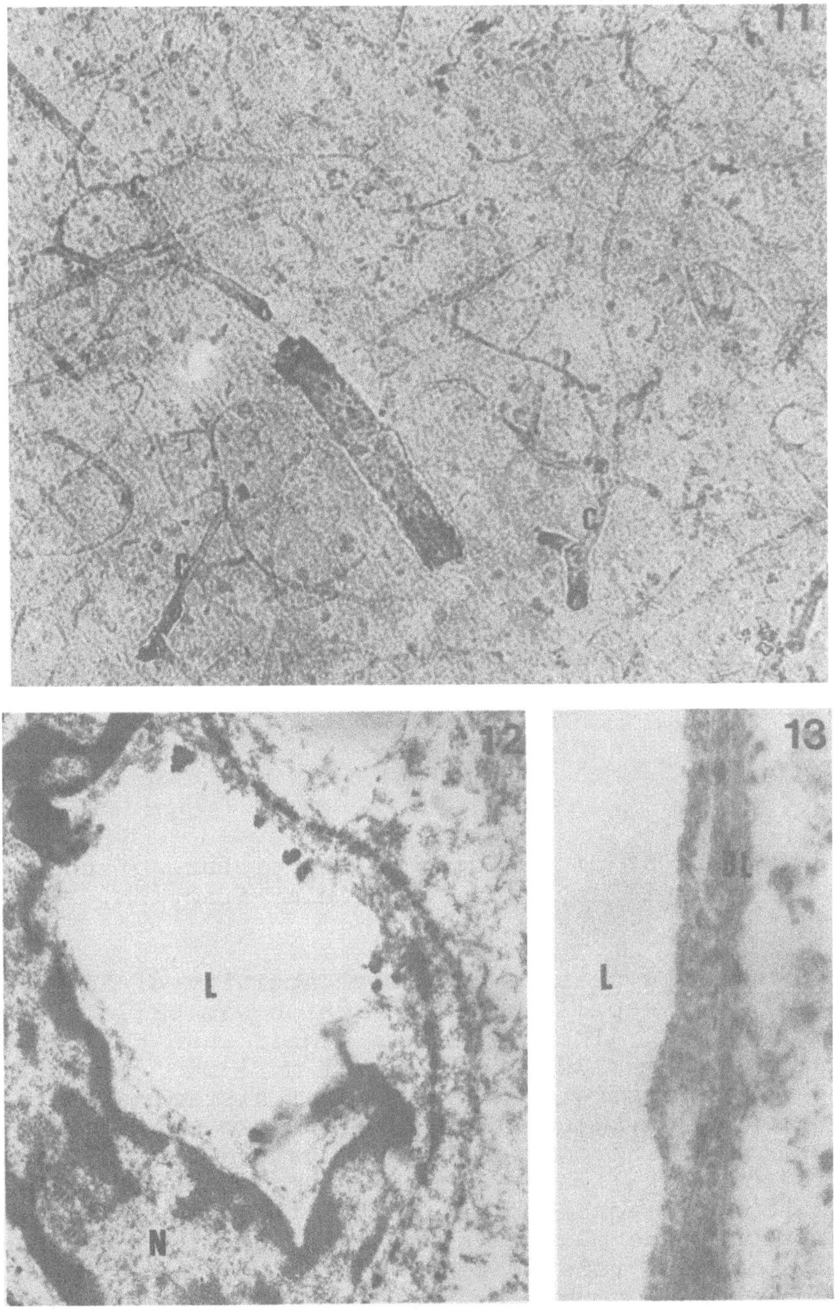

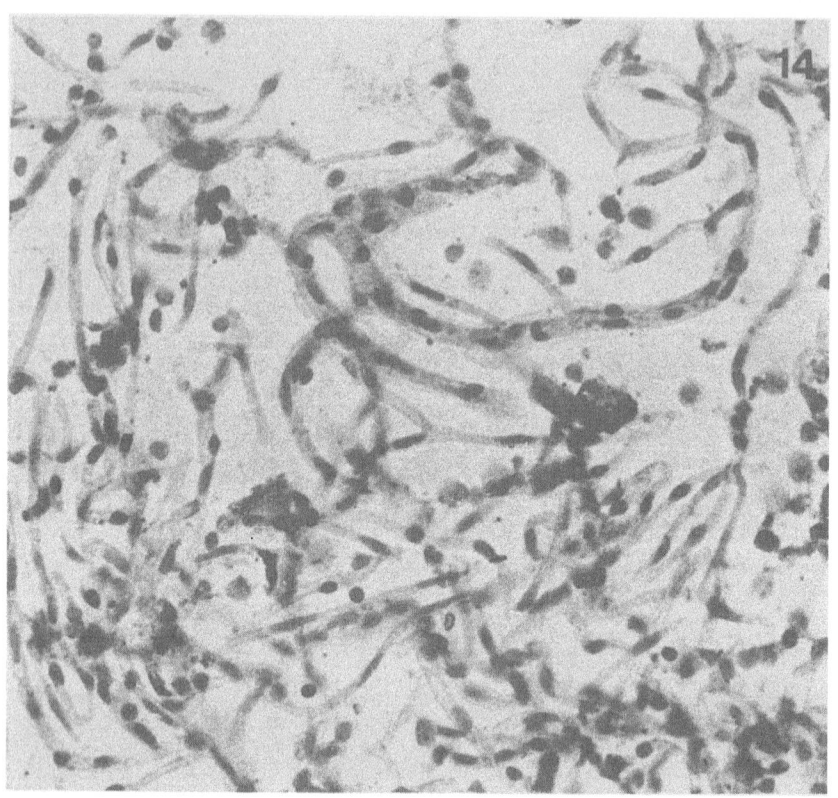

Figure 14.   Light microscopic survey picture of the
             capillary-rich subcellular fraction, X 400

  Adenylate cyclase. The basal activity of adenylate
cyclase in the capillary-rich fraction was 63.2 ± 8.0
pmol mg^{-1} min^{-1}. This value was higher than that of the
initial homogenate (22.7 ± 4.4 pmol mg^{-1} min^{-1}). The
effect of increasing concentration of histamine on the
activation of capillary adenylate cyclase was shown
earlier (30). The concentration required for half-maximal
stimulation (EC_{50}) was around $5.10^{-6}M$, whereas maximal
stimulation was achieved at about $10^{-4}M$. In various
preparations, the maximal stimulation by histamine ranged
from 1.4- to 2.4-fold of basal enzyme activity. Concentration-response curves of two H_1-receptor agonists and
two H_2-receptor agonists are compared to that of histamine
(Fig. 16). Mepyramine and cimetidine were employed as
H_1- or H_2-receptor antagonists, respectively. Figure 17
shows the inhibition of mepyramine on the histamine-
stimulated adenylate cyclase in capillaries and the

Table 1. Specific Activities of ATPases in the Homogenate and Capillary-Rich Fraction of the Guinea Pig Cerebral Cortex

Enzyme	Homogenate	Capillary-Rich Fraction	Enrichment Factor
$Na^+ - K^+$ ATPase	1988 ± 202	3125 ± 52	1 : 1.57
Mg^{++} - ATPase	5574 ± 815	5423 ± 465	1 : 0.97
$Ca^{++} - Mg^{++}$ - ATPase	48 ± 26	1269 ± 180	1 : 26.4

Mean ± SD
Specific activities = nmol Pi mg^{-1} 30 min^{-1}

inhibitory constant (Ki) was determined to be 1.1×10^{-6}M. On the other hand, the Ki for cimetidine, an H_2-receptor antagonist, was much lower, 7.5×10^{-7}M. Figure 18 shows the pattern of inhibition, which is competitive in nature, like a true antagonist. Further details of these results will be published elsewhere (31).

Guanylate cyclase. The basal activity of guanylate cyclase in the capillary-rich fraction was 20.1 ± 1.7 pmol mg^{-1} min^{-1}. The cyclic GMP accumulation was found to be linear in the course of incubation up to 15 minutes and was linear with protein concentration from 50 µg/tube to 200 µg/tube in the incubated samples. The K_m value of the enzyme for GMP was about 0.25 mM (Fig. 19 B). Acetylcholine, histamine and enkaphalin (10^{-4}M) failed to activate the enzyme. Only Triton X-100 (Fig. 19 A) was found to increase the enzyme activity up to 6-fold. The K_m value was unaffected by the Triton.

DISCUSSION

The development of methods for isolating capillaries from brain has made it possible to investigate the transport processes of the blood-brain barrier directly at cellular level. Several enzymes that participate in transport processes were found to be confined mainly to the brain capillaries (2, 9, 10, 12, 14, 20, 27, 28). For the uptake of amino acids, the presence of Na^+-independent L system and the Na^+-dependent A system was also evident in isolated brain microvessels (3, 19, 48).

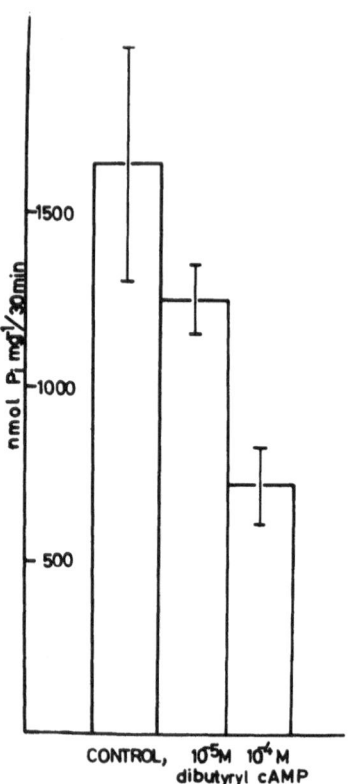

Figure 15.   Effects of dibutyryl cyclic AMP on the
              Ca^{++} - Mg^{++} - ATPase of isolated brain
              capillaries.

Adenosine triphosphatase. ATPase is a component of
the plasma membranes of animal cells (49). It is well
established that this enzyme takes an important part in
the active ion transport across the cell membrane. Certain neurotransmitters were shown to influence the membrane-bound Na^+ - K^+ ATPase activity in brain synaptosomes
(8). Na^+ - K^+ ATPase was found to be present in the isolated brain capillaries (12, 52); the activity (9.07 ± 1.98
μmol mg^{-1} hr^{-1}) was greater than in cultured glial cells.
Recent histochemical evidence (13) using p-nitrophenyl-
phosphate as substrate also indicated the presence of
Na^+ - K^+ ATPase in the endothelial cytoplasm. Thus, it
seems to be the case that the brain capillaries, like the

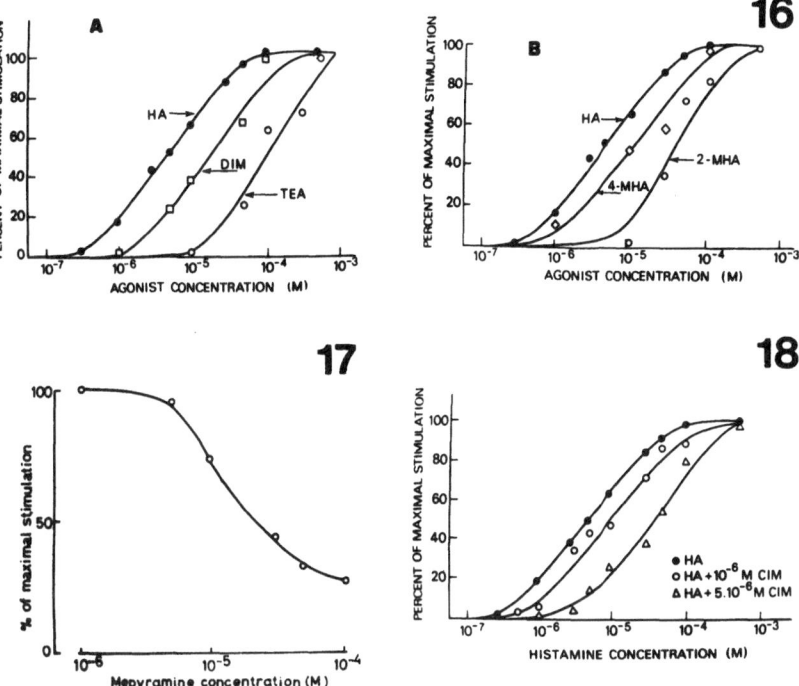

Figure 16:   Concentration-response curves are compared
             to that of histamine (HA).
             A - TEA = 2-thiazolethylamine, H_1 agonist;
                 DIM = Dimaprit, H_2 agonist.

             B - 2-MHA = 2-methylhistamine, H_1 agonist;
                 4-MHA = 4-methylhistamine, H_2 agonist.

Figure 17:   Inhibition of different concentrations of
             mepyramine on the adenylate cyclase activity
             stimulated by histamine in the capillary-
             rich fraction.

Figure 18:   Inhibition of histamine-activation of capil-
             lary adenylate cyclase by different concen-
             trations of cimetidine.

glial cells, have the capacity to participate in the ionic
maintenance of homeostasis of the central nervous system.
Our biochemical data, obtained from measurements on isolated brain capillaries, are in good agreement with the

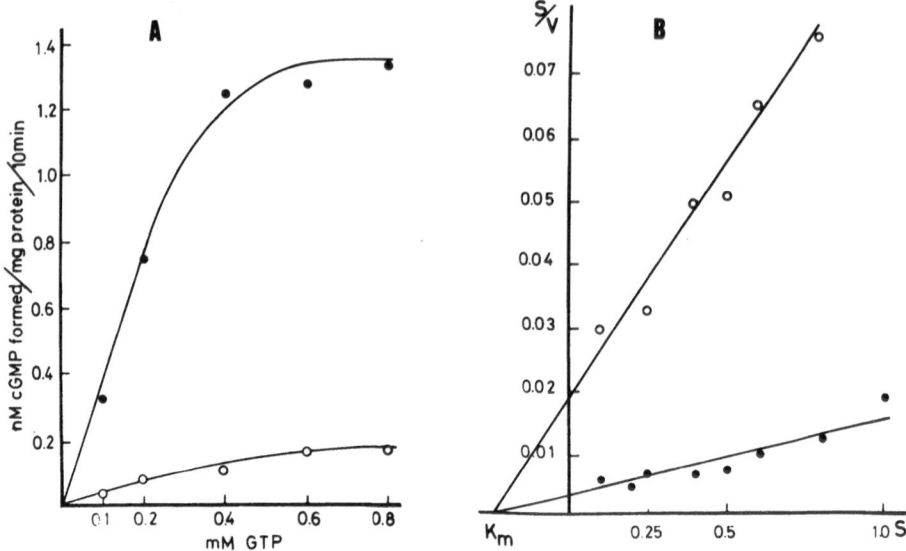

Figure 19.   A - Guanylate cyclase activity could be
                 activated by Triton X-100 treatment
                 (•——•).   Basal activity: o——o.

             B - The K_m value of the activated enzyme
                 (•——•) is the same as that of the con-
                 trol (o——o).

above-mentioned findings. Our determinations showed a
little lower enzyme activity (3.12 ± 0.52 μmol mg^{-1}
30 min^{-1}), but this could be explained by differences in
preparations. To avoid the contamination of blood components, we perfused the animals with sodium chloride to
wash out the blood. Thus, in our data, the ATPase activity measured refers only to the capillary compartment. It
is important to point out that, in addition to the Na$^+$ -
K$^+$ ATPase, the brain capillaries contain a high amount of
Ca^{++} - Mg^{++} - ATPase. The 26-fold enrichment of this
enzyme in the capillary fraction after isolation is consistent with earlier histochemical data (23, 54). Our
results have shown that, when the transendothelian macro-
molecular transport was enhanced by ATPase inhibition,
collagen-like fibers were formed at the same time in the
basal lamina of brain capillaries (21, 22). This structural change in the molecular organization of basal lamina
suggests a close relationship between the tropocollagen
filaments and the Ca^{++} - Mg^{++} - ATPase of the basal
lamina (29).

Adenylate cyclase. The adenylate cyclase, together with the effects of cyclic AMP, on the permeability of brain capillaries has been extensively studied in our laboratory (24, 27, 28, 30, 31). Hormone receptors, in addition to histamine receptors characterized in our studies, have been reported to be present in isolated brain capillaries. These data suggest the presence of mixed populations of receptors for adenylate cyclase in brain capillaries. β-adrenergic-sensitive adenylate cyclase has been shown to occur in intracranial blood vessels (38). Baca and Palmer (2) have shown that, in the 10,000 g particulate fraction of homogenized capillary-rich fractions, the adenylate cyclase was activated by noradrenaline, adrenaline and dopamine. In his latest work, Palmer (39) was able to establish that the adrenergic receptors mediating adenylate cyclase activity in cerebral capillaries are predominantly of the beta$_2$-type. The presence of adenosine (20) and prostaglandin receptors (14) in microvessel preparations was also evident. Westergaard (56) reported that serotonin perfused through the cerebral ventricles and produced an increase in transendothelial transport of circulating exogenous peroxidase. The results of our studies strongly indicate that, at least in rat and guinea pig, the brain capillaries contain adenylate cyclase, which is in close functional relationship to the histamine receptors of H$_2$-type (27, 30, 31). This finding may be of clinical importance, since with metiamide, a non-toxic histamine H$_2$-receptor blocker, we could reduce the extent of brain edema evoked by Yttrium 90 irradiation (7, 28). In addition, the therapeutic use of ergot alkaloids in migraine headaches via dopamine-sensitive adenylate cyclase of cerebral vessels was also demonstrated (45). Finally, evidence was recently published (17) that an action of noradrenaline is responsible for controlling the outflow of water from brain to blood stream.

Guanylate cyclase. The presence of guanylate cyclase activity was described in the brain microvessels for the first time. Bioactive substances tested to stimulate the capillary guanylate cyclase proved to be ineffective. In general, the failure of activating the capillary guanylate cyclase by transmitter substances, together with the demonstration that extracellular calcium is required in order for various neurotransmitters and hormones to increase cyclic GMP in intact target tissue, has led to the suggestion (46) that hormones activate the guanylate cyclase by elevating intracellular calcium level in the target tissue. It is worth mentioning that both cyclic AMP and cyclic GMP have been implicated in responses in vascular tissue (11). It was shown recently (35) that cyclic GMP can increase Na$^+$ permeability, possibly by phosphorylating a membrane protein, and regulate the membrane potential in this way. Guanylate cyclase may have a similar function in brain capillaries, as well.

To date little information is available with regard to the central interactions of drugs as related to the cyclic nucleotide systems, in particular those associated with the microcirculation. This information can be gained only from further studies of molecular interactions taking place at the hormone receptors.

ACKNOWLEDGMENTS

The authors are grateful to collaborators with whom certain parts of this paper have been already published. We are thankful to Miss Gabriella Gazdagh, Mrs. Zsizsanna Horváth, Mrs. Krisztina Mohácsi and Mr. János Szeles for their skilled technical assistance.

REFERENCES

1.   Arnold, W.P., Mittal, C.K., Katsuki, S. and Murad, F. (1977): Nitric oxide activates guanylate cyclase and increases guanosine 3':5'-cyclic monophosphate levels in various tissue preparations. Proc. Nat. Acad. Sci. USA 74: 3203-3207.

2.   Baca, G.M. and Palmer, G.C. (1978): Presence of hormonally-sensitive adenylate cyclase receptors in capillary-enriched fractions from rat cerebral cortex. Blood Vessels 15: 286-298.

3.   Betz, A.L. and Goldstein, G.W. (1978): Polarity of the blood-brain barrier: Neutral amino acid transport into isolated brain capillaries. Science 202: 225-227

4.   Bodenheimer, T.S. and Brightman, M.W. (1968): A blood-brain barrier to peroxidases surrounded by perivascular spaces. Am. J. Anat. 122: 249-268.

5.   Brendel, K., Meezan, E. and Carlson, E.C. (1974): Isolated brain microvessels: A purified, metabolic- ally active preparation from bovine cerebral cortex. Science 185: 953-955.

6.   Brown, B.L., Albano, J.D.M., Ekins, R.P. and Sgherzi, A.M. (1971): A simple and sensitive satu- ration assay method for the measurement of adenosine 3':5'-cyclic monophosphate. Biochem. J. 121: 561-562.

7. Csanda, E., Joó, F., Somogyi, I., Szücs, A., Saal, M., August, A. and Komoly, S. (1977): Structural, ultra-structural and functional reactions of the brain after implanting Yttrium 90 rods used in stereotactic neurosurgery. Acta Neurochirurg. Suppl. 24: 139-147.

8. Deshaiah, D. and Ho, I.K. (1977): Kinetics of catecholamine sensitive $Na^+ - K^+$ ATPase activity in mouse brain synaptosomes. J. Neurochem. 26: 2029-2035.

9. Djuricic, B.M. and Mrsulja, B.B. (1977): Enzymic activity of the brain: Microvessels vs. total fore-brain homogenata. Brain Res. 138: 561-564.

10. Djuricic, B.M. and Mrsulja, B.B. (1979): Brain microvessel hexokinase: Kinetic properties. Experientia 35: 169-171.

11. Dunham, E.W., Haddox, M.K. and Goldberg, N.D. (1974): Alteration of vein cyclic 3':5' nucleotide concen-trations during changes in contractility. Proc. Nat. Acad. Sci. USA 71: 815-819.

12. Eisenberg, H.M. and Suddith, R.L. (1977): Sodium-potassium ATPase in brain capillaries. Soc. for Neurosci. Abstract Vol. 3: 217.

13. Firth, J.A. (1977): Cytochemical localization of the K^+ regulation interface between blood and brain. Experientia 33: 1093-1094.

14. Gerristen, M.E., Parks, T.P. and Printz, M.P. (1979): Prostaglandin E_2 is the major enzymatic product of endoperoxide/PGH_2/metabolism in isolated bovine cerebral microvessels. Fed. Proc. 38: 752.

15. Goldstein, G.W., Wolinsky, J.S., Csejtey, J. and Diamond, I. (1975): Isolation of metabolically active capillaries from rat brain. J. Neurochem. 25: 715-717.

16. Hamberger, A. and Hamberger, B. (1966): Uptake of catecholamines and penetration of trypan blue after blood-brain barrier lesions. Z. Zellforsch. 70: 386-392.

17. Hartman, B.K. (1973): The innervation of cerebral blood vessels by central noradrenergic neurons. In: Frontiers in Catecholamine Research, E. Usdin and S.H. Snyder (eds.), pp. 91-96, Pergamon Press, New York.

18.   Hegstrand, L.R., Kanof, P.D. and Greengard, P. (1976)
      Histamine-sensitive adenylate cyclase in mammalian
      brain.   Nature 260: 163-165.

19.   Hjelle, J.T., Baird-Lambert, J., Cardinale, G.,
      Spector, S. and Udenfriend, S. (1978):   Isolated
      microvessels: The blood brain barrier in vitro.
      Proc. Nat. Acad. Sci. USA 75: 4544-4548.

20.   Huang, M. and Drummond, G.I. (1979):   Adenylate
      cyclase in cerebral microvessels.   Fed. Proc. 38: 532.

21.   Joó, F. (1968):   The effect of the inhibition of
      adenosine triphosphatase activity on the fine struc-
      tural organization of brain capillaries.   Nature 219:
      1378-1379.

22.   Joó, F. (1969):   Changes in the molecular organiza-
      tion of the basement membrane after inhibition of
      adenosine triphosphatase activity in the rat brain
      capillaries.   Cytobios 3: 289-301.

23.   Joó, F. (1969):   Electron histochemical structure of
      capillaries in the rat brain.   Acta Biol. Szegediense
      15: 79-88.

24.   Joó, F. (1971):   Increased production of coated
      vesicles in the brain capillaries during enhanced
      permeability of the blood-brain barrier.   Brit. J.
      Exp. Pathol. 52: 646-649.

25.   Joó, F. (1972):   Effect on N^6O^2-dibutyryl cyclic
      $3':5'$-adenosine monophosphate on the pinocytosis of
      brain capillaries of mice.   Experientia 28: 1470.

26.   Joó, F. and Karnushina, I. (1973):   A procedure for
      the isolation of capillaries from rat brain.
      Cytobios 8: 41-48.

27.   Joó, F., Rakonczay, Z. and Wollemann, M. (1975):
      cAMP-mediated regulation of the permeability in the
      brain capillaries.   Experientia 31: 582-583.

28.   Joó, F. (1979):   Significance of adenylate cyclase
      in the regulation of permeability of brain capillarie
      In: Pathophysiology of Cerebral Energy Metabolism,
      B.B. Mrsulja, Lj.M. Rakic, I. Klatzo and M. Spatz
      (eds.), pp. 211-237, Plenum Publ. Corp., New York.

29. Joó, F. (1979): The role of adenosine triphosphatase in the maintenance of molecular organization of the basal lamina in the brain capillaries. In: Frontiers in Matrix Biology, A.M. Robert, R. Boniface and L. Robert (eds.), pp. 166-182, S. Karger, New York.

30. Karnushina, I.L., Palacios, J.M., Barbin, G., Dux, E., Joó, F. and Schwartz, J.C. (1979): Histamine-related enzymes and histamine receptors in isolated brain capillaries. Agents and Actions 9: 89-90.

31. Karnushina, I.L., Palacios, J.M., Barbin, G., Dux, E., Joó, F. and Schwartz, J.C. (1979): Studies on a capillary-rich fraction isolated from brain: Histaminergic components and characterization of the histamine receptors linked to adenylate cyclase. J. Neurochem. (submitted for publication).

32. Kolber, A.R., Bagnell, C.R., Krigman, M.R., Hayward,J. and Morell, P. (1979): Transport of sugars into microvessels isolated from rat brain: A model for the blood-brain barrier. J. Neurochem. 33: 419-432.

33. Lowry, O.H., Rosenbrough, N.J., Farr, A.L. and Randall, R.J. (1951): Protein measurements with the folin phenol reagent. J. Biol. Chem. 193: 265-275.

34. Mason, R.G. and Saba, S.R. (1969): Platelet ATPase activities. I. Ecto-ATPases of intact platelets and their possible role in aggregation. A. J. Path. 55: 215-223.

35. Miller, W.H. and Nicoll, G.D. (1979): Evidence that cyclic GMP regulates membrane potential in rod photoreceptors. Nature 280: 64-66.

36. Millonig, G. (1969): Advantages of a phosphate buffer for OsO_4 solutions in fixation. J. Appl. Physics 32: 1637.

37. Mrsulja, B.B., Mrsulja, B.J., Fujimoto, T., Klatzo, I. and Spatz, M. (1976): Isolation of brain capillaries: A simplified technique. Brain Res. 110: 361-365.

38. Nathanson, J.A. and Glaser, G.H. (1979): Identification of β-adrenergic-sensitive adenylate cyclase in intracranial blood vessels. Nature 278: 567-569.

39. Palmer, G.C. (1979): Beta adrenergic receptors
    mediate adenylate cyclase response in rat cerebral
    cortex. Neuropharmacol. (in press).

40. Rathbun, W.B. and Betlach, M.V. (1969): Estimation
    of enzymatically produced ortophosphate in the pres-
    ence of cystein and adenosinetriphosphate. Anal.
    Biochem. 28: 436-445.

41. Reese, T.S. and Karnovsky, M.J. (1967): Fine struc-
    tural localization of a blood-brain barrier to exo-
    genous peroxidase. J. Cell Biol. 34: 207-217.

42. Reik, L., Petzold, G.L., Higgins, J.A., Greengard, P.
    and Barrnett, R.J. (1970): Hormone-sensitive adenyl
    cyclase: Cytochemical localization in rat liver.
    Science 168: 382-384.

43. Rodriguez, L.A. (1955): Experiments on the histo-
    logical locus of the haematoencephalic barrier.
    J. Comp. Neurol. 102: 27-45.

44. Saito, T. and Keino, H. (1976): Ultrastructural
    demonstration of guanylate cyclase activity in rat
    kidney. Fifth Internat. Congress of Histochemistry,
    pp. 303; Bucharest-Roumania.

45. Schmidt, M.J. and Hill, L.E. (1977): Effects of
    ergots on adenylate cyclase activity in the corpus
    striatum and pituitary. Life Sci. 20: 789-798.

46. Schultz, G., Hardman, J.G., Schultz, K., Baird, C.E.
    and Sutherland, E.W. (1973): The importance of
    calcium ions for the regulation of guanosine 3':5'-
    cyclic monophosphate levels. Proc. Nat. Acad. Sci.
    USA 70: 3889-3893.

47. Selivonchick, D.P. and Roots, B.I. (1977): Lipid
    and fatty acyl composition of rat brain endothelia
    isolated by a new technique. Lipids 12: 165-169.

48. Sershen, H. and Lajtha, A. (1976): Capillary trans-
    port of amino acids in the developing brain. Exp.
    Neurol. 53: 465-474.

49. Skou, J.C. (1957): The influence of some cations
    on an adenosine triphosphatase from peripheral
    nerves. Biochim. Biophys. Acta 23: 394-401.

50. Steiner, A.L., Wehman, R.E., Parker, C.W. and Kipnis, D.M. (1972): Radioimmunoassay for the measurement of cyclic nucleotides. In: Advances in Cyclic Nucleotide Research, Vol. 2., P. Greengard and G.A. Robinson (eds.), pp. 51-61, Raven Press, New York.

51. Suddith, R.L., Crawford, J.S. and Eisenberg, H.M. (1978): Isolated brain microvessels: Sodium-potassium ATPase and potassium transport. J. Cell Biol. 79: 103a.

52. Sun, A.Y. and Samorajski, T. (1970): Effects of ethanol on the activity of adenosine-tripholphatase acetylcholinesterase in synaptosomes isolated from guinea pig brain. J. Neurochem. 17: 1365-1372.

53. Torack, R.M. and Barrnett, R.J. (1964): The fine structural localization of nucleoside phosphatase activity in the blood-brain barrier. J. Neuropath. exp. Neurol. 23: 46-59.

54. Webb, J.L. (1966): Enzyme and metabolic inhibitors, Vol. II. Academic Press, New York.

55. Westergaard, E. (1975): Enhanced vesicular transport of exogenous peroxidase across cerebral vessels, induced by serotonin. Acta Neuropathol. 32: 27-42.

METABOLIC PATTERNS IN CEREBRAL CAPILLARIES

B. B. Mršulja, B. M. Djuričić and
D. V. Micic

Laboratory for Neurochemistry
Institute of Biochemistry
Faculty of Medicine
Belgrade, Yugoslavia

Brain capillaries and parenchyma are morphologically and metabolically separate and distinct compartments; the function of endothelial cells in transporting substances from blood to brain requires some specialized properties. The simplicity in isolating cerebral capillaries permits the investigation on their biochemical properties. Such an approach has revealed that the enzymatic organization of brain capillaries differs from that of the parenchyma (4-7, 11, 15). The brain capillaries which we have been studying are most probably a part of the blood-brain barrier (BBB); hence it is tentatively proposed that the specific biochemical organization of the brain microvasculature regulates the transport of substrates across the blood to the brain.

Knowledge of the metabolism and structural composition of isolated brain capillaries is of the utmost importance in order to elucidate the cellular mechanisms which might regulate the transport of substances from blood to the brain. Our previous investigations on the biochemical organization of cerebral capillaries (4, 6, 11) indicated: (a) some specific metabolic patterns of glucose-6-P metabolism in capillaries; (b) possible relevance of NADP-linked enzymes for brain capillary metabolism and function; and, (c) the existence of enzymatic barrier for monoamines in endothelial cells.

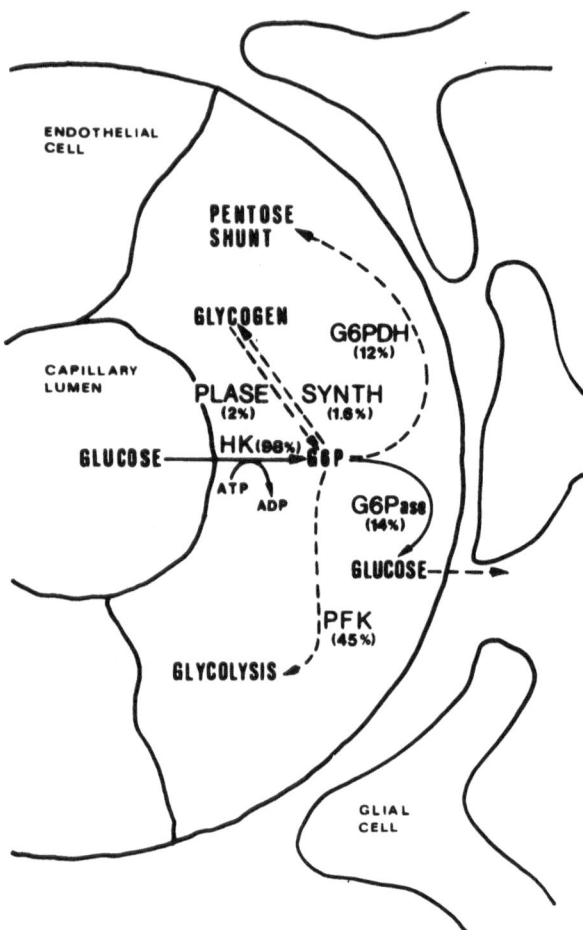

Fig. 1.   Glucose-6-phosphate metabolism in endothelial
           cell.   Abbreviations: G6P - glucose-6-phosphate;
           G6Pase - glucose-6-phosphatase; G6PDH -
           glucose-6-phosphate dehydrogenase; HK -
           hexokinase; Plase - glycogen phosphorylase;
           Synth - glycogen synthetase.   The numbers
           indicate percentage participation of given
           enzymatic reaction in glucose-6-P formation or
           consumption.

A)   GLUCOSE-6-P METABOLISM IN BRAIN CAPILLARIES
     (Figure 1).

     In the endothelial cells, glucose is phosphorylated
in the presence of ATP by hexokinase to form glucose-6-P.
The characteristics of the capillary hexokinase are different from those of parenchymal enzyme (5). The kinetic
studies of hexokinase in brain capillaries and parenchyma
have revealed the following: (a) the parenchymal enzyme
is more susceptible to glucose inhibition; and (b) the
parenchymal hexokinase can tolerate greater variations
in the ATP concentration. Also, there are kinetic differences at different pH values and in the presence of
fructose, sodium, potassium. Therefore, there appear to
be two kinetically distinct enzymes, one in the brain
parenchyma and the other in the endothelial cell (5). It
is most likely that these differences are related to the
specialized functions of these cells.

     Based on our results (Table 1), one can conclude
that (a) the glycolytic pathway is present in the endothelial cells; (b) glycolysis takes about 50% of the
glucose-6-P formed in capillaries as well as in parenchyma; (c) the glucose-6-P metabolized by the pentose
shunt is greater in endothelial than in parenchymal
cells (the respective values are 12% and 4.5% of maximal
glucose-6-P producing capacities) and, (d) glycogen
synthetase is more active in the microvasculature. The
glycogenolytic activity in the endothelial cell is
negligible since 98% of glucose-6-P is formed by the
hexokinase reaction. Since the activity of energy producing enzymes is low (11) and glycolysis appears to be
the main source of ATP for the intrinsic metabolism of
capillaries, it is very possible that the energy demands
of the endothelial cells are low.

     In the capillaries 14% of the glucose-6-P is hydro-
lyzed by glucose-6-Pase, while in parenchyma the value is
negligible. The glucose-6-Pase activity is 4.2 times
greater in the capillary than in the parenchyma. Previously we postulated that hexokinase and glucose-6-Pase
participate in the glucose transport system through the
endothelial cell (5, 6); the excess glucose-6-P not
metabolized by the endothelial cell is hydrolyzed by
glucose-6-Pase to glucose and released. The nature of
the glucose carrier through membranes of the endothelial
cells remains to be elucidated, but there is evidence
that the transport of glucose into capillaries is saturable with a Km value (12) very close to that for the
capillary hexokinase Km (5).

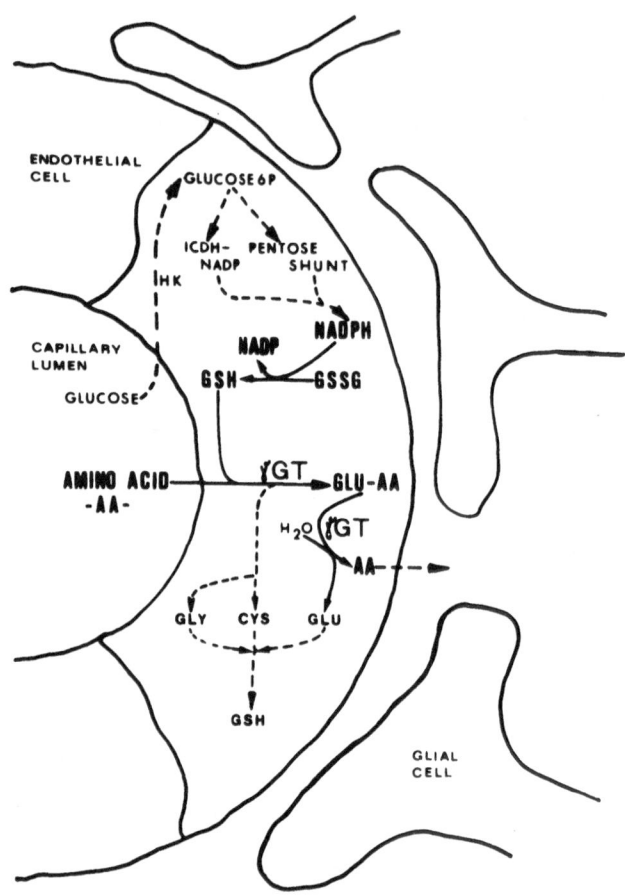

Fig. 2.   Possible role of NADPH, glutathione and gamma-
           glutamyl transpeptidase in amino acid transport
           in capillaries. Abbreviations: Cys - cysteine;
           Glu - glutamate; Glu-AA - glutathione-amino
           acid complex; GSH - reduced glutathione;
           GSSG - oxidized glutathione; γ GT - gamma-
           glutamyl transpeptidase; Gly - glycine; ICDH-
           NADP - NADP-linked isocitrate dehydrogenase.

Table 1. Glucose-6-phosphate related enzymes in brain
          capillaries and parenchyma and activity ratios.

| | Activity | | |
Enzyme	In Capillaries	In Parenchyma	Ratio
Hexokinase	1320 + 120	3960 + 240	0.33
Glycogen phospho- rylase	32 + 3	164 + 7	0.20
Glucose-6-P dehydrogenase	159 + 13	196 + 10	0.81
Glucose-6-Pase	184 + 18	44 + 2	4.18
Glycogen synthe- tase	22 + 3	14 + 0.7	1.52

Enzyme activities are given in nMoles of substrate converted per hr per mg protein, 37°C, pH 7.1. S.E.M. for
n = 5 is given.

B) POSSIBLE RELEVANCE OF NADPH-PRODUCING ENZYMES FOR
    CAPILLARY AMINO ACID TRANSPORT (Figure 2).

     All the evidence indicates that the plasma membrane
and intracellular tight junctions of brain capillary
endothelial cells are responsible for the proper function
of BBB (14). The studies which measured unidirectional
uptake of amino acids from the blood readily demonstrated
a BBB transport system for large amino acids while that
for small neutral amino acids was minimal; the specific
entry of neutral amino acids into the isolated cerebral
capillaries is a saturable, oxygen- and sodium-independ-
ent process (16).

     Orlowski (13) proposed the possible role of gluta-
thione in the transport of amino acids through cell
membranes; this system involves gamma-glutamyl transpeptidase and reduced glutathione, the levels of which

Table 2.   NADPH-producing enzymes in brain capillaries
            and parenchyma and activity ratios.

Enzyme	Activity		
	In Capillaries	In Parenchyma	Ratio
Glucose-6-P dehydrogenase	159 ± 13	196 ± 10	0.81
6-P-gluconate dehydrogenase	83 ± 6	111 ± 6	0.75
Isocitrate dehydrogenase (NADP-linked)	315 ± 32	415 ± 18	0.76

Enzyme activities are given as nMoles of substrate
converted per hr per mg protein, 37°C, pH 7.1. S.E.M.
for n = 5 is given.

are determined by the availability of NADPH for glutathione reductase reaction.   All the NADP-linked enzymes
are fairly active in the endothelial cells (Table 2).
In addition, the capillary/parenchyma activity ratio for
gamma-glutamyl transpeptidase is 5.3.   NADP-dependent
dehydrogenases catalyze the formation of NADPH which is
then available for the reduction of oxidized glutathione
(GSSG): $NADPH + H^+ GSSG \rightarrow NADP^+ + 2 GSH$ ....1).

The main pathway of glutathione degradation is the
reaction catalyzed by the gamma-glutamyl transpeptidase.
This membrane-bound enzyme catalyzes the transfer of
gamma-glutamyl group of glutathione to amino acids
according to the following reaction: Glutathione + amino
acid (AA) ⇄ gamma-glutamyl-AA + cysteinylglycine ......
..... 2).

The gamma-glutamyl-AA complex formed in transpep-
tidation can be converted to glutamate with the con-
sequent liberation of amino acid [for details see Orlowski
(13)]. Therefore, gamma-glutamyl transpeptidase reaction

could be involved in amino acid transport in brain
capillaries. Since the NADPH-producing capacity and the
gamma-glutamyl transpeptidase activity are relatively
high in the brain capillaries, these enzymatic activities
may be linked to amino acid transport. However, some
cells have little if any gamma-glutamyl transpeptidase
and yet are still capable of amino acid transport (13).
Therefore, the gamma-glutamyl transport system is limited
only to selective sites where gamma-glutamyl transpeptidase activity is highly concentrated, (such as the
brain capillaries). The gamma-glutamyl system is probably
not solely responsible for the transport of all amino
acids. The transport of the non-metabolizable amino
acids (3) probably cannot be mediated by the gamma-
glutamyl transpeptidase reaction, since the amino acids
in which the alpha hydrogen is replaced by some other
group are not substrates for the enzyme. With respect
to NADPH requirement in the capillary amino acid transport, it is worthwhile to point out that the uptake of
neutral amino acids is increased in ischemia (10) at a
time when the activities of NADPH-producing enzymes are
slightly enhanced (Djuričić and Mršulja, unpublished).

C) ENZYMATIC BARRIER FOR MONOAMINES IN BRAIN CAPILLARIES (Figure 3).

The BBB is generally regarded as a mechanism whereby
harmful substrates circulating in the blood are excluded
from entering the central nervous system. Monoamines
and some amine precursors formed and released from
peripheral organs are prevented from entering the brain,
and histochemical techniques have provided some evidence
that the BBB for the monoamines is enzymatic in nature
(2, 8, 9). The activities of the monoamine degradating
enzymes are fairly high in brain capillaries (Table 3).
However, isolated brain capillaries took up the ^{3}H-nor-
epinephrine and the labeled substance increased with the
duration of the incubation (1). In addition, authors have
found in capillaries a large amount of methylated metabolites of norepinephrine, namely metanephrine and
normetanephrine. Brain capillaries are probably unable
to retain the norepinephrine after the inhibition of
degradative enzymes, since the inhibition of COMT and
MAO inhibited also the uptake of norepinephrine (Spatz,
personal communication). It is our tentative conclusion
that monoamines can enter the endothelial cell and
thereafter are degradated by the enzymes. In addition,
the same system probably plays a role in preventing an

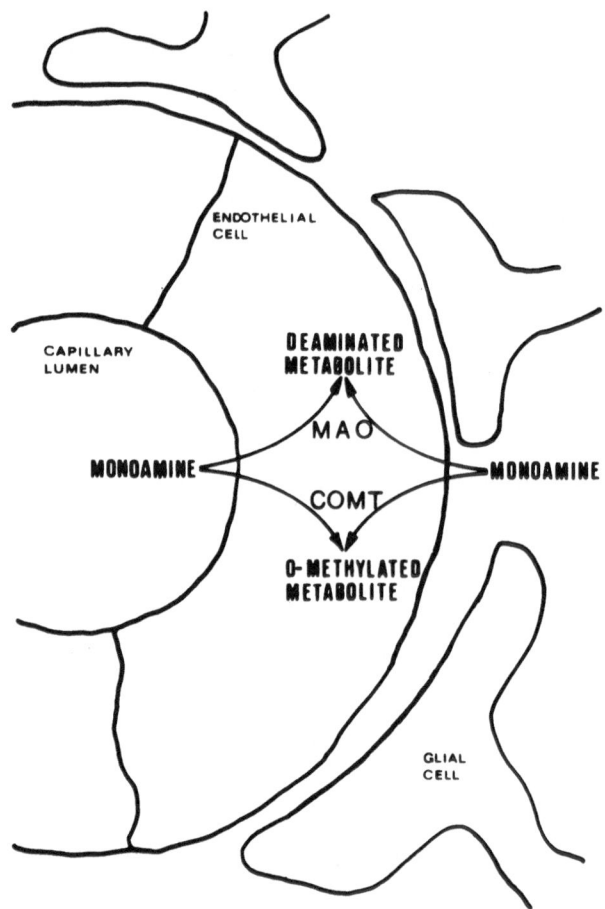

Fig. 3.   Enzymatic barrier for monoamines in capillaries.
           Abbreviations: MAO - monoamine oxidase;
           COMT - catechol-O-methyl transferase.

Table 3. Monoamine-metabolizing enzymes in brain
          capillaries and parenchyma and activity
          ratios.

Enzyme	Activity		Ratio
	in capillaries	in parenchyma	
Monoamine oxidase (kynuramine)	32 ± 5	14 ± 0.3	2.27
Monoamine oxidase A	37 ± 8	4.1 ± 1.2	9.02
Monoamine oxidase B	0.82 ± 0.3	0.034 ± 0.01	24.12
Catechol-O-methyl transferase	2.15 ± 0.14	1.61 ± 0.10	1.33

Enzyme activities are given as nMoles of substrate
converted per hr per mg protein, 37°C. S.E.M. for
n = 5 is given.

outpouring of the vasoactive amines from the brain to
the peripheral tissues during the time of their enhanced
formation or release.

CONCLUSION

For many years the complexity of BBB was not fully
realized until ultrastructural and biochemical studies
revealed that the capillary endothelial cells differ
morphologically and biochemically from other organs. In
addition, it can be concluded that brain endothelial cells
not only differ morphologically but also biochemically
from the cells of brain parenchyma. These differences
are most likely related to BBB function and maintenance
of the brain milieu.

ACKNOWLEDGMENT

These studies were supported by grant from the Union
of Sciences of Republic Serbia (No. 40404-14).

REFERENCES

1. Abe, T., Abe, K_3, Swink, M.E., Klatzo, I. and Spatz,
    M. (1978): The ^{3}H-norepinephrine uptake and fate in
    the isolated cerebral capillaries. Soc. Neuroscience,
    8th Ann. Meet. (Abstracts), Vol. 4, p. 441.

2. Bertler, A., Falck, B., Owman, Ch. and Rosengren, E.
    (1966): The localization of monoaminergic blood-brain
    barrier mechanism. Pharmacol. Rev. 18: 369-385.

3. Christensen, H.N., Aspen, A.J. and Rice, E.G. (1956):
    Metabolism in the rat of three amino acids lacking
    alpha-hydrogen. J. Biol. Chem. 220: 287-294.

4. Djuričić, B.M. and Mršulja, B.B. (1977): Enzymatic
    activity of the brain: microvessels vs. total brain
    homogenate. Brain Res. 138: 561-564.

5. Djuričić, B.M. and Mršulja, B.B. (1979): Brain micro-
    vessel hexokinase: kinetic properties. Experientia
    35: 169-171.

6. Djuričić, B.M., Rogač, Lj., Spatz, M., Rakić, Lj.M.
    and Mršulja, B.B. (1978): Brain microvessels. I.
    Enzymic activities. In: Adv. Neurology, Vol. 20,
    J. Cervos-Navarro, et al. (eds.), pp. 197-205,
    Raven Press, New York.

7. Goldstein, G.W., Wolinsky, J.S., Csejtey, J. and
    Diamond, I. (1975): Isolation of metabolically
    active capillaries from rat brain. J. Neurochem. 25:
    715-717.

8. Hardebo, J.E., Edvinson, L., Falck, B., Lindvall, M.,
    Owmen, Ch., Rosengren, E. and Svengaard, N.A.
    (1976): Experimental models for histochemical and
    chemical studies of the enzymatic blood-brain
    barrier for amine precursors. In: The Cerebral
    Vessel Wall, J. Cervos-Navarro et al., (eds.),
    pp. 233-244, Raven Press, New York.

9.   Kaplan, G.P., Hartman, B.K. and Creveling, C.R.
     (1979): Immunohistochemical demonstration of
     catechol-O-methyl transferase in mammalian brain.
     Brain Res. 167: 241-250.

10. Mićić, D.V., Swink, M., Mićić, J., Klatzo, I. and
     Spatz, M. (1979): The ischemic and postischemic
     effects on the uptake of neutral amino acids in
     isolated cerebral capillaries. Experientia 35:
     625-626.

11. Mršulja, B.B. and Djuričić, B.M. (1979): Biochemical
     characteristics of cerebral capillaries. Symposium:
     The Cerebral Microvasculature, Galveston (in press).

12. Mršulja, B.B., Mršulja, B.J., Fujimoto, T., Klatzo,
     I. and Spatz, M. (1976): Isolation of brain capil-
     laries: a simplified technique. Brain Res. 110:
     361-365.

13. Orlowski, M. (1976): Possible role of glutathione
     in transport processes. In: Transport Phenomena in
     the Nervous System, G. Levy et al., (eds.) pp. 13-28,
     Plenum Press, New York.

14. Rapoport, S.I. (1976): Blood-Brain Barrier in
     Physiology and Medicine, Raven Press, New York.

15. Sessa, G., Orlowski, M. and Green, J.P. (1976):
     Isolation from bovine brain of a fraction containing
     capillaries and fraction containing membrane frag-
     ments of the choroid plexus. J. Neurobiol. 7: 51-61.

16. Spatz, M., Mićić, D.V., Fujimoto, T., Mršulja, B.B.
     and Klatzo, I. (1979): Transport phenomena in
     cerebral ischemia. In: Pathophysiology of Cerebral
     Energy Metabolism. B.B. Mršulja, et al., (eds.)
     pp. 143-153, Plenum Press, New York.

STUDIES ON THE BLOOD-BRAIN BARRIER (BBB) TO MONOAMINES

T. Abe, K. Abe, D. Mićić, B.M. Djuričić,
B.B. Mršulja and M. Spatz

Laboratory of Neuropathology and
Neuroanatomical Sciences
National Institute of Neurological and
Communicative Disorders and Stroke
National Institutes of Health
Bethesda, Maryland 20205, U.S.A.

Laboratory for Neurochemistry
Institute of Biochemistry
Faculty of Medicine
Belgrade, Yugoslavia

INTRODUCTION

Cerebral ischemia leads to a postischemic modification of the blood-brain barrier (BBB) permeability to
substances which under normal conditions do not easily
cross the BBB. The BBB change appeared to be selective
since the leakage of various molecules tested was related
to their size and/or their specific transport system (2,
12, 13). A distinct alteration of BBB permeability has
also been observed for the monoamines (4). The enhanced
passage of 5-hydroxytryptamine (5-HT) occurred prior to
that of norepinephrine (NE) in gerbils subjected to bilateral common carotid artery occlusion and release (2).

An integral part of the BBB is the microvasculature
which remains metabolically active even after separation
from the rest of the brain tissue (1, 9). It has been
shown that the isolated cerebral microvessels contain
the monoamines' catabolic enzymes and are able to take
up and metabolize 5-HT and NE (1, 6). Thus, the microvessels are capable of regulating flux of the amines by

enzymatic inactivation. In view of these observations, the changes in the uptake of the amines and the activity of their degradating enzymes were investigated in capillaries isolated from ischemic and postischemic brains. Experiments were designed to coincide with the observed presence and absence of increased BBB permeability to 5-HT and NE (2). The correlative studies comprised a) the uptake of the metabolizable amines (5-HT and NE) and the nonmetabolizable NE analogue, metaraminol (MA), and b) the determination of catechol-O-methyl transferase (COMT) and monoamine oxidase [total MAO, (A) and (B) forms] levels in the cerebral microvessels.

This report will demonstrate a postischemic disturbance in the capillary barrier to amines which most likely is responsible for the observed selective increase in the monoamines' passage across the BBB.

MATERIAL AND METHODS

The cerebral microvessels were separated from the brains of gerbils subjected to bilateral common carotid artery occlusion for 15 minutes only, or after 24 and 72 hours of clip release. The control samples were obtained from sham-operated animals. The capillary fraction was prepared by tissue homogenization, centrifugation and discontinuous sucrose gradient (1-1.08 M) (9).

The uptake of the tritiated amines (L-NE, DLM and 5-HT) was individually determined in aliquots of the same capillary suspension (0.1 ml containing 20-40 γ of protein) in a buffered Ringer albumin solution (0.5 ml) pH 7.4 for 2-15 minutes at 36°C. The specificity of the uptake was tested by addition of respective unlabeled (cold substrates (5-10 μmoles) to the incubating medium containing the labeled substance. The reaction was stopped with 0.32 M sucrose and the samples were rinsed prior to the addition of Bray's solution. The level of radioactivity of each sample was counted in the Beckman LS 250 liquid scintillation spectrophotometer. The protein content of the tissues was determined by Lowry's method (8).

For the enzymatic assays, the microvessels were disrupted by repeated freezing and thawing. However, the addition of 0.5% triton x-100 was required for the analysis of COMT. The determination of COMT was based on Nagatsu's technique using S-adenosyl-L-methionine as the donor and the epinephrine as the acceptor for the methyl group (11). The concentration of metanephrine

separated from the epinephrine by toluene-isobutanol
mixture (68/38 v/v) and reextracted into 0.1 N HCl was
measured fluorospectrophotometrically. Appropriate concentrations of metanephrine were processed in the same
way as a measure of recovery.

The total MAO was assayed by microfluorometric techniques which measure the conversion of kynuramine to
4-hydroxyquinoline (5). Serotonin was used for MAO (A)
while p-nitrophenethylamine served as MAO (B) specific
substrate (3, 7, 10 14).

MAO (A). The samples were incubated in phosphate
buffer (75 mm, pH 7.4) containing semicarbazide-HCl (9 mm)
and serotonin (0.05 mm) for 60 minutes at 37°C. The reaction was stopped with $HClO_4$ (1.5 N) standard solution
of serotonin (5-50 µg) (without tissue) were treated
similarly. The fluorescence (the remainder of serotonin)
was measured in spectrophotofluorometer at 310 nm and
400 nm (excitation/emission).

MAO (B). The reaction mixture for the samples incubated for 60 minutes at 37°C contained phosphate buffer
(75 mm, pH 7.4), non-ionic detergent [OP-10, 1% (v/v)],
Triton-X 100 (.5% v/v), and p-nitrophenethylamine (8 mm).
The PEA (p-nitrophenylethylaldehyde) for the controls
was added just before the readings of the absorbance.
The degraded PEA was determined spectrophotometrically
at 450 nm.

The following labeled substances were used in these
investigations: norepinephrine Dl-[1-^{3}H (N)], spec.
activity 10.43 Ci/mmol, norepinephrine L-[7,8-^{3}H (N)],
spec. activity 25.4 Ci/mol, metaraminol, DL [7-^{3}H (N)],
spec. activity 6.93 Ci/mmol, and 5-hydroxytryptamine
creatine sulfate 5[1,2-^{3}H (N)], spec. activity 25.1
Ci/mmol, and were purchased from New England Nuclear,
Boston, Mass. All the cold substrate except for metaraminol and S-adenosyl-L-methionine were obtained from
Sigma Chemical Co., St. Louis, Mo., while metaraminol
and S-adenosyl-L-methionine were purchased from Regis
Chemical Co., Chicago, Ill., and Boehringer Mannheim Co.,
Indianapolis, Ind., respectively.

RESULTS AND COMMENTS

The changes in the uptake of 5-HT, NE and MA as
well as in the concentration of COMT and MAO activity
observed in the ischemic and postischemic cerebral microvessels are summarized in Tables 1 and 2, respectively.

Table 1. Uptake of labeled monoamines in isolated cerebral microvessels

Substrate	Sham	Ischemia (15 min)	Recovery	
			24 hours	72 hours
		Relative Uptake in Percent		
^{3}H 5-HT	100	102.9 + 7.7	140.5 + 12.0*	133.8 + 4.6*
^{3}H NE	100	106.0 + 9.0	104.4 + 4.3	137.0 + 2.0*
^{3}H MA	100	100.8 + 7.4	103.9 + 4.2	154.2 + 12.8

The values represent the mean + S.E. in 8-12 assays determined either in duplicate or quadruplicate. Controls (100%) 5-HT 66379 + 3726; NE 63338 + 1563; MA 18493 + 1040 CPM/mg-1P

The yield of the total capillary proteins in all experimental groups was similar to their respective sham-operated controls

* Period of BBB substrate's leakage observed in vivo experiments

Table 2. Catechol-0-methyltransferase and monoamine oxidase activities in cerebral microvessels

	Controls	Ischemia (15 min)	Recovery	
			24 hours	72 hours
COMT	1.93 ± 0.4	27.6 ± 1.0	4.7 ± 1.3	28.1 ± 4.8
MAO (kynuramine)	32.5 ± 5.3	7.1 ± 0.8	6.2 ± 0.3	12.5 ± 0.8
MAO A (serotonin)	37.6 ± 8.2	14.2 ± 1.7	5.9 ± 0.5	10.3 ± 0.6
MAO B (p-nitrophenyl-ethylamine)	0.82 ± 0.3	0.48 ± 0.04	0.28 ± 0.01	0.53 ± 0.04

The values are expressed as nMoles/mg prot./hr M ± S.E.M (4)

As may be seen from Table 1, the increase in the capillary uptake of 5-HT occurred prior to that of NE and MA,
in concurrence with the findings of selective postischemic
BBB leakage. At the same time (24 hours recovery), the
depression of MAO activity was maximal and the elevation
of COMT activity was considerably smaller than that of
the other experimental periods (Table 2). Thus, the
most marked reduction of the MAO activity coincided with
the increased capillary 5-HT uptake and the increased
passage of 5-HT from blood to brain. Subsequently, at
72 hours of postischemia when the BBB permeability was
altered for 5-HT and NE, the uptake of each amine, the
metabolizable (5-HT and NE) and the nonmetabolizable
(MA), in the isolated microvessels was above the normal
levels (Table 1). At this point, it should be mentioned
that the increase in the amines' capillary uptake was
almost always seen with 5-HT, while the increase of NE
uptake was not found unless the MA showed a similar
change, as was the case in the 72-hour postischemic
microvessels. Likewise, the incidence of 5-HT increased
BBB permeability was higher than that of NE in this period
(2). The ischemic effect on the capillary NE uptake was
the same whether the L- or the DL-form of the labeled
substance was used for these experiments. The rise of
each amine's capillary uptake was a specific one, since
it could be inhibited to the same degree as the normal
uptake by addition of the respective cold substrates (80%
inhibition with 10 μm of each amine) to the incubating
medium. Furthermore, a fourteen-fold higher activity
of COMT, with a 30% reduction in the concentration of
MAO, was seen in the microvessels from 72-hour post-
ischemic brains over those from the controls. A similarly altered activity of both enzymes was observed in
cerebral microvessels after ischemia only (Table 2).

These findings indicate that the enzymes involved
in the capillary monoamines' inactivation are conspicuously modified by cerebral ischemia. However, the
mechanism responsible for the ischemic and postischemic
elevation of COMT activity is unknown. Our preliminary
observations suggest that ischemia alters the kinetic
properties of the enzyme since Co^{++} stimulated the NE
methylation to a greater extent in the experimental
than in the control microvessels (unpublished observations). As far as MAO is concerned, this enzyme's
activity is oxygen-dependent; therefore, the observed
reduction of MAO could be the result of either the
inadequate supply or the inefficient utilization of the
oxygen. In any case, the changes in the capillary
enzymes cause an increased methylation and decreased

deamination of the exogenous amines in ischemic and post-ischemic microvessels, as indicated by the recovered levels of the natural substrates used for the enzymatic assays.

On the other hand, the increased microvascular methylating ability apparently had no direct relationship to the capillary NE uptake, since the enzyme's levels were similarly altered in ischemia and postischemia but the NE uptake was not found changed before 72 hours of recovery. The greater accumulation of not only NE but also of MA above the normal levels strongly suggests that at this particular time the disturbance of both the transport and metabolism might be responsible for the observed increased passage of NE across the BBB (Table 1).

Under normal conditions, most of the NE taken up by the isolated cerebral capillaries is methylated and re-leased into the incubating medium (1). It is therefore possible that the BBB leakage of NE does not occur until the microvascular uptake exceeds the catabolism of the amine. Moreover, the isolated cerebral microvessels were found to have a greater affinity for 5-HT than for NE uptake and metabolism (5-HT Km 2.1 and NE Km 14.7 nmoles/mg P). Most of the 5-HT taken up by the capil-laries was normally deaminated and released into the incubating medium (1).

Thus, the high level of 5-HT in the cerebral micro-vessels observed at an earlier time than that of the NE (Table 1) is most probably due to the capillary higher affinity for 5-HT than for NE and the reduced capability of the microvessels to deaminate the 5-HT. Both factors therefore may play the decisive role in the appearance of 5-HT before NE leakage of BBB in postischemia.

CONCLUSION

Selectively altered cerebral microvascular accu-mulation of monoamines was found in postischemia while the enzymes involved in the degradation of the amine were affected by ischemia and postischemia. The modi-fied function of the isolated cerebral microvessels was closely related to the BBB permeability changes for monoamines observed in parallel studies (1).

REFERENCES

1. Abe, T., Abe, K., Rausch, W.D., Klatzo, I. and
    Spatz, M. (1979):   Characteristics of some monoamines
    uptake systems in isolated cerebral capillaries.   In:
    Symposium on Cerebral Microvasculature: Investigation
    of the Blood Brain Barrier, Galveston, July 1979.
    Plenum Press, New York (in press).

2. Abe, K., Abe, T., Klatzo, I. and Spatz, M. (1979):
    The effect of endogenous central nervous system
    depressants in ischemic cerebral edema of gerbils.
    In: International Symposium: Brain Edema - Pathology
    and Therapy, Berlin, September 1979.   Raven Press,
    New York (in press).

3. Gallagher, B.M. (1977): Multiple monoamine oxidase
    activities in heterogenous populations of mouse lung
    mitochondria.   Biochem. Pharmacol. 26: 935-938.

4. Hervonen, H., Steinwall, O., Spatz, M. and Klatzo, I.
    (1979): Behaviour of the blood-brain barrier (BBB)
    toward biogenic amines in experimental cerebral
    ischemia.   In: Symposium on Cerebral Microvas-
    culature: Investigation of the Blood Brain Barrier,
    Galveston, July 1979.   Plenum Press, New York (in
    press).

5. Krajl, M. (1965): A rapid microfluorimetric determi-
    nation of monoamine oxidase.   Biochem. Pharmacol.
    14: 1684-1685.

6. Lai, F.M. and Spector S. (1978): Studies on the
    monoamine oxidase and catechol-O-methyltransferase
    of the rat cerebral microvessels.   Arch. Int.
    Pharmacodyn. 233: 227-234.

7. Lewinsohn, R., Böhm, K.H., Glover, V. and Sandler,
    M. (1978): A benzylamine oxidase distinct from
    monoamine oxidase B - widespread distribution in
    man and rat.   Biochem. Pharmacol. 27: 1857-1863.

8. Lowry, O.H., Rosebrough, N.J., Farr, A.L. and
    Randall, R.J. (1951): Protein measurement with the
    folin phenol reagent.   J. Biol. Chem. 193: 265-275.

9. Mrsulja, B.B., Mrsulja, B.J., Fujimoto, T., Klatzo,
    I. and Spatz, M. (1976): Isolation of brain capil-
    laries: a simplified technique.   Brain Res. 110:
    361-365.

10. Murphy, D.L., Donnelly, C.H., Richelson, E. and
     Fuller, R.W. (1978): N-substituted cyclopropylamines
     as inhibitors of MAO-A and B forms. Biochem. Pharma-
     col. 27: 1767-1769.

11. Nagatsu, T. (1973): Assay of catechol-O-methyltrans-
     ferase activity. In: Biochemistry of Catecholamines,
     The Biochemical Method, pp. 181-183, University of
     Tokyo Press, Tokyo.

12. Nishimoto, K., Wolman, M., Spatz, M. and Klatzo, I.
     (1978): Pathophysiologic correlations in the blood-
     brain barrier damage due to air embolism. Adv.
     Neurol. 20: 237-244.

13. Spatz, M., Fujimoto, T. and Go, G.K. (1976):
     Transport studies in ischemic cerebral edema. In:
     Dynamics of Brain Edema, H.M. Pappius and W. Feindel
     (eds.), pp. 181-186, Springer Verlag, Berlin-
     Heidelberg.

14. Yang, H.Y.T. and Neff, N.H. (1973): N-phenethylamine:
     a specific substrate for type B monoamine oxidase of
     brain. J. Pharmacol. Exp. Therapeut. 187: 365-371.

DYNAMICS OF POSTISCHEMIA:

ENZYMES IN BRAIN CAPILLARIES

B. M. Djuričić and B. B. Mršulja

Laboratory for Neurochemistry
Institute of Biochemistry
Faculty of Medicine
Belgrade, Yugoslavia

The isolated brain microvessels have been described as the morphological and functional correlate of the blood-brain barrier (9) and represent a separate and comparatively unique metabolic compartment (3, 5, 6). Previous investigations on the metabolism of brain capillaries have indicated that the brain microvasculature is more resistant to ischemia than the brain parenchyma (18). However, the metabolism of the microvessels does change during recirculation regardless of the intensity of the ischemic insult (5). Just as in the brain parenchyma, the pathophysiological events related to ischemia are evident in brain microvessels even after the blood circulation has been re-established (12). The severity of an ischemic insult is manifested primarily in the postischemic period (15). Recirculation appears to trigger its own set of events, including alterations in metabolite levels as well as in enzyme activities (5, 7, 19, 24). For example, some derangements in the levels of certain neurotransmitters of the brain were exhibited even after 7 days of recirculation following 15 minutes of bilateral ischemia in gerbils (2). In contrast, the enzymatic changes seem to be more pronounced in earlier stages of postischemia.

The report that brain edema develops during the early stages of reflow may be related to disturbances in the blood-brain barrier (17). A group of enzymes (I) have been implicated in the function of the blood-brain barrier, because their activities are somewhat greater in

the brain capillaries than in the parenchyma (6, 16).
The enzymes included glucose-6-phosphatase (G6Pase),
phosphodiesterase (PDE), gamma-glutamyl transpeptidase
(GGT) and adenosine deaminase (ADA), and these enzymes
were measured at various periods of recirculation following 15 minutes of bilateral ischemia. Another group of
enzymes (II), which include the pentose pathway enzymes
and hexokinase (HK), also exhibit some enrichment in the
capillaries. The results indicate that the metabolic
changes in brain capillaries are more dramatic during the
early stages of postischemia.

EXPERIMENTAL

Mongolian gerbils (Meriones unguiculatus) of both
sexes were used; animals weighing 60-70 g had free access
to food and water.

Brain ischemia was induced by clipping both common
carotid arteries in the neck region with Heifetz aneurysm
clips (20). The ischemia of the cerebral hemispheres
after bilateral occlusion has been attributed to an
anomaly of the circle of Willis (8). The animals were
sacrificed by decapitation either after 15 minutes of
bilateral ischemia or after the designated time of recirculation. Forebrain capillaries were isolated and prepared for enzymatic assays as described by Djuricić and
Mrsulja (3).

Standard spectrophotometric and fluorometric procedures were used for the determination of enzymatic activities. The enzyme substrate for the GGT reaction was
gamma-glutamyl-p-nitroanilide (25), for the PDE it was
nitrophenyl-p-thymidine 5'-phosphate (23) and for the ADA
it was adenosine. The other enzymes were assayed with
their respective substrates, using NAD(P)-coupled reactions (13). The GGT and PDE were assayed at a pH of 8.9,
whereas the rest of the enzymes were measured at a pH of
7.1. All incubations were performed at 37°C. Enzymatic
activities were expressed on a protein basis (14).
Statistical significance was estimated with the Student's
t-test; $P < 0.05$ was considered significant.

To facilitate the explanation of the changes in
enzyme activity during postischemia, a differential enzyme
velocity (ΔV) was calculated by the following equation:

$$\Delta V = S_y - S_x$$, where S represents the enzyme
activities at time x and y of postischemia. In order to

compare the changes that occurred between each interval, the values were reduced to their lowest common denominator, as shown by the following equation:

$$\Delta V \text{ / 5 min of postischemia} = \frac{\Delta V \text{ / } n}{\text{control activity}} \times 100,$$

where $n = \frac{y - x}{5 \text{ min}}$ describes the number of five minute periods between the sampling at times x and y. Hence, each value for ΔV / 5 min represents the percent of the control value for which the enzyme activity either increased (+) or decreased (-).

RESULTS

Ischemia

I.   In the first group of enzymes, the PDE and GGT increased with a rate of 63 and 55% per 5 minutes, respectively, and ADA decreased at a rate of 12% (Figs. 1 and 2). The changes in G6Pase activity were not significant.

II.   The relative activities of HK and G6PDH significantly decreased at about the same rate during ischemia, whereas the activity for 6GPDH was not significantly different (Figs. 3 and 4). All the enzymes of the non-oxidative branch of the pentose pathway, with the exception of transketolase (TK), were increased after 15 minutes of ischemia. Transaldolase (TA) was the most severely affected, and the ribose-5-phosphate isomerase the least (Figs. 5 and 6). The TK activity is reduced after the ischemic episode. It should be noted that the enzymes of the non-oxidative branch of the pentose pathway exhibit greater changes than those for the oxidative (i.e., G6PDH and 6PGDH).

Postischemia

I.   The most profound changes in the enzymatic activities in brain capillaries were observed in the first five minutes of recirculation, the rate of change reaching as high as 400% of the control activity per 5 minutes for TK (Fig. 6).

The changes in those enzymes that are thought to play a role in capillary function did not exhibit a uniform pattern during postischemia. G6Pase activity decreased significantly during the first 5 minutes of post-

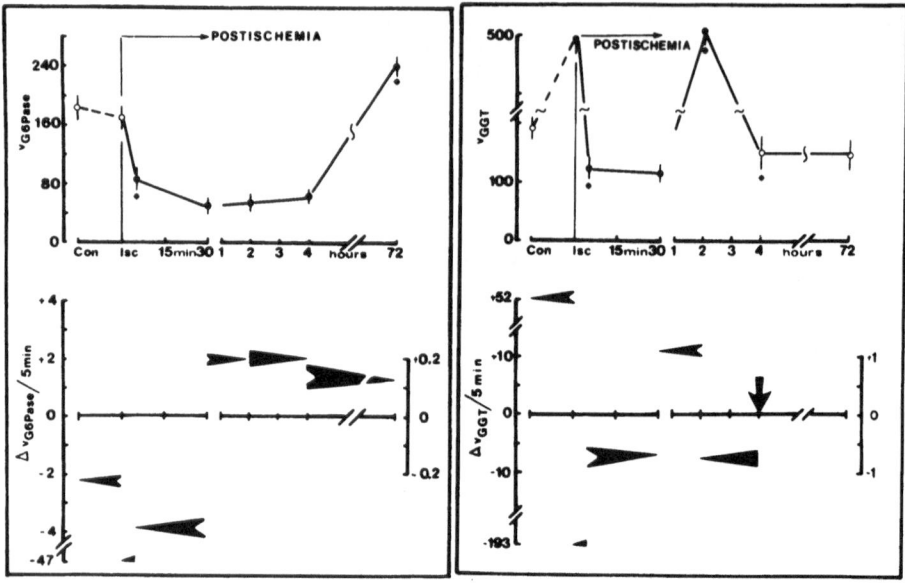

Fig. 1.   Changes in the activities of glucose-6-phospha-
           tase (G6Pase) and gamma-glutamyl transpeptidase
           (GGT) during postischemia in brain capillaries.
           Closed circles indicate significant differences
           when compared with the preceding value. Verti-
           cal bars indicate the SEM for 4 determinations.
           The Δ V/5 min was calculated as described under
           the experimental methods. The control activity
           for G6Pase and GGT was 184 ± 18 and 194 ± 12
           nmoles/hour^{-1}/mg^{-1} protein, respectively. Con
           is an abbreviation for controls and Isc for 15
           min of ischemia.

ischemia, remained depressed for up to 4 hours of reflow
and thereafter slowly increased. At 72 hours of recirculation, the activity was significantly greater than that
for control. In general, the changes in G6Pase activity
after the initial depression were rather slow (Fig. 1).

        During the first five minutes of reflow, the GGT
activity decreased rapidly, but did not change further
at 30 minutes of recirculation. Subsequently, there was
a secondary rise in activity comparable to that observed
during ischemia. By 4 hours of reflow, the values were
back to control and remained there at 72 hours.

        After the large increase in PDE activity during
ischemia, the levels drop off rapidly and approximate

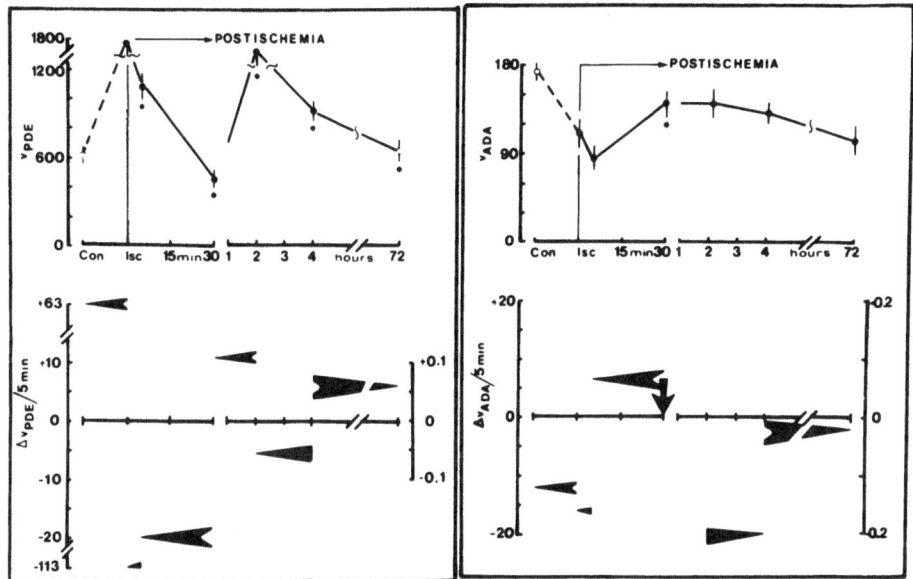

Fig. 2.   Changes in the activities of phosphodiesterase
          (PDE) and adenosine deaminase (ADA) during post-
          ischemia in the capillaries. For details, see
          legend to Fig. 1. The control activity for PDE
          and ADA was 616 ± 39 and 172 ± 8 nmoles/hour^{-1}/
          mg^{-1} protein, respectively.

those of control by 30 minutes of reflow (Fig. 2). Of
this particular group of enzymes, this is the largest
change to be observed. Similar to the response with GGT,
the activity once again rises at 2 hours of reflow to a
peak value similar to that for ischemia and thereafter
decreases to control values by 72 hours (Fig. 2).

    The changes in ADA are less dramatic than for the
others in this group. The only significant change was
the increase in activity between 5 and 30 minutes of recirculation. Full recovery of this enzyme to control
level was not evident even after 72 hours of recirculation.

    II. The HK activity remained low during the first
2 hours of reflow and then increased toward control level
(Fig. 3). A possible control point for the pentose pathway was indicated by the opposite changes in the activities of the two NADP-linked dehydrogenases (G6PDH and
6PGDH). When the activity of G6PDH is low during the
early stages of postischemia, 6PGDH activity is increasing
to a peak at 30 minutes of reflow. After 30 minutes of

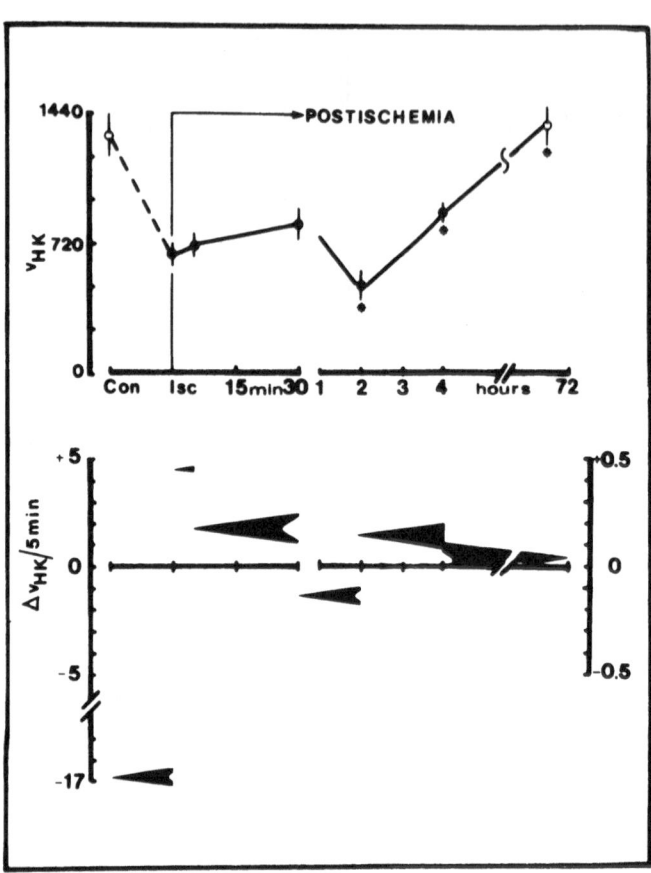

Fig. 3. Changes in the activity of hexokinase (HK)
         during postischemia in brain capillaries. For
         details, see legend to Fig. 1. The control
         activity for HK was 1320 ± 120 nmoles/hour^{-1}/
         mg^{-1} protein.

recirculation, the G6PDH activity increases to a maximum
at 4 hours, as that for 6PGDH decreases to near control
values. At 72 hours, the levels of G6PDH were somewhat
lower than control, but those for 6PGDH were not significantly different from control.

    The most dramatic changes in activity during recir-
culation were observed in the 4 enzymes of the nonoxidative portion of the pentose pathway. With the exception
of TK, there is a reversal of the ischemic changes during
the first 5 minutes of reflow. The TK activity increases
to a peak 30 minutes after reflow and by 2 hours had decreased to control values. All 4 of the nonoxidative

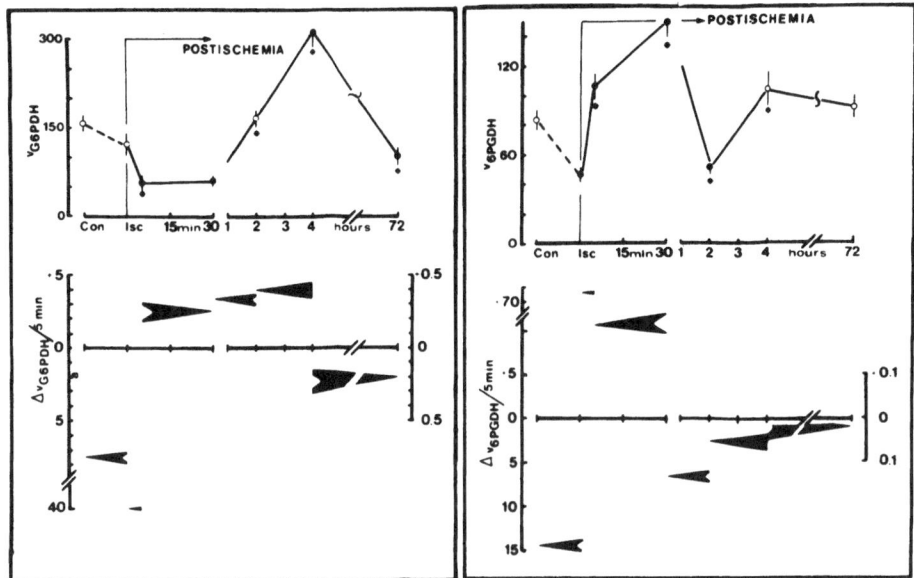

Fig. 4.   Changes in the activities of glucose-6-phosphate
          dehydrogenase (G6PDH) and 6-phosphogluconate de-
          hydrogenase (6PGDH) during postischemia in brain
          capillaries. For details see legend to Fig. 1.
          The control activity for G6PDH and 6PGDH was
          159 ± 13 and 83 ± 6 nmoles/hour^{-1}/mg^{-1} protein,
          respectively.

enzymes measured increased to a peak at 4 hours of recirculation and were back to or below control at 72 hours.

DISCUSSION

     The metabolic effects of brain ischemia are not
solely restricted to the duration of blood deprivation.
Although there is generally a rapid restoration of energy
metabolites in the brain tissue (15), changes in electro-
lytes, water content, neurotransmitters and enzymes persist during postischemia (17). Ischemia severely trauma-
tizes the brain, and no compartment of the tissue remains
unaffected. Brain capillaries are not an exception; their
metabolic alterations are probably caused not only by the
ischemia but also secondarily by the derangements in the
brain environment.

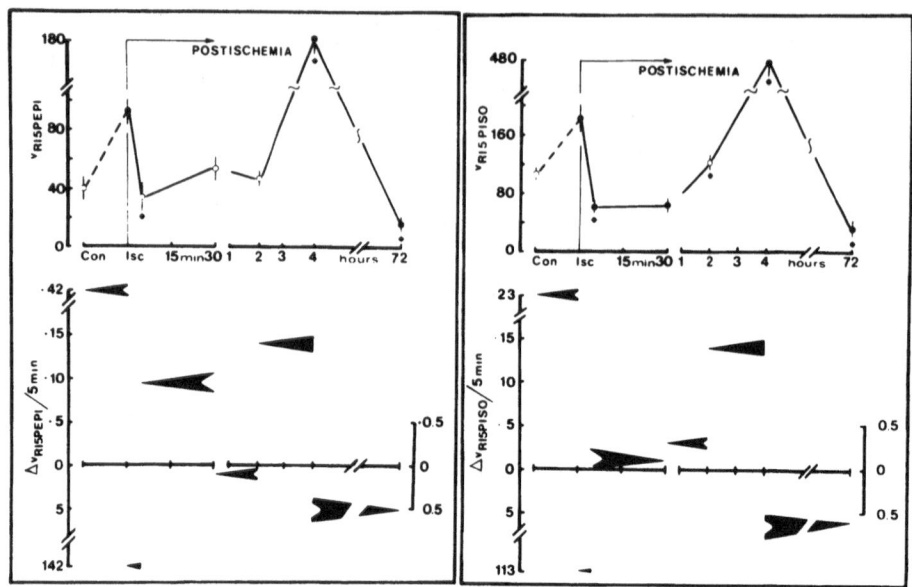

Fig. 5.   Changes in the activities of ribulose-5-phosphate
epimerase (Ri-5-P Epi) and ribose-5-phosphate
isomerase (Ri-5-P Iso) during postischemia in
brain capillaries. For details, see legend to
Fig. 1. The control activity for Ri-5-P Epi and
Ri-5-P Iso was 40 ± 8 and 107 ± 7 nmoles/hour^{-1}/
mg^{-1} protein, respectively.

Brain microvessels represent the morphological and
functional basis of the blood-brain barrier. The transport of substances from blood to brain is a specific
function of the endothelial cells and their metabolism
most probably plays a role in this process. It is possible that certain of the microvessel enzymes have a
specific function in the mechanism of the blood-brain
barrier. Although proof of such a relationship is lacking,
certain similarities exist between the response of HK
and G6Pase (drop of activities in ischemia and early postischemia) and the decrease in the 2-deoxy-D-glucose uptake in capillaries isolated from an ischemic brain (26).
Similarly, the observation that uptake of amino acids in
brain capillaries from an ischemic brain is increased
(26) may be related to the change in GGT activity, an
enzyme implicated in amino acid transport (22).

A higher specific activity of certain enzymes in
capillaries may indicate that enzymes are "markers" of
capillaries and/or that these enzymes are related to

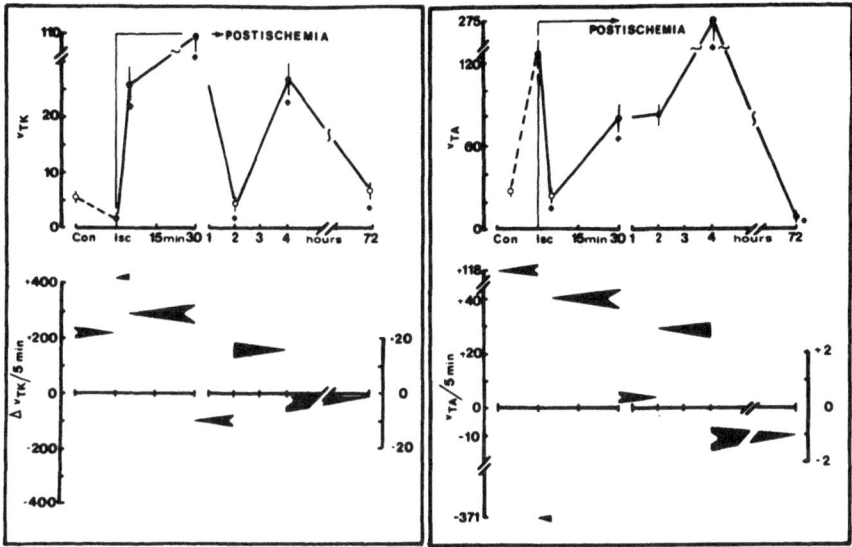

Fig. 6.   Changes in the activities of transketolase (TK)
and transaldolase (TA) during postischemia in
brain capillaries. For details, see legend to
Fig. 1. The control activity for TK and TA was
5.8 ± 0.5 and 28 ± 2 nmoles/hour^{-1}/mg^{-1} protein,
respectively.

capillary function. With this in mind, it is evident
that brain capillaries do partially recover their functional metabolism within 4 hours of postischemia. However, the earlier changes during postischemia are quite
pronounced. As the alterations in the enzymes associated
with capillaries differ, the regulation of their activities appears to be different.

HK is the only enzyme which was more affected by
ischemia than by postischemia. In contrast to the other
enzymes, HK activity did not change during the early
periods of reflow. Since numerous metabolic factors have
been shown to affect capillary HK activity (4), these
metabolic alterations may partially account for the
changes in the enzyme activity.

The pentose shunt seems to be more active in brain
capillaries than in the brain parenchyma (16). However,
even within the pentose pathway the oxidative and nonoxidative branches react differently to reflow. In capillaries, G6PDH is the least reactive enzyme among the
pentose pathway enzymes, and this finding could be

explained by this enzyme's being under separate control
from the other shunt enzymes (10). The time course of
the 6PGDH activity during postischemia more closely resembles that of HK; the changes are more pronounced during
later phases of postischemia. In addition, the opposite
pattern of the G6PDH and 6PGDH changes during reflow further indicate a control point in the pentose pathway (11).

The non-oxidative branch of the pentose shunt is
very sensitive to ischemia but much more so to postischemia. Two rate-limiting enzymes of the shunt, TA and
TK (21, 27), exhibit the greatest changes, increasing the
overall metabolic capacity of the pentose pathway. Moreover, the pentose pathway can also serve in a non-oxidative capacity, since the TA and TK reactions are reversible. In this situation the pentose phosphates are formed
from fructose phosphate and triose phosphates, without a
reduction of either $NADP^+$ or NAD^+, which are essential for
anaerobic glycolysis. Two pentose-5-P related enzymes
(epimerase and isomerase) responded similarly during postischemia, and the changes were more pronounced between the
second and fourth hour of postischemia. However, a full
explanation of these changes in the enzymes in the pentose
pathway is still lacking, since a number of factors that
influence the enzyme activities [e.g., norepinephrine,
serotonin, acetylcholine, ions (1)] are also altered during
ischemia and postischemia (2).

CONCLUSION

Postischemia supposedly represents a period of recovery. However, "recovery" is questionable, because the
number of metabolic derangements continues to occur or
appear. Since the brain capillaries are an integral part
of the brain, they too are susceptible to the noxious
effects of ischemia. But when the capillary enzymes and
the enzymes of the non-oxidative branch of the pentose
pathway are considered, the most significant changes are
observed not during ischemia but in the early stages of
postischemia (up to the fourth hour). Most of the
enzymic changes are evident during the period that coin-
cides with the development of ischemic brain edema (17).
Moreover, some of the altered enzyme activities are not
restored even after 3 days of reflow. Therefore, the
"stress" of metabolism that occurs in ischemia is not
readily reversed in postischemia.

ACKNOWLEDGMENT

This study was supported by a grant from the Union of Sciences of Republic Serbia (No. 40404-14).

REFERENCES

1. Appel, S.H. and Parrot, B.L. (1970): Hexose monophosphate pathway in synapses. J. Neurochem. 17: 1619-1626.

2. Cvejić, V., Djuričić, B.M. and Mršulja, B.B. (1980): Oscillatory pattern of catecholamine metabolism following transient cerebral ischemia in gerbils. This volume.

3. Djuričić, B.M. and Mršulja, B.B. (1977): Enzymic activity of the brain: Microvessels vs. total forebrain homogenate. Brain Res. 138: 561-564.

4. Djuričić, B.M. and Mršulja, B.B. (1979): Brain microvessel hexokinase: Kinetic properties. Experientia 35: 169-171.

5. Djuričić, B.M. and Mršulja, B.B. (1979): Brain microvessels: Glucose metabolizing enzymes in ischemia and subsequent recovery. In: Pathophysiology of Cerebral Energy Metabolism, B.B. Mršulja, Lj.M. Rakić, I. Klatzo and M. Spatz (eds.), pp. 239-252, Plenum Press, New York

6. Djuričić, B.M., Rogač, Lj., Spatz, M., Rakić, Lj.M. and Mršulja, B.B. (1978): Brain microvessels. I. Enzymic activities. Adv. neurol. Vol. 20, pp. 197-206, Raven Press, New York.

7. Djuričić, B.M., Simić, M. and Mršulja, B.B. (1979): Experimental brain ischemia and postischemia: Lactate dehydrogenase isoenzymes. Expt. neurol. (Submitted for publication).

8. Gaudet, R.J. and Levine, L. (1979): Transient cerebral ischemia and brain prostaglandins. Biochem. Biophys. Res. Comm. 86: 893-901.

9. Katzman, R. (1976): Workshop summary. In: Dynamics of Brain Edema, H.M. Pappius and W. Feindel (eds.), pp. 373-377, Springer, Berlin.

10. Kauffman, F.C. (1972): The quantitative histochemistry of the pentose phosphate pathway in the central nervous system of the rat. J. Neurochem. 19: 1-9.

11. Kauffman, F.C., Brown, J.G., Passonneau, J.V. and
     Lowry, O.H. (1969): Effects of changes in brain
     metabolism on levels of pentose phosphate pathway
     intermediates. J. biol. Chem. 244: 3647-3653.

12. Klatzo, I. (1979): Cerebral oedema and ischaemia.
     In: Recent Advances in Neuropathology, Vol. 1,
     W.T. Smith and J.B. Cavanagh (eds.), pp. 27-39,
     ChurchillLivingstone, Edinburgh, London and New York.

13. Lowry, O.H. and Passonneau, J.V. (1972): A flexible
     system óf enzymatic analysis. Academic Press, New
     York.

14. Lowry, O.H., Rosebrough, N.J., Farr, A.L. and
     Randal, J.R. (1951): Protein measurement with the
     Folin phenol reagent. J. biol. Chem. 193: 265-275.

15. Mršulja, B.B. (1979): Some new aspects of the patho-
     chemistry of the postischemic period. In: Pathophysi-
     ology of Cerebral Energy Metabolism, B.B. Mrsulja,
     Lj.M. Rakić, I. Klatzo and M. Spatz (eds.), Plenum
     Press, New York.

16. Mršulja, B.B. and Djuričić, B.M. (1979): Biochemical
     characteristics of cerebral capillaries. Symposium on
     the Cerebral Microvasculature, Galveston (In press).

17. Mršulja, B.B., Djuričić, B.M., Cvejić, V., Mršulja,
     B.J., Abe, K., Spatz, M. and Klatzo, I. (1980): Bio-
     chemistry of experimental ischemic brain edema. Pro-
     ceedings of First International Ernst Reuter Sympo-
     sion on Brain Edema, Raven Press, New York (In press).

18. Mršulja, B.B., Djuričić, B.M., Mršulja, B.J., Rogač,
     Lj., Spatz, M. and Klatzo, I. (1978): Brain micro-
     vessels. II. Effects of ischemia and dihydroergotoxine
     in enzymic activities. Adv. neurol. Vol. 20, pp.207-
     213, Raven Press, New York.

19. Mršulja, B.B., Lust, W.D., Mršulja, B.J., Passonneau,
     J.V. and Klatzo, I. (1976): Postischemic changes in
     certain metabolites following prolonged ischemia in
     the gerbil cerebral cortex. J. Neurochem. 26: 1099-
     1103.

20. Mršulja, B.B., Mršulja, B.J., Cvejić, V., Djuričić,
     B.M. and Rogač, Lj. (1978): Alterations of putative

neurotransmitters and enzymes during ischemia in gerbil cerebral cortex. J. Neural. Transm., Suppl. 14: 23-30.

21. Novello, F. and McLean, P. (1968): The pentose phosphate pathway of glucose metabolsim. Biochem. J. 107: 775-791.

22. Orlowski, M. (1976): Possible role of glutathione in transport processes. In: Adv. Expt. Med. Bioll., Vol. 69, pp. 13-28, Plenum Press, New York.

23. Razzel, W.E. and Kherana, H.G. (1959): Studies on polynucleotides. I. J. biol. Chem. 234: 2105-2113.

24. Schwartz, J.P., Mršulja, B.B., Mršulja, B.J., Passonneau, J.V. and Klatzo, I. (1976): Alterations of cyclic nucleotide-related enzymes and of ATPase during unilateral ischemia and recirculation in gerbil cortex. J. Neurochem. 27: 101-107.

25. Sessa, G., Orlowski, M. and Green, J.P. (1976): Isolation from bovine brain of a fraction containing capillaries and a fraction containing membrane fragments of the choroid plexus. J. Neurobiol. 7: 51-61.

26. Spatz, M., Mićić, D., Fujimoto, T., Mršulja, B.B. and Klatzo, I. (1979): Transport phenomena in cerebral ischemia. In: Pathophysiology of Cerebral Energy Metabolism, B.B. Mršulja, Lj.M. Rakić, I. Klatzo and M. Spatz (eds.), pp. 143-153, Plenum Press, New York.

27. Srivastava, L.M. and Hübscher, G. (1966): Glucose metabolism in the mucosa of small intestine. Enzymes of the pentose phosphate pathway. Biochem. J. 101: 48-55.

ONTOGENY OF MEMBRANE-BOUND PROTEIN PHOSPHORYLATING SYSTEMS IN THE RAT

Richard Rodnight and Helen Holmes

Department of Biochemistry
Institute of Psychiatry
British Postgraduate Medical Federation
London University
De Crespigny Park, London, SE5 8AF UK

ABSTRACT

The ontogeny of the major protein phosphorylating systems in membrane fragments from the cerebral cortex was studied in the rat, a species in which synaptogenesis predominantly occurs after birth. The phosphorylation of certain proteins associated with kinase activity largely dependent on either cyclic AMP or Ca^{2+} (in the presence of calmodulin) increased. The activity changed markedly during the period of 10-15 days after birth, i.e. coinciding with the onset of major synaptogenesis. However, the activity of other systems controlled by these factors increased more gradually from birth to adulthood. In contrast, the activity of a phosphorylating system transferring to a protein of 47K daltons in a reaction dependent on Ca^{2+} only, was abundant at birth and until 15 days of age, but had decreased some 5-fold by the time adulthood was reached.

INTRODUCTION

Considerable circumstantial evidence (reviewed in references 5, 14) points to roles for the process of membrane-located protein phosphorylation in synaptic transmission in the mammalian CNS. Thus, preparations of synaptic plasma membrane fragments constitute a remarkably rich source of protein phosphorylating

systems through which integral membrane proteins are
cyclically phosphorylated and dephosphorylated in
reactions dependent on cyclic AMP or calcium ions, or
apparently independent of both these factors (5, 14).
Very little is known of the functions of the numerous
phosphate acceptor proteins typical of synaptic membranes
except that they are located in the terminal plasma
membranes as well as in the post-synaptic membrane and
its associated densities (6, 8, 13, 15, 16). Many are
no doubt membrane-bound enzymes whose activity is regulated by a phosphorylation - dephosphorylation cycle;
others may be 'structural' proteins concerned in such
processes as the control of membrane permeability.
Other possible functional roles for the phosphorylating
system include (1) neurotransmitter release (for Ca^{2+}-
regulated systems 3, 9), (2) regulation of the
polarization state of the post-synaptic cell (5)
(possibly for neurotransmitters which bind to receptors
linked to adenylate cyclase), and (3) modulation over
relatively long time intervals of the 'receptivity'
of the post synaptic cell.

A striking feature of the protein phosphorylating
systems present in synaptic plasma membrane fractions
made from adult animals, compared with other subcellular
fractions and membrane preparations from other sources,
is the complexity of the pattern of endogenous proteins
that accept phosphate from ATP. Are all of these acceptors typical of synapses or are some of them less
specialized membrane components? To study this
question we have examined the ontogeny of protein
phosphorylating systems in the rat, an animal in which
the main spurt of synaptogenesis is delayed until the
10-26 days after birth (1, 2).

METHODS

Preparation of membrane fragments

These were prepared from rat cerebral cortex
essentially by the procedure of Jones and Matus (7),
except that, for animals up to 5 days old, cerebral
cortices were homogenized in 4.5 vol. of 0.32 M-sucrose,
instead of 9 vol. Three fractions were obtained from
the gradient: a 'light' fraction (over 0.925M-sucrose),
an 'intermediate' fraction (over 1.12M-sucrose), and
a 'heavy' fraction (pellet in 1.12M-sucrose). In
adult animals, it was shown by Jones and Matus (7) that
the 'light' fraction consists of myelin fragments, the
'intermediate' fraction of synaptic plasma membrane

fragments, and the 'heavy' fraction of crude mitochondria. In the present work, we have used the
'intermediate' fraction only. Preparations in pellet
form were stored at -30°C overnight. The pellets were
resuspended in 4 mM-imidazole/HCl buffer (pH 7.4) to
give a final protein concentration of 10 mg/ml approximately one hour before use.

Protein Determination

Protein was determined by the method of Miller (12)
with bovine plasma albumin as standard.

Phosphorylation of membrane proteins

The incubation mixture for the labelling reaction
(final volume, 82.5 µl) contained $20\mu M$-^{32}P-[ATP] (600
cpm pmol^{-1}), 30 mM-Tris/HCl (pH 7.4 at 37°C), 1 mM-MgSO$_4$,
with the addition where indicated of 60 M-cAMP and
1 mM EGTA, or 0.5 mM-CaCl$_2$, 0.2 mM-EGTA and a partially
purified preparation of calmodulin. The incubation
tubes were shaken at a rate of 10-20 oscillations per
second with a 'Microid' flask shaker (A. Gallenkamp &
Co Ltd; Technico House, London EC2P 2ER, UK). The
reaction was started by the rapid injection from a
microsyringe of 10 µl of membrane preparation (100 µg
of protein) and terminated by the injection of 17.5 µl
of 'stopping solution'. The latter contained 286 mM-
Tris HCl (pH 8.7 at 20°C), 266 mM-glycine, 5.7% (w/v)
sodium dodecyl sulphate (SDS), 5.7% (v/v) 2-mercapto-
ethanol, 28.5% (w/v) sucrose and 0.03% bromophenol
blue. Each preparation was labelled for 10s in
duplicate.

Polyacrylamide gel electrophoresis

A discontinuous buffer system was used. The
temperature conditions and the composition of buffers,
stacking (4%) and resolving (10%) gel mixtures were the
same as used in previous work (4). Exponential
gradient polyacrylamide slab gels (6%-17%) were made
using a stock solution of 38% (w/v) of acrylamide and
2% (w/v) NN'-methylenebisacrylamide.

The solubilized membrane proteins (20 µg of each
sample) were applied to the slabs - each slab took 20
samples. A starting current of 10 mA per slab was
applied until the bromophenol blue dye entered the
resolving gel, and then the current was increased to
25 mA and was retained at that value until the dye

front reached the bottom of the slab. After electrophoresis, the slabs were stained with Coomassie Blue,
destained, dried under vacuum, and exposed to Kodak
"No-Screen" X-ray film, and the film was developed and
scanned in a Chromoscan densitometer, all by procedures
described before (4).

RESULTS AND DISCUSSION

      At least 17 major phosphate acceptors were found
in synaptic membrane fragments prepared from adult rat
cerebral cortex (Fig. 1). The phosphorylation of 6 of
these acceptors (designated 0_{3a}, 0_{3b}, α_5, β_3, β_4,
and γ_2) was markedly stimulated by cyclic AMP. One
acceptor (γ_5) required Ca^{2+} alone for its phosphorylation, and the phosphorylation of 4 others (α_3, β_6,
β_7 and γ_4) was stimulated by Ca^{2+} in the presence of
calmodulin. This pattern of phosphorylation was then
compared with that observed using the 'intermediate'
fraction prepared from the cerebral cortex of rat, at
1, 5, 10 and 15 days after birth. In 1-day-old rats
the cyclic AMP-dependent phosphorylating systems were
very low; thereafter they increased with age reaching
a maximum in the adult animal (Fig. 1). For some
systems (β_3 and β_4) the major increase coincided with
the onset of synaptogenesis. In contrast, the activity
of a phosphorylating system transferring to a protein
of 47K daltons in a reaction dependent on Ca^{2+} only,
was abundant at birth and until 15 days of age, but had
decreased some 5-fold by the time adulthood was reached.
Of the four proteins which required Ca^{2+} plus calmodulin
for maximal phosphorylation, two (β_6, and β_7) were
present in the 1-day-old rat at somewhat decreased
quantities, and two (α_3 and γ_4) did not appear until
the 15th day after birth. Of the cyclic AMP-independ-
ent systems, only one (γ_3) was present at birth in
amounts similar to that of the adult rat.

      Our findings with regard to protein β_3 and β_4
confirm the observations of Lohmann et al. (10)
on the ontogeny of the equivalent phosphorylating
systems Ia and Ib, which are believed to be enriched in
the synaptic membrane and synaptic vesicle fractions of
adult rat. These authors also reported that membranes
prepared from cerebral cortex of newborn rat contained
a protein of 50K daltons (II) in somewhat greater
amounts than in adult rat. The phosphorylation of this
protein appeared to be heavily dependent on cyclic AMP.
In the present study, it is evident from Fig. 1 that
no cyclic AMP-dependent systems were found in this

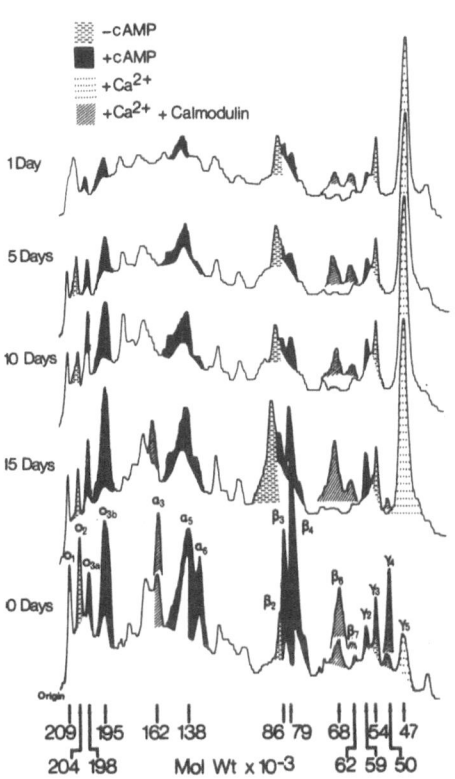

Fig. 1. Typical scans of radioautographs of ^{32}P-
labelled 'intermediate' membrane fragments
from animals of various ages fractionated on
polyacrylamide gels. Each sample of labelled
fragments was fractionated on an exponential
gradient gel (6%-17% acrylamide), and in
parallel on a 10% single concentration gel.
The gradient gel gave the best separation of
proteins above about 90K daltons, whereas
the 10% gels were superior for the lower MW
proteins. In the diagram the relevant sections
of the scans from each gel have been joined in
the trough to the left of the peak labelled
β_3.

region in new born animals. So far we are unable to
explain this discrepancy.

Since the phosphorylation of substrate proteins
in membrane fragments is catalysed by endogenous
membrane-bound protein kinases, the low phosphorylation
of proteins 0_{3a}, 0_{3b}, α_5, β_3 and β_4 in immature animals
could be due either to deficiency of substrate or
enzyme. The present work offers no evidence on this
point. However, Lohmann et al. (11) demonstrated that
protein kinase activities I and II (believed to phos-
phorylate the corresponding proteins Ia and Ib) in
newborn rat cerebral cortex did not differ significantly
from that of the adult.

Examination of other subcellular fractions (data
not shown) revealed that, in rats up to 15 days of age,
the system γ_5 is present in the 'light' membrane
fraction at concentrations which are approximately twice
those seen in the 'intermediate' membrane fraction,
whereas 'heavy' fractions contained negligible amounts
of any of the phosphorylated proteins seen in the
'intermediate' fraction.

The increase in protein phosphorylating systems
β_3 and β_4 and possibly α_5 in rat cerebral cortex
correlates roughly with the morphological development
of synaptic structures (1). It has been demonstrated
by means of electron microscopy that the number of
synapses in various regions of the rat brain increases
markedly between 10 to 26 days postnatally. Synaptic
vesicles have been shown to increase with a time
course similar to that for synaptic membranes (2).
The demonstration by this study that the increase in
certain cyclic AMP-dependent and Ca^{2+} + calmodulin
dependent protein phosphorylating systems in rat
cerebral cortex is concomitant with synapse formation
provides support for the possibility that phosphorylation
of proteins by cyclic AMP-dependent and Ca^{2+} + calmodulin
dependent protein kinases may be important in the
physiology of the synapse.

We are grateful to the Medical Research Council
of the U.K. for a grant and to Mr. H. Russell and
Miss J. Sankar for technical help.

REFERENCES

1. Aghajanian, G.K. and Bloom, F.E. (1967): The
    formation of synaptic junctions in developing rat
    brain: a quantitative electron microscopic study.
    Brain Res. 6: 716-727.

2. Armstrong-James, M. and Johnson, R. (1970):
    Quantitative studies of postnatal changes in
    synapses in rat superficial motor cerebral cortex.
    An electron microscopical study. Z. Zellforsch.
    Mikrosk. Anat. 110: 559-568.

3. De Lorenzo, R.J., Freedman, S.D., Yohe, W.B. and
    Maurer, S.C. (1979): Stimulation of Ca^{2+}-depen-
    dent neurotransmitter release and presynaptic nerve
    terminal protein phosphorylation by calmodulin and
    a calmodulin-like protein isolated from synaptic
    vesicles. Proc. Natl. Acad. Sci. USA 76: 1838-
    1842.

4. Dunkely, P.R., Holmes, H. and Rodnight, R. (1977):
    Phosphorylation of synaptic-membrane proteins from
    ox cerebral cortex in vitro: Preparation of
    fractions enriched in phosphorylated proteins by
    using extraction with detergents and urea and gel
    filtration. Biochem. J. 163: 369-378.

5. Greengard, P. (1978): Cyclic nucleotides,
    phosphorylated proteins and neuronal function.
    Raven Press, New York.

6. Holmes, H. and Rodnight, R. (1977): Phosphoryla-
    tion of synaptic membrane proteins from rat cerebral
    cortex in vitro: localization of phosphorylated
    proteins in the synaptic junction and post-synaptic
    lattices. Proc. Int. Soc. Neurochem. 6: 508.

7. Jones, D.H. and Matus, A.L. (1974): Isolation of
    synaptic plasma membranes from brain by combined
    flotation - sedimentation density gradient centri-
    fugation. Biochim. Biophys. Acta. 356: 276-287.

8. Kelly, P.T., Cotman, C.W. and Largen, M. (1979):
    Cyclic AMP-stimulated protein kinases at brain
    synaptic junctions. J. Biol. Chem. 254: 1564-
    1575.

9.   Krueger, B.K., Forn, J. and Greengard, P.   (1977):
      Depolarization-induced phosphorylation of specific
      proteins, mediated by calcium ion influx in rat
      brain synaptosomes.   J. Biol. Chem.   252: 2764-
      2773.

10.   Lohmann, S.M., Ueda, T. and Greengard, P.   (1978):
      Ontogeny of synaptic phosphoproteins in brain.
      Proc. Natl. Acad. Sci. USA   75: 4037-4041.

11.   Lohmann, S.M., Walter, U. and Greengard, P.   (1978):
      Protein kinases in developing rat brain.   J. Cyclic
      Nucl. Res.   4: 445-452.

12.   Miller, G.L.   (1959):   Protein determination for
      large numbers of samples.   Analyt. Chem. 31: 964.

13.   Ng, M. and Matus, A.   (1979):   Protein phosphory-
      lation in isolated plasma membranes and postsynaptic
      junctional structures from brain synapses.   Neuro-
      science   4: 169-180.

14.   Rodnight, R.   (1979):   Cyclic nucleotides as
      second messengers in synaptic transmission.   In-
      ternational Review of Biochemistry edited by
      K. F. Tipton.   University Park Press, Baltimore,
      1-80.

15.   Weller, M.   (1977):   Evidence for the presynaptic
      location of adenylate cyclase and the cyclic AMP-
      stimulated protein kinase which is bound to synaptic
      membranes.   Biochim. Biophys. Acta.   469:   350-
      354.

16.   Weller, M. and Morgan, I.G.   (1976):   Localization
      in the synaptic junction of the cyclic AMP stimu-
      lated intrinsic protein kinase activity of synapto-
      somal plasma membranes.   Biochim. Biophys. Acta.
      433:   223-228.

TUBULIN IN DEVELOPING RAT BRAIN: REGIONAL

DISTRIBUTION AND EFFECT OF GLUCOCORTICOIDS

R. Mileusnić, S. Kanazir and Lj. M. Rakić

Institute of Biochemistry
Faculty of Medicine
Institute for Biological Research
Belgrade, Yugoslavia

The relationship between the various anatomical and chemical compartments of the developing brain changes continuously as a result of many processes. They reach their "optimum" level during early adulthood but thereafter do exhibit some effects with age (1, 19, 42).

Special interest has been given to the study of protein synthesis, since the proteins in neural tissue could be directly concerned with the encoding, storage and transmission of information (32). The structural and functional changes in the properties of the specialized protein synthesizing systems which are caused by genetic, developmental and environmental factors might be expected to alter neural function (4, 28, 31, 39).

In this report, we will focus on experiments concerned with the regional changes in tubulin content and the effect of cortisol on protein synthesis in the brain. The heterogeneity of the brain makes it extremely difficult to compare one region with another one, even at the same age (8, 31). It is therefore necessary to measure carefully and compare timetables of changes of soluble and particulate tubulin content in different brain regions. Bearing in mind that adrenal glucocorticoids bind to (25, 40) and influence (9, 27) the various proteins contained in the brain cells, and that the treatment of neural cells with cortisol increases the rate of axon elongation (21), we decided to investigate the effect of glucocorticoid treatment on the content of tubulin, which has previously been shown to be the major soluble protein of the axon (7, 11).

MATERIALS AND METHODS

The rats of the Lewis inbred strain were used for all the experiments. Bilateral adrenalectomized rats were maintained on drinking water containing 0.9% saline for 5 days. The brains were quickly removed and the following areas were dissected on ice: frontal cortex (Cx.), nucleus caudate (N.Cd.), thalamus (Th.), hypothalamus (HTh.), septum (S) and hippocampus (Hippo.). The effect of glucocorticoid was studied in intact or adrenalectomized (ADX) animals injected i.p. with 0.5 ml of a saline solution containing 100 µCi of ^{35}S-methionine (NEN, spec. act. 400 Ci/mmole) 3 hours prior to the hormone administration. Hydrocortisone acetate (Sigma, 25 mg/kg) dissolved in 50% ethanol was injected i.p., while control animals received only 50% ethanol. All rats were sacrificed 1 hour later (always between 10-12 a.m.).

Tissue samples were homogenized in (1:10/w:v) 0.32 M sucrose containing 20 mM sodium phosphate (pH 6.8), 5 mM MgCl$_2$ and 0.1 mM GTP with a Dounce homogenizer. The samples were centrifuged at 105,000 xg for 60 minutes to yield a soluble and a particulate fraction. Aliquots of the soluble and particulate fractions were taken for the determination of protein, SDS-polyacrylamide gel electrophoresis, TCA-precipitation, colchicine-binding activity and DE81 filter discs-retained radioactivity. Proteins were determined by the method of Lowry et al. (20) using BSA (fraction V) as the standard. The high speed supernatants were subjected to 10% SDS-polyacrylamide gel electrophoresis according to Laemmli and Favre (16). Gels were stained with Commasie Brilliant Blue, dried and then exposed to x-ray film (Kodak RPI 54) according to Laskey and Mills (17). Cold 10% TCA was used for TCA-precipitation. Samples were filtered under vacuum on Whatman GF/C discs and washed four times with 5 ml of 5% TCA. Discs were dried and counted in PPO/POPOP/toluene scintillator.

Each fraction was incubated at a final concentration of 2.5 x 10^{-6} M ^{3}H-colchicine (NEN, spec, act. 16-20 Ci/mmole)(0.01 µCi/sample) at 37°C for 60 minutes as described Weisenberg (41). The reaction for the soluble fraction was terminated by adding 1 ml of unlabeled 10^{-5} M colchicine and 5 ml of washing buffer (67 mM sodium phosphate pH 6.8

containing 100 mM KCl) according to Borisy (3). The
samples of soluble fraction were filtered under vacuum
through a stack of 4 prewashed Whatman DE81 filter discs
and washed 4 times with 5 ml of washing buffer. Discs
were dried and counted in a PPO/POPOP/methoxyethanol/
Triton X-100 scintillator. For the particulate fraction,
the pellets were resuspended in the homogenizing medium
and 0.1 ml aliquots were incubated as described previously.
After the termination of the reaction 30 volumes of homogenizing buffer containing 3×10^{-5} M unlabeled colchicine
was added, and the postmitochondrial pellet was collected
at 16,000 xg for 30 minutes. The pellet was resuspended
and radioactivity and protein concentration were determined as above.

RESULTS

Developmental Changes in Tubulin Concentration

      Tubulin concentrations in adult animals were mea-
sured by SDS-polyacrylamide gel electrophoresis and colchicine-binding assay. As has been shown (36), these
two methods gave almost identical results. During the
early postnatal development, however, the sequential rate
of appearance of brain specific proteins (D_3, synaptin,
GFA) (2) and their co-migration with tubulin, as well as
denaturation of beta tubulin subunit into protein with
smaller molecular weight (13) prevented correct determination of tubulin. Consequently, SDS-polyacrylamide gel
electrophoresis appears not to be sufficiently accurate
for tubulin determination in the early postnatal development.

      The soluble and particulate tubulin concentrations
isolated from total rat brain telencephalon are given in
Figure 1 for the various stages of development, as
described by McIlwain (4). There is some decrease of
tubulin concentration in the soluble and increase in the
particulate fraction during the period of growth and
differentiation of nerve cells (stage II of development).
One-day-old rat telencephalon contains 300-350 μg of
tubulin/mg of protein in the soluble fraction and 60-70 μg
of tubulin/mg of protein in the particulate fraction. In
the period of myelinization (stage III of development) the
tubulin concentration continues to decrease in the soluble
fraction and to increase in the particulate fraction.
Stabilization of tubulin concentration in both fractions
occurs during maturation (stage IV of development) and
aging.

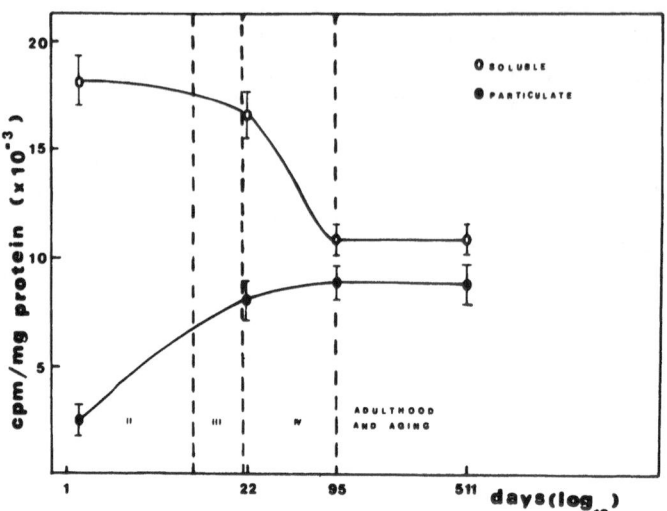

Figure 1.   Tubulin concentration during different
             stages of development.

        In the fourth stage, the changes in tubulin content
of each fraction were similar in all regions (Figs. 1 and
2).   In the course of aging, tubulin redistribution between soluble and particulate fractions showed a different
trend compared to the total telencephalon; that is, tubulin concentration increased in the soluble and decreased
in the particulate fraction in the caudate nucleus and
thalamus.   In the frontal cortex the distribution was
reversed: decreased in the soluble and increased in the
particulate fraction; whereas in the hippocampus it was
increased in both fractions.

Effect of Cortisol on Protein Synthesis in Different
Brain Regions

        The treatment with cortisol did not change the per-
centage of ^{35}S-methionine incorporated into TCA-precipitable proteins of cytosol from intact animals in any
region except the cortex.   In adrenalectomized animals,
the same treatment produced significant changes in the
septum, caudate nucleus and hypothalamus (Fig. 3).   However, the incorporation ratio between hormone-treated and
non-treated animals, whether intact or adrenalectomized,
showed that: a) the amount of ^{35}S-methionine incorporated
into cytosol of adrenalectomized animals was lower than
in the intact animals, in all regions except the septum;
and b) stimulation of incorporation occurred only in the

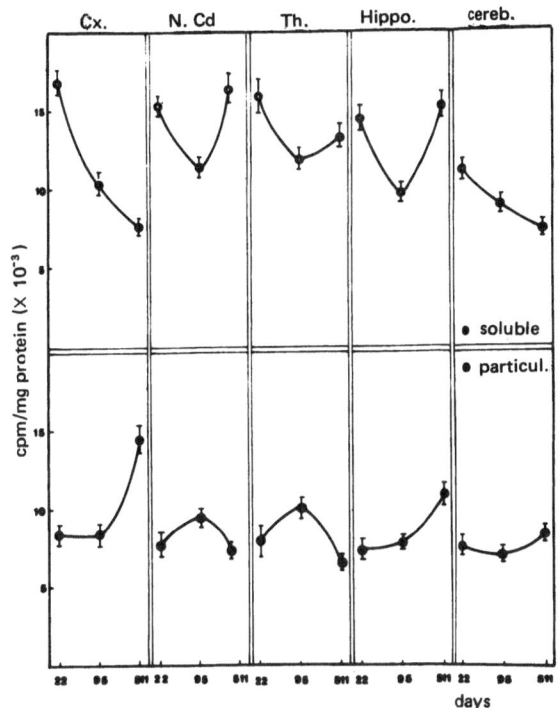

Figure 2.   Tubulin distribution in different brain
            regions during maturation and aging.

septum (Fig. 4).   Further determinations were done on
intact animals because of the differences found in the
amount of labeled amino acid found in the soluble fraction.

     TCA-precipitated soluble fraction isolated from dif-
ferent brain regions showed no difference in the amount
of ^{35}S-methionine incorporated in the cortex, hypothalamus,
thalamus and hippocampus of control versus hormone-treated
group.   However, incorporation of ^{35}S-methionine was found
to be increased in septum and decreased in caudate nucleus
(Table).   The acidic proteins retained on DE81 filter discs
showed no difference in the amount of ^{35}S-methionine in the
control as compared to hormone-treated group in cortex,
caudate nucleus and thalamus, but an increased amount of
radioactivity was seen in septum, hypothalamus and hippocampus (Table).   Determination of tubulin dimers showed
that the tubulin concentration was affected in regions
which are by definition primary target sites for glucocorticoids (14, 15, 23), namely septum, hypothalamus and
hippocampus.   In our experiments, there was also an

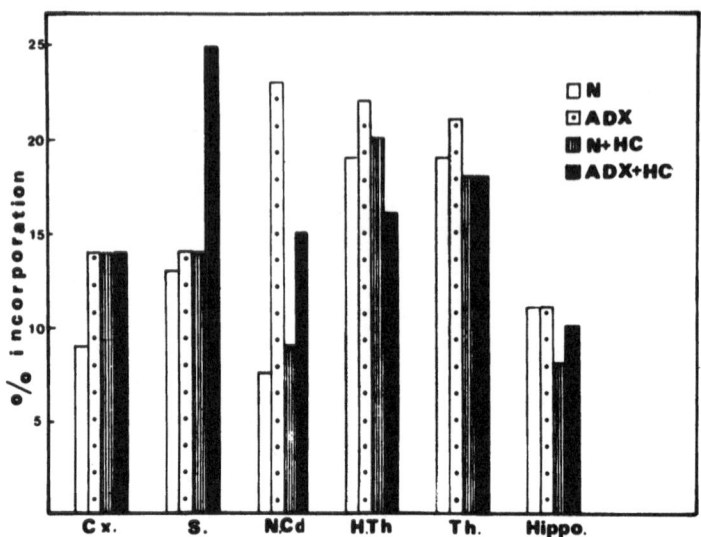

Figure 3.   The percentage of ³⁵S-methionine incorpora-
            tion into the TCA-precipitable proteins of
            cytosol. N - intact; ADX - adrenalectomized;
            N + HC - intact hormone-treated; ADX + HC -
            adrenalectomized hormone-treated.

increase in the tubulin dimers in response to the cortisol in the thalamus (Table).

        Analysis of SDS-polyacrylamide gels shown in Figure 5
revealed that hormone treatment increased ³⁵S-methionine
incorporation into TCA-precipitable proteins of soluble
fraction isolated from the septum, thalamus, hypothalamus
and hippocampus. In all these regions the same group of
proteins was affected and the most prominent changes were
found in zones of 33,000, 45,000, 53,000-56,000 and
130,000 daltons.

DISCUSSION

        Tubulin appears to possess an important role as a
regulatory protein in the brain (2, 6, 7, 29, 33, 35, 37).
We have already demonstrated in a previous series of experiments (28, 29) that, in the developing rat brain,
tubulin undergoes characteristic changes during fetal,

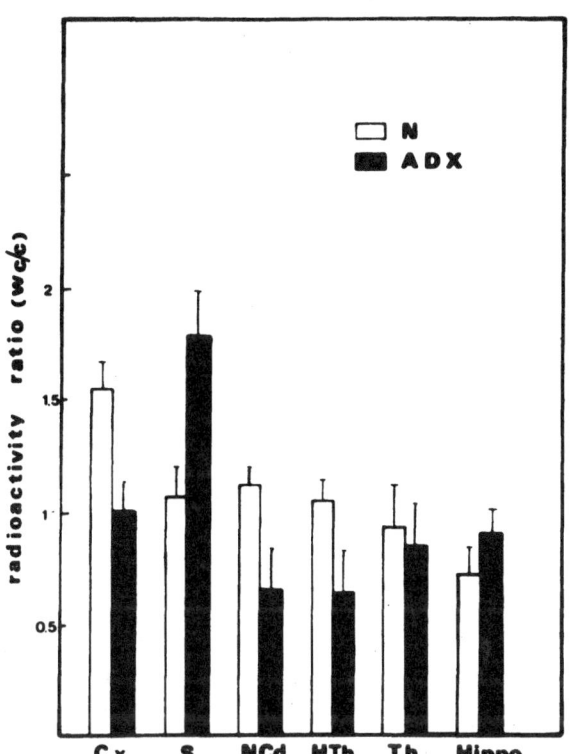

Figure 4. ^{35}S-methionine incorporation ratio between the hormone-treated and untreated animals. N - intact; ADX - adrenalectomized.

Table. Effect of Hydrocortisone-acetate on ^{35}S-methionine Incorporation and ^{3}H-colchicine Binding in Cytosol from Different Brain Regions

Brain Regions	Ratio:hormonal-treatment/control (cpm/mg ± SEM)		
	TCA-precipitable	DE81-retained	^{3}H-colchicine
Frontal cortex	0.99 ± 0.12	1.19 ± 0.05	0.99 ± 0.12
Septum	1.47 ± 0.20	1.38 ± 0.20	1.28 ± 0.11
Caudate nucleus	0.83 ± 0.11	0.91 ± 0.14	1.07 ± 0.19
Hypothalamus	1.06 ± 0.04	1.39 ± 0.34	1.04 ± 0.15
Thalamus	1.15 ± 0.06	1.17 ± 0.09	1.48 ± 0.24
Hippocampus	1.22 ± 0.09	1.27 ± 0.05	1.33 ± 0.22

postnatal and aging periods.   The amount of tubulin in the soluble fraction was found to be increased from the fourteenth day of gestation until birth.   From birth to the tenth postnatal day, there is some decrease in tubulin content but from the tenth to the twenty-first postnatal day (a critical period of development in this species),· there is a sudden drop in its concentration. This decline continues at a slow rate throughout adolescence and adulthood; thus, at the five-hundred-eleventh day, the concentration of tubulin is the same as on the fourteenth day of gestation.   Similar changes in the level of brain tubulin were reported for the first to thirtieth days of life by Schmitt et al. (36).

In the present study we have demonstrated that, during the critical period of brain development, there is a change in the tubulin distribution and content,as indicated by a decrease in the soluble fraction and an increase in the particulate fraction isolated from whole telencephalon or its different regions. The stabilization of this process occurs in the whole telencephalon, when the process of biochemical maturation is completed (95th day). The changes in the tubulin concentration occurring during the critical period of brain development may represent an important factor in the regulation of growth and differentiation.

The differences in the distribution of tubulin between the soluble and the particulate fractions in the various regions of telencephalon indicate that the amount of tubulin in these two fractions may be correlated with the processes of aging.   During this period, there is a striking redistribution of the tubulin between the two fractions obtained from the frontal cortex and the hippocampus. Accretion of tubulin in the particulate fraction of both regions may be correlated with specific age-dependent changes.   Iqbal (13) has described accumulation of beta subunit of tubulin in the hippocampus in Alzheimer disease. It is known that these regions are the "hot spots" for the processes related to learning and memory formation.   Bearing in mind our findings and the known importance of tubulin in the nerve functions, including the learning processes (33, 34, 35), we postulate that the temporal similarities between the ontogenetic evolution of the brain and the tubulin changes could have a physiological implication.

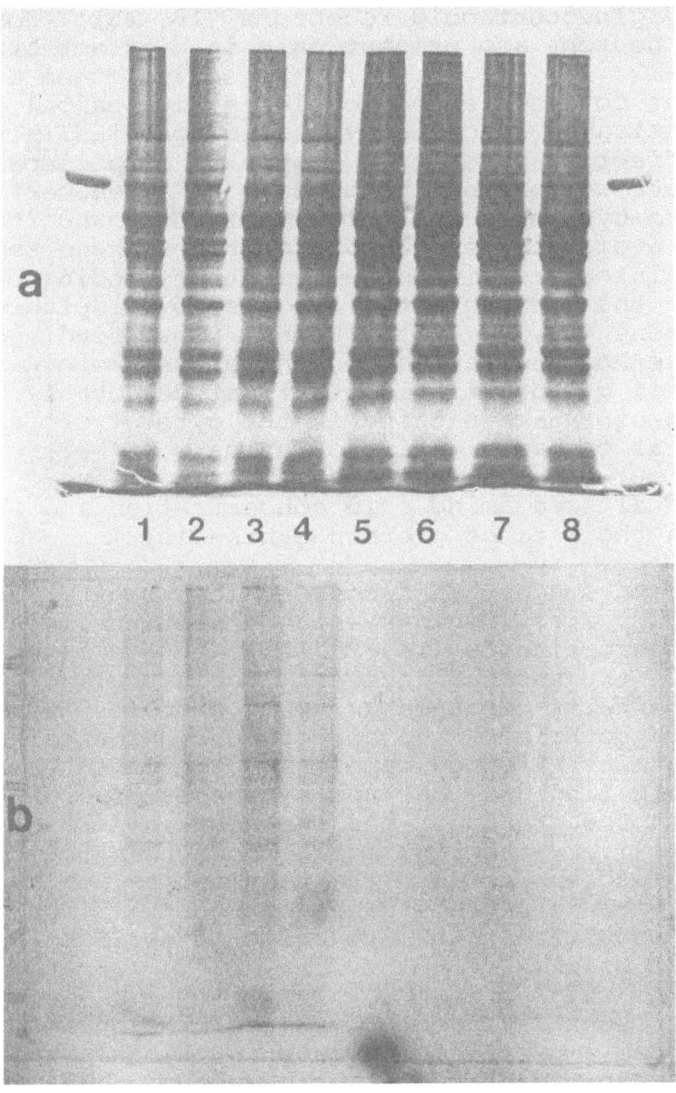

Figure 5.   SDS-polyacrylamide gel electrophoresis (a)
            and autoradiography (b) of cytosol proteins
            from different brain regions.   1 and 5 -
            hypothalamus; 2 and 6 - thalamus; 3 and 7 -
            septum; 4 and 8 - hippocampus; 1-4 - hormone-
            treated; 5-8 - untreated.

There is a growing amount of data that the cytosol of
different target tissues contains very similar, if not
identical, glucocorticoid receptors (18, 43). Hormone-
sensitive neurons are present only in distinct brain regions (5, 12, 15, 21, 22, 38). Recently, Etgen et al. (10)
showed that corticosterone modulates hippocampal synthesis
or degradation of specific cytosolic proteins in vitro
without effecting an overall change in hippocampal metabolism. Data concerning the binding of adrenocortical
hormones to cytosolic and nuclear sites showed that adre-
nalectomy avoids the endogenous corticosterone secretion
which can interfere with measurement of binding by com-
peting for the binding sites and also by displacing the
bound hormone (22, 24, 26). However, our findings concerning the hormonal effect on the brain protein synthesis
suggest that glucocorticoid treatment detectably altered
the amount of ^{35}S-methionine in the cytosol of adrenalectomized as compared to intact animals. Therefore, we
continued our work using intact animals to avoid the
problem of altered amino acid concentration and its influence on the rate of protein synthesis.

Our data suggest that besides the primary target
sites of CNS, i.e., regions which accumulate, bind and
retain glucocorticoids (14, 21, 25), the thalamus also
responds to the hormonal action. The glucocorticoid
treatment affects the same group of proteins in septum,
thalamus, hypothalamus and hippocampus. The full identity
of all proteins affected cannot be determined from autoradiographic findings, but the synthesis of proteins
involved in the transport and the neural plasticity (6,
30, 33, 34)(i.e., tubulin, actin and myosin) may be modulated by glucocorticoids. The tubulin results are in
agreement with in vitro studies of Etgen (10) on glucocorticoid modulation of the tubulin metabolism in the
hippocampal slices.

The present data seem to suggest that hormonal treatment produces a parallel increase in the turnover rates
for a select group of proteins in septum, thalamus, hypothalamus and hippocampus, and in the amount of tubulin
dimers, indicated by colchicine-binding assay. The evidence suggests that the rate of synthesis and degradation
of both tubulin and other acidic proteins investigated
could be modulated by glucocorticoid treatment.

In conclusion, the changes in tubulin concentration
coincide with certain critical stages of development,
while the glucosteroids may have an effect on the synthesis and degradation of tubulin and other acidic proteins.

REFERENCES

1. Bamburg, J.R., Shooter, E.M. and Wilson, L. (1973):
    Developmental changes in microtubule protein of
    chick brain. Biochem. 12: 1476-1482.

2. Bock, E. (1978) Brain specific proteins. J. Neuro-
    chem. 30: 7-14.

3. Borisy, G.G. (1972): A rapid method for quantitative
    determination of microtubule protein using DEAE cellu-
    lose filters. Anal. Biochem. 50: 373-385.

4. Brattgard, S.O., Endström, J.E. and Hydén, H. (1957):
    The chemical changes in regenerating neuron. J.
    Neurochem. 1: 316-325.

5. Chytil, F. and Toft, D. (1972): Corticoid binding
    component in rat brain. J. Neurochem. 19: 2877-2880.

6. Cronly-Dillon, J.R. and Perry, G.W. (1976): Tubulin
    synthesis in developing rat visual cortex. Nature
    261: 581-583.

7. Dahlström, A. (1971): Axoplasmic transport with
    particular respect to adrenergic neurons. Phylos.
    Trans. R. Soc., London (Biol.) 261: 325-358.

8. Davison, A.N. (1977): Biochemical, morphological
    and functional changes in developing brain. In:
    Biochemical Correlates of Brain Structure and Function,
    A.N. Davison (ed.), pp. 1-13, Academic Press, New York.

9. Dunn, A.J., Gildersleeve, N.B. and Gray, H.E. (1978):
    Mouse brain tyrosine hydroxylase and glutamic acid
    decarboxylase following treatment with adrenocortico-
    tropic hormone, vasopressin or corticosterone. J.
    Neurochem. 31: 977-982.

10. Etgen, A.M., Lee, K.S. and Lynch, G. (1979): Gluco-
    corticoid modulation of specific protein metabolism
    in hippocampal slices maintained in vitro. Brain
    Res. 165: 37-45.

11. Feit, H.G., Dutton, G.R., Barondes, S.H. and
    Shelanski, M.L. (1971): Microtubule protein. Iden-
    tification and transport to nerve endings. J. Cell
    biol. 51: 138-147.

12. Gerlach, J.L. et al. (1976): Cells in regions of
     rhesus monkey brain and pituitary retain radioactive
     estradiol, corticosterone and cortisol differentially.
     Brain Res. 103: 603-612.

13. Iqbal, K., Grundel-Iqbal, I., Wisniewski, H.M. and
     Terry, R.D. (1977): On neurofilament and neuro-
     tubule proteins from human autopsy material. J.
     Neurochem. 29: 417-424.

14. Knizley, H., Jr. (1972): The hippocampus and septal
     area as primary target sites for corticosterone. J.
     Neurochem. 19: 2737-2745.

15. Kraulis, I., et al. (1975): Distribution, metabolism
     and biological activity of deoxycorticosterone in the
     central nervous system. Brain Res. 88: 1-14.

16. Laemmli, U.K. and Favre, M. (1973): Maturation of
     the head of bacteriophage T_4.I.DNA packing events.
     J. Mol. Biol. 80: 575-599.

17. Laskey, R.A. and Mills, D.A. (1975): Quantitative
     film detection of ^{3}H + ^{14}C on polyacrylamide gels by
     fluorography. Europ. J. Biochem. 56: 335-341.

18. Lehrer, G.M., Maker, H.S. and Weisserbath, S. (1973):
     Brain uptake of cortisol and corticosterone from CSF
     and systemic sites. Neurology 23: 63-68.

19. Lim, L. (1977): Regulation of RNA metabolism in the
     developing brain. In: Biochemical Correlates of
     Brain Structure and Function, A.N. Davison (ed.),
     pp. 15-41, Academic Press, New York.

20. Lowry, O.B., Rosebrough, N.S., Farr, A.L. and Randal,
     R.J. (1951): Protein measurement with folin-phenol
     reagent. J. Biol. Chem. 193: 265-285.

21. McEwen, B.S. (1976): Interactions between hormones
     and nerve tissue. Sci. American, June: 48-58.

22. McEwen, B.S., DeKloet, R. and Wallach, G. (1976):
     Interactions in vivo and in vitro of corticosteroids
     and progesterone with cell nuclei and soluble macro-
     molecules from rat brain regions and pituitary.
     Brain Res. 105: 129-136.

23. McEwen, B.S. and Wallach, G. (1973): Corticosterone binding to hippocampus: Nuclear and cytosol binding in vitro. Brain Res. 57: 373-386.

24. McEwen, B.S., Wallach, G. and Magnus, C. (1974): Corticosterone binding to hippocampus: Immediate and delayed influence of adrenal secretion. Brain Res. 70: 321-334.

25. McEwen, B.S., Weiss, J.M. and Schwartz, L.S. (1969): Uptake of corticosterone by rat brain and its concentration by certain limbic structures. Brain Res. 16: 227-241.

26. McEwen, B.S., Weiss, J.M. and Schwartz, L.S. (1969): Retention of Corticosterone by cell nuclei from brain regions of adrenalectomized rats. Brain Res. 17: 471-482.

27. Meyer, J., Lnine, V.N., Khylchevskaya, R.I. and McEwen, B.S. (1979): Glucocorticoids and hippocampal enzyme activity. Brain Res. 166: 172-175.

28. Mileusnić, R. (1978): Changes in Tubulin Content during Development and Aging of Rat Brain. Doctoral Dissertation, Medical Faculty, University of Belgrade, June, 1978.

29. Mileusnić, R. and Rakić, Lj. M. (1979): Brain tubulin in the function of pre- and postnatal development and aging. Developmental Neurosci., submitted.

30. Perry, G.W. and Cromly-Dillon, J.R. (1978): Tubulin synthesis during a critical period in visual cortex development. Brain Res. 142: 374-378.

31. Rakić, Lj. M., Mileusnić, R., Rogać, Lj. and Veskov, R. (1979): Some biochemical aspects of electroconvulsive seizure. In: Pathophysiology of Cerebral Energy Metabolism, B. B. Mršulja, Lj. M. Rakić, I. Klatzo and M. Spatz (eds.), pp. 281-311, Plenum Press, New York.

32. Roberts, S. (1973): Alterations in cerebral protein-synthesizing system during maturation. In: Progress in Brain Research, D. H. Ford (ed.), Vol. 10, Elsevier Scientific Publ. Co., Amsterdam.

33.   Rose, S.P.R.:   Neurochemical correlates of
      early learning in the chick.   In: Neurobiological
      Basis of Learning and Memory, B.W. Agranoff and Y.
      Tsukuda (eds.), John Wiley, in press.

34.   Rose, S.P.R. and J. Haywood (1977):   Experience,
      learning and brain metabolism.   In: Biochemical
      Correlates of Brain Structure and Function, A.N.
      Davison (ed.), pp. 249-292, Academic Press, New York.

35.   Rose, S.P.R., Sinha, A.K. and Jones-Lecointe, A.
      (1976):   Synthesis of tubulin-enriched fraction in
      rat visual cortex is modulated by dark-rearing and
      light-exposure.   FEBS Letters 65, 2: 135-139.

36.   Schmitt, H., Gozes, I. and Littauer, U.Z. (1977):
      Decrease in levels and rates of synthesis of tubulin
      and actin in developing rat brain.   Brain Res. 121:
      327-342.

37.   Shelanski, M.L. (1973):   Microtubules.   In: Proteins
      of the Nervous System, D.J. Schneider (ed.), pp. 227-
      241, Raven Press, New York.

38.   Stith, R.D., Pearson, R.J. and Dana, R.C. (1976):
      Uptake and binding of ^{3}H-hydrocortisone by various
      pig brain regions.   Brain Res. 117: 115-124.

39.   Vesco, R. and Guiditta, A. (1968):   Disaggregation
      of brain polysomes induced by electroconvulsive
      treatment.   J. Neurochem. 15: 81-85.

40.   Warembourg, M. (1975):   Radioautographic study of
      the rat brain after injection of 1,2,-^{3}H-cortico-
      sterone.   Brain Res. 89: 61-70.

41.   Weisenberg, R.C., Borisy, G.G. and E.W. Taylor (1968):
      The colchicine binding protein of mammalian brain
      and its relation to microtubules.   Biochem. 7: 4466-
      4479.

42.   Weiss, B. (1971):   Ontogenetic development of adenyl
      cyclase and phosphodiesterase in rat brain.   J.
      Neurochem. 18: 469-477.

43.   Wrange, O. (1979):   A comparison of the glucocorti-
      coid receptor in cytosol from rat liver and hippo-
      campus.   Biochim. Biophys. Acta 582: 346-357.

TRANSFER RIBONUCLEIC ACIDS IN THE DEVELOPING BRAIN:

THE EFFECT OF THE CONVULSANT METHIONINE SULFOXIMINE

ON tRNA FUNCTION

Orchid Der and Otto Z. Sellinger

Laboratory of Neurochemistry
Mental Health Research Institute
University of Michigan Medical Center
Ann Arbor, Michigan 48109

SUMMARY

Effects of the convulsant L-methionine-dl-sulfoximine (MSO) on brain transfer ribonucleic acids (tRNAs) were discovered through a comparative biochemical analysis of tRNAs from control and MSO-treated (150 mg/kg, single dose) 3-day old rats. The study of the in vitro formation of aminoacyl-tRNA, the key intermediate in translation, coupled with benzoylated-DEAE cellulose (BCD) chromatography of control and MSO-tRNAs, revealed a relationship between the molecular structure of tRNA, as affected by MSO, and its function in amino acid acceptance. Post-charging of tRNA following BDC chromatography revealed 3 tRNA^{phe} and 3 tRNA^{lys} iso-acceptors in control brains. The administration of MSO resulted in a significant alteration of the mobility of all 6 isoacceptors on BDC and in a marked reduction of their aminoacylating capacity when compared to control molecules. The MSO-elicited impairment of aminoacyl-tRNA formation was studied using both liver (heterologous) and brain (homologous) aminoacyl-tRNA synthetase preparations. Acceptance of lysine and phenylalanine by MSO-tRNA was uniformly lower than acceptance by control tRNA and was reflected by a drop in the V_{max} values for the two amino acids. Different optima for $[K^+]$ and $[CTP]$ were also found when phenylalanine acceptance by control and MSO-tRNA was compared.

The possibility that changes in brain tRNA methylation (6, 29), previously noted after MSO, may account

ror their impaired performance in aminoacylation is
briefly discussed.

INTRODUCTION

Although relatively little research has addressed
the question of how neural development is controlled
at the molecular level, there is no doubt that the
mechanisms involved are not only sophisticated but
also exquisitely fine-tuned, both temporally and
topographically. For nerve cells to make specific
proteins and for these to be endowed with functionally
relevant roles for later life, there must exist
appropriate controlling mechanisms to assure the
correctness of their structural makeup and the timeli-
ness of their appearance, i.e. during development,
proteins must be synthesized in strict synchrony to be
fully operational (18). One level of molecular organization at which controls are most likely to possess
cell specific features is the post-transcriptional
level, and it is at this level that potentially
competent molecules gain their final signals which
transform them into operationally competent molecules.
Transfer ribonucleic acids (tRNAs) are such molecules,
and certain aspects of their synthesis (3, 25) and
processing (1, 4, 7, 11, 14, 17-19, 21, 28) in the
early post-natal rat brain and of the developmental
dysfunction which they may be subjected to are the
theme of this contribution (6-9, 12, 28, 29, 31).

The tRNAs are of interest mainly because they lie
squarely astride the crossroads linking polynucleotides
to polypeptides and because their principal, though
not exclusive role, is that of making cell-specific
protein synthesis possible. This role relies heavily
on the ability of tRNAs to undergo structural modifications and hence diversify into subpopulations of
molecules, termed isoacceptor tRNAs (23-24, 31-32).
The search for a link-up between tRNAs and their
modifiable and developmentally adaptive isoacceptors,
(2-3, 10) with the synthesis, during a narrow developmental time-band, of key neural proteins is the object
of the cellular brain research effort ongoing in our
laboratory. To prime and sustain this, as of late,
lagging effort (18, 31), we are attempting to characterize and purify sample brain tRNAs (9) and their precursors (11) and isoacceptors (9) in the very young
rat brain. We also wish to be able to interfere with
the synthesis, the processing and the modification of
tRNA in the immature brain and hence, possibly with its

function. The findings presented in this paper may be a useful link in the global chain of facts required to understand how tRNAs participate in and shape the molecular events in neural development, for they document the presence, in the 3-day old rat brain, of 3 isoacceptors each of tRNA^{lys} and tRNA^{phe}. We also show that the acute treatment of the 3-day old animals with a single convulsant dose of L-methionine-dl-sulfoximine (MSO) leads to significant qualitative alterations of the 3 tRNA^{lys} and tRNA^{phe} isoacceptors. These alterations, in turn, appear to lead to changes in their functional competence, i.e., the aminoacylation of tRNA^{lys} and tRNA^{phe} isolated from brains of MSO-treated animals becomes significantly impaired compared to that noted with control brain tRNA. Since in previous studies we demonstrated that MSO: a) stimulates two base specific cerebral tRNA methyltransferases, N^2-methyl and N_2^2-dimethylguanine -tRNA methyltransferase (29), and b) alters the specificity of brain tRNA methyltransferases and the pattern of in vivo brain tRNA methylation (6), the present findings infer that the observed reduced functional competence of tRNA^{lys} and tRNA^{phe} is somehow due to specific changes in their methylation and/or that of any one of their isoacceptors.

EXPERIMENTAL PROCEDURES

Animals.

The animals were 3 day-old Sprague-Dawley rats.

The administration of MSO.

MSO was dissolved in saline and injected intraperitoneally at a dose of 150 mg/kg. The animals were killed 3 h later. Controls received an equal volume of saline.

The extraction of tRNA.

The cerebral cortices of control and MSO-treated rats were homogenized in 0.25 M sucrose, 0.035 M Tris-HCl, pH 7.4, 0.025 M KCl, 0.025 M $MgCl_2$, 0.001 M EDTA, 0.01 M β -mercaptoethanol and 0.1% (v/v) diethylpyrocarbonate. The homogenate was centrifuged for 20 min at 30,000 x g and the tRNA extracted from the resulting supernatant by the procedure of Ortwerth & Der (24). For further purification, the tRNA fraction was chromatographed on DEAE-cellulose according to Jank & Gross (16).

Preparation of Aminoacyl-tRNA synthetase.

Livers or brains (ca. 10 g) were homogenized in
2 volumes of buffer containing 0.25 M sucrose, 0.05 M
KCl, 0.01 M Tris-HCl, pH 7.5, 0.01 M $MgCl_2$, 0.01 M
β-mercaptoethanol and 10% (v/v) glycerol. The homogenate was centrifuged as above for tRNA extraction and
the supernatant recentrifuged for 75 min at 160,000 x
g. The resulting supernatant (10-30 ml) was passed
through a column of Sephadex G-75 (2.5 x 75 cm) previously equilibrated with Tris-HCl (0.05 M, pH 7.5),
containing 0.01 M KCl and $MgCl_2$, 0.01 M β-mercaptoethanol
and 10% glycerol. Elution was at 1 ml/3 min, and
fractions of 3.5 ml were collected. Fractions 23-25
contained maximal aminoacyl-tRNA synthetase activity.
They were stored at -20°C following addition of glycerol
to 50%.

Chromatography on benzoylated diethylaminoethyl-
cellulose (BDC).

BDC (Biorad Labs., Richmond, California) was
washed with 0.05 M sodium acetate, pH 5.0, containing
0.01 M $MgCl_2$ (column buffer), 2.0 M NaCl and 40% (v/v)
ethanol, and a column (1 x 20 cm) was prepared and
equilibrated with the above solution (27). Samples
containing tRNA (under 200 A_{260} units) were eluted with
120 ml of a 0.3 - 1.0 M NaCl gradient in column buffer,
followed by 120 ml of a 0-30% ethanol gradient in
column buffer, containing 1.0 M NaCl (27). The flow
rate was adjusted to 2 ml/min and 5 ml fractions were
collected.

Aliquots (0.1 ml) of each fraction were mixed with
9 ml of ACS scintillation fluid and the radioactivity
determined. When BDC fractions were to be post-charged
with radioactive lysine or Phenylalanine, 0.3 ml were
taken to perform the aminoacylation assay.

tRNA aminoacylation.

tRNA was routinely aminoacylated in a reaction
volume of 1 ml containing, in the order of addition:
1 μ curie of ^{14}[C] or ^{3}[H] -labelled amino acid or amino
acid mixture, as indicated (^{14}[C] lysine (U): 270
mCi/mmole; ^{3}[H] -lysine (4,5): 60-80 Ci/mmole
^{14}[C] -phenylalanine (U): 490 mCi/mmole;
^{3}[H] -phenylalanine: 60-80 Ci/mmole; ^{14}[C] amino
acid mixture: 10 mCi/mmole), 0.4 M Tris-HCl, pH 7.5,
10 (lysine and mix) or 20 mM (phenylalanine) $MgCl_2$,

20 (lysine and mix) or 50 mM (phenylalanine) KCl, 1 (lysine and mix) or 4 mM (phenylalanine), ATP, 0.6 mM CTP, 2 mM β-mercaptoethanol, varying amounts of tRNA, as indicated and ca. 1 mg of aminoacyl-tRNA synthetase protein. Following incubation for 10 min at $37^{\circ}C$ with gentle shaking, the reactions were stopped by placing the tubes on ice and adding 5 ml of ice-cold 10% TCA. After 15 min the radioactive aminoacyl-tRNA was collected on Whatman GF/A filters and processed for counting according to Yang and Novelli (33). In later experiments, the concentrations of Mg^{++}, K^+ and CTP for lysine and phenylalanine aminoacylation were those listed in Table 3.

Protein determination.

The procedure of Lowry et al. (22) was used with crystalline bovine serum albumin as standard.

RESULTS

Aminoacylation of the phenol-extracted and DEAE-cellulose purified tRNAs was effected using a preparation of aminoacyl-tRNA synthetases isolated from control livers, as well as a similar preparation of this enzyme complex isolated from livers of MSO-treated rats. Their specific activities, reflecting the acceptance capacity of control vs MSO-tRNAs for a mixture of amino acids and, separately, for lysine and phenylalanine, are shown in Table 1. Acceptance by MSO-tRNAs was uniformly reduced, irrespective of the source of the synthetases. Quantitatively, the reductions were lowest with the amino acid mixture, greatest with lysine, and intermediate with phenylalanine. MSO-tRNAs were also poorer substrates for aminoacylation than were control tRNAs in a homologous aminoacylation system, again irrespective of whether the aminoacyl-tRNA synthetases were from control or MSO-brains (Table 2). The findings shown in Tables 1 and 2 thus demonstrate that both the control and the MSO, hepatic and cerebral, lysyl and phenylalanyl-tRNA synthetases effectively discriminate between control and MSO-tRNAs.

In an effort to determine whether the aminoacylation of control and MSO-tRNAs requires different concentra-tions of some of the assay components to become optimal, samples of control and MSO-tRNAs were incubated in the presence of control liver aminoacyl-tRNA synthetases and the concentrations of Mg^{++}, K^+ and CTP varied, one at a time, each in a separate experiment. Table 3

Table 1.   Amino acid acceptance by brain tRNA of
            control and MSO-treated rats:
                 Heterologous Aminoacylation

Source of tRNA	Source of Liver Synthetases	Amino Acid Mixture pmoles/A_{260}/mg	Lysine of protein	Phenyl-alanine
Control	Control	107.9	77.9	4.34
MSO	Control	93.9	54.5	3.46
Difference, %		−13	−30	−20
Control	MSO	138.8	92.8	7.36
MSO	MSO	115.7	73.4	5.90
Difference, %		−17	−21	−20

MSO: 150 mg/kg, 3 h before death. The rats were
        3-days old. All amino acids were [14C]-
        labelled.

Table 2. Amino acid acceptance by brain tRNA of
control and MSO-treated rats:
Homologous Aminoacylation

Source of tRNA	Source of Brain Synthetases	Amino Acid Mixture	Lysine	Phenyl-alanine
		pmoles/A_{260}/mg of protein		
Control	Control	.31.1	22.4	1.78
MSO	Control	26.9	15.0	1.62
Difference, %		−14	−33	−10
Control	MSO	35.8	21.3	2.88
MSO	MSO	29.6	15.8	2.38
Difference, %		−17	−26	−17

MSO: 150 mg/kg, 3 h before death. The rats were
3-days old. All amino acids were [14C]-labelled.

Table 3.   Optimal assay concentrations of Mg^{++}, K^{+}, and cytidine triphosphate (CTP) for the aminoacylation of brain tRNA of control and MSO-treated rats.

Amino Acid	Source of Brain tRNA	Mg^{+} (mM)	K^{+} (mM)	CTP (mM)
Lysine	Control	10	20	0.6
	MSO	10	20	0.6
Phenylalanine	Control	10	50	> 2
	MSO	10	80	2

MSO: 150 mg/kg, 3 h before death. The rats were 3-days old. Aminoacyl-tRNA synthetase was from control livers.

indicates that the aminoacylation of control and MSO-tRNAs by lysine became optimal in the presences of identical concentrations of Mg^{++}, K^+ and CTP, while that of MSO-tRNAs by phenylalanine appeared to require more K^+ and CTP than that of control tRNAs. When the concentrations of lysine and phenylalanine were similarly varied, no differences in K_m values between the control and the MSO-tRNAs were noted. The V_{max} values for lysine, however, were found to be 19%, and those for phenylalanine, 32% lower with MSO-tRNAs as substrate.

Following phenol extraction and passage through a column of DEAE-cellulose (see Methods), tRNA-containing fractions from brains of control and MSO-treated rats were aminoacylated with ^{14}[C] -phenylalanine (control) and ^{3}[H] -phenylalanine (MSO) and the double-labelled mixture loaded on a column of BDC. Figure 1 shows very little difference in the elution patterns of the two differentially labelled (precharged) phe-tRNAs. When this experiment was repeated, but instead of chromatographing precharged tRNAs, these were chromatographed first and were then post-charged with lysine and phenylalanine, the results depicted in Figure 2 (A-C) were obtained. In panel B are shown the tRNA^{lys} and the tRNA^{phe} profiles derived from control brain tRNA. Several points of interest may be noted: a) the left portion (salt gradient) of the panel contains the predominant tRNA^{lys} isoacceptor and b) the right portion (EtOH-salt gradient) contains two additional tRNA^{lys} and three tRNA^{phe} isoacceptors. Figure 2, panels A and C, shows the elution profiles of tRNA^{lys} (A) and tRNA^{phe} (C) derived from brains of MSO-treated rats. The salient features of these panels and their differences relative to panel B are: a) the elution of all three tRNA^{lys} isoacceptors were markedly retarded, so that a significant portion of the predominant tRNA^{lys} isoacceptor eluted astride the salt-EtOH-salt gradient interface (Panel A); b) the position of the three tRNA^{phe} isoacceptors also shifted toward higher EtOH concentrations; and c) the heights of the MSO-tRNA^{lys} and tRNA^{phe} isoacceptor peaks were significantly reduced compared to those of the control molecules (panel B).

Using BDC-chromatography of the uncharged tRNA molecules, but not the technique employed in Figure 1, it thus became possible to detect major differences between the chromatographic mobility of control and MSO-tRNA, both in terms of the elution order of their respective tRNA isoacceptors and of their functional competence.

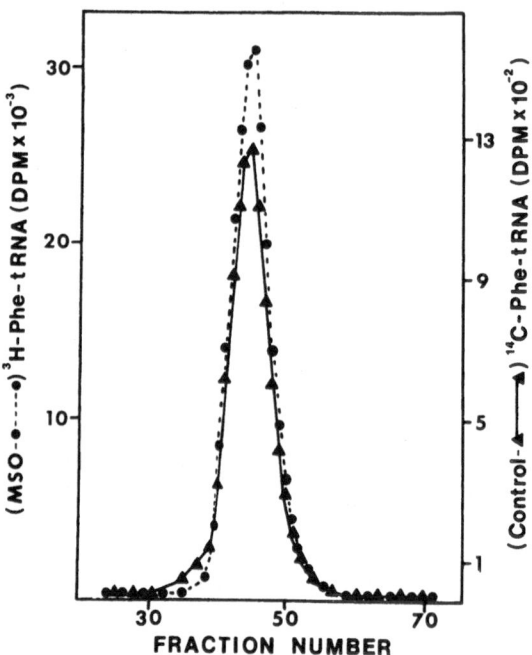

Fig. 1.   Chromatography of tRNA on benzoylated DEAE-
           cellulose.   Precharged [14][C] -Phe-control-
           tRNA was co-chromatographed with precharged
           [3][H] -Phe-MSO-tRNA on a 1 x 8 cm column of
           BDC as described under Methods.   The column was
           run at 4°C, and 5 ml fractions were collected
           at a flow rate of 2 ml/min.

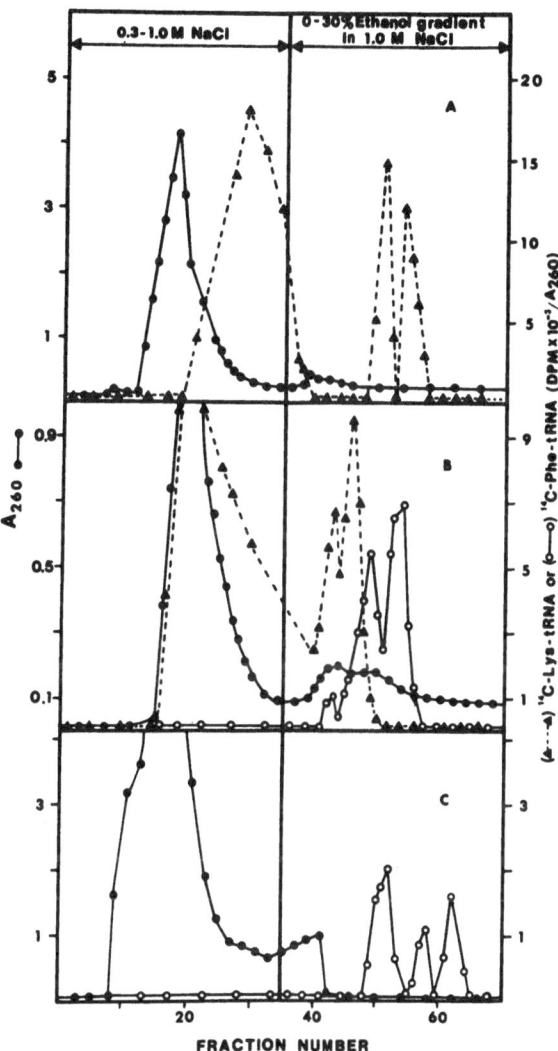

Fig. 2.   Chromatography of tRNA on benzoylated DEAF-
cellulose. Approximately 90 A$_{260}$ units of
unfractionated tRNA were chromatographed.
Amino acid acceptance was assayed as described.
(A)   MSO-tRNA, [^{14}C]-phenylalanine.
(B)   Control-tRNA, [^{14}C]-phenylalanine.
(C)   MSO-tRNA, [^{14}C]-lysine.

DISCUSSION

Comparisons of aminoacylating capacities of brain
tRNAs have been made rather infrequently and asystemati-
cally in a number of different test systems. Johnson
and his associates (17-19) and Barra et al. (1) studied
the homologous process as a function of mouse or rat
age and, inter alia, found that the aminoacylation of
brain tRNA by lysine and phenylalanine requires
different concentrations of Mg^{++}, ATP and CTP. The
results shown in Table 3 confirm these observations in
a heterologous aminoacylating system. More recently,
Harris and Maas (14) compared the aminoacylation of
5 and 55-day old rat brain tRNA by 18 amino acids,
including lysine and phenylalanine, using the aminoacyl-
tRNA synthetases of E. coli to achieve maximum acceptance (pmoles/A_{260}). No difference in the acceptance
of these two amino acids was noted under any of the
conditions tested, a finding in disagreement with our
results (Tables 1 and 2), which show markedly higher
acceptance values for lysine, relative to phenylalanine,
in both the heterologous and the homologous systems.

Attempts to interfere with cerebral aminoacylation
of tRNAs in vivo have also been reported, but to date
most involve the chronic administration of drugs, such
as ethanol or morphine, to rodents (8, 12) or exposure
to an extensive "training" routine in the goldfish (20).
In the studies of Tewari and associates (12), chronic
ethanol ingestion was shown to inhibit cerebral leucyl
tRNA synthetase, leaving the cognate brain tRNA unscathed.
In the case of the chronic administration of morphine
(8), this treatment failed to result in significant
alterations of brain tRNAs, but caused an apparent
reduction of the ability of cerebral arginine-,
leucine- and phenyl- alanine-tRNA synthetases to effect
the aminoacylation of cognate E. coli tRNAs, with no
effect on brain tRNA structure or function. In the
single study describing exogenously elicited changes
in tRNAs, Kaplan and Sirlin (20) noted an apparent increase in the accepting capacity of a minor species of
$tRNA^{leu}$ in goldfish brain trained to criterion. The
present results demonstrate the feasibility of achieving
rapid (3 h) changes in brain tRNA function via the
administration of a single dose of the convulsant MSO
to 3-day old rats. The data conclusively show that
brain tRNAs are the target of MSO action. However, as
noted in Table 1, aminoacylation of control and MSO-tRNA
by the hepatic synthetases of the MSO-treated animals

resulted in higher acceptance values and, as noted
further in Table 2, cerebral phenylalanine-tRNA
synthetase of MSO-treated animals aminoacylated tRNA^{phe}
of control and MSO-brains more efficiently than did the
control cerebral enzyme. No adequate explanation of
this apparently specific, but secondary, effect of MSO
on synthetase activity is available at this time.

The effect of MSO on brain tRNA function could be
detected using a mixture of amino acids, a clear indication of its general nature, but the effect became
larger when tRNA^{lys} and tRNA^{phe} were tested (Table 1).
The decrement in lysine acceptance by MSO-tRNA was
more marked using control synthetases, irrespective of
their tissue of origin, while, with phenylalanine,
liver synthetases appeared to recognize the cognate
MSO-tRNA less efficiently than brain synthetases,
irrespective of the treatment of the animals. It
should be emphasized, however, that in both systems
(Table 1 and 2), acceptance of lysine and phenylalanine
by MSO-tRNA was inferior to that effected by control
tRNA. It is also important to note that, since only
the V_{max} and not the K_m values for lysine and phenylalanine became altered as a result of the administration
of MSO (Table 4), the affinity of these amino acids for
the cognate aminoacyl-tRNA synthetase remained unchanged,
but the extent of aminoacylation of MSO-tRNA relative
to control-tRNA became significantly reduced. Moreover,
in view of no differences in the optima for Mg^{++} and
CTP between the control and the MSO-tRNAs (Table 3),
it is unlikely that the administration of MSO brought
about alterations of their interaction with the
synthetase complex at the latter's active site(s) (23,
30).

Further and more direct evidence that brain tRNA^{lys}
and tRNA^{phe} represent targets of MSO action was adduced
by the results of BDC chromatography (Figures 2A-C).
Inspection of panel B reveals the presence of 3 tRNA^{lys}
and 3 tRNA^{phe} isoacceptor species in the 3-day old rat
cerebral cortex. Whereas the predominant tRNA^{lys}
isoacceptor species eluted with the main A_{260} peak,
the minor tRNA^{lys} species eluted inside the EtOH-salt
portion of the chromatographic run, overlapping with
2 tRNA^{phe} isoacceptor species. Following the administration of MSO (panel A), the predominant tRNA^{lys}
species eluted away from the main A_{260} peaks, while
the two minor tRNA^{lys} species shifted towards higher
EtOH concentrations. Panel C shows that the 3 tRNA^{phe}

Table 4. Kinetic constants for lysine and phenylalanine
          acceptance by brain tRNA of control and MSO-
          treated rats

Source of Brain tRNA	Lysine		Phenylalanine	
	K_m	V_{max}	K_m	V_{max}
	(uM)	(pmole/A_{260}/min)	(uM)	(pmole/A_{260}/min)
Control	2.2	208	1.4	3.7
MSO	2.2	169	1.4	2.5
Difference, %		−19	0	−32

MSO: 150 mg/kg, 3 h before death. The rats were
      3-days old.

isoacceptor species also became retarded in their eluting position, yet, as in the control situation (Panel B), the last tRNA^{phe} peak remained free of overlap with any of the tRNA^{lys} peaks.

Future work will be directed at a thorough and comparative characterization of the slowest control and MSO-tRNA^{phe} species. Experiments to purify and characterize relevant tRNA^{lys} isoacceptors are also planned (32). It is hoped that these future efforts may reveal that the different mobilities of MSO-tRNA^{lys} and tRNA^{phe} underlie changes in content and/or position of methyl groups and that the observed loss of their functional capacity stems directly from such post-transcriptional, MSO-elicited modifications of their primary structure (13, 15, 26).

ACKNOWLEDGEMENT

The research was supported by the United States Public Health Service, grant NS-06294.

REFERENCES

1. Barra, H.S., Unates, L.E., Sayavedra, M.S. and Caputto, R. (1972): Capacities for binding amino acids by tRNAs from rat brain and their changes during development. J. Neurochem. 19: 2289-2298.

2. Carpousis, A., Christner, P. and Rosenbloom, J. (1977): Preferential usage of glycyl-tRNA iso-accepting species in collagen synthesis. J. Biol. Chem. 252: 2447-2449.

3. Christner, P.J. and Rosenbloom, J. (1976): A comparison of transfer RNA isoaccepting species between collagenous and noncollagenous tissues in the embryonic chick. Arch. Biochem. Biophys. 172: 399-409.

4. Cummins, C.J., Salas, C.E. and Sellinger, O.Z. (1975): The homologous methylation of tRNA in rat brain. Brain Res. 96: 407-412.

5. Dainat, J., de Balbian Verster, F., Zand, R. and Sellinger, O.Z. (1979): Age-dependent changes in the specificity of tRNA methyltransferases in the cerebellum of the icteric and nonicteric Gunn rat. Neurochem. Res. 4: 557-565.

6. Dainat, J., Salas, C.F., and Sellinger, O.Z. (1978): Alteration of the specificity of brain tRNA methyltransferases and of the pattern of brain tRNA methylation in vivo by methionine sulfoximine. Biochem. Pharmacol. 27: 2655-2658.

7. Dainat, J. and Sellinger, O.Z. Cerebellar tRNA methyltransferases: a developmental study. Brain Res. (in press)

8. Datta, R.K. and Antopol, W. (1973): Effect of chronic administration of morphine on mouse brain aminoacyl-tRNA synthetase and tRNA-amino acid binding. Brain Res. 53: 373-386.

9. Der, O. and Sellinger, O.Z. (1979): Changes in methylation in tRNA^{lys} and tRNA^{phe} isoacceptor populations in the methionine sulfoximine epilepto-genic rat brain. Abstracts, VII Intern. Meeting, International Soc. Neurochem., Jerusalem.

10. Drabkin, H.F. and Lukens, L.N. (1978): Preferential use in collagen synthesis of the same glycyl-tRNA species that is elevated in collagen-synthesizing tissues. J. Biol. Chem. 253: 6233-6241.

11. Elahi, E. and Sellinger, O.Z. (1979): The post-natal methylation of transfer ribonucleic acid in brain. Evidence for the methylation of precursor transfer ribonucleic acid. Biochem. J. 177: 381-384.

12. Fleming, E.W., Tewari, S. and Noble, E.P. (1975): Effects of chronic ethanol ingestion on brain aminoacyl-tRNA synthetase and tRNA. J. Neurochem. 24: 553-560.

13. Ginzburg, I., Cornelis, P., Giveon, D. and Littauer, U.Z. (1979): Functionally impaired tRNA from ethionine treated rats as detected in injected Xenopus oocytes. Nucl. Acids Res. 6: 657-672.

14. Harris, C.L. and Maas, J.W. (1974): Transfer RNA and the regulation of protein synthesis in rat cerebral cortex during neural development. J. Neurochem. 22: 741-750.

15. Harris, I.S. and Randerath, K. (1978): Amino-
     acylation of undermethylated mammalian transfer
     RNA. Biochim. Biophys. Acta 521: 566-575.

16. Jank, P. and Gross, H.   (1974): Methyl-deficient
     mammalian transfer RNA: II. Homologous methylation
     in vitro of liver tRNA from normal and ethionine-
     fed rats: ethionine effect on 5-methylcytidine
     synthesis in vivo. Nucl. Acids. Res. 1: 1259-1267.

17. Johnson, T.C. (1969): Aminoacyl-tRNA synthetase
     and transfer RNA binding activity during early
     mammalian brain development. J. Neurochem. 16:
     1125-1132.

18. Johnson, T.C. (1976): Regulation of protein
     synthesis during postnatal maturation of the brain.
     J. Neurochem. 27: 17-23.

19. Johnson, T.C. and Chou, L.   (1972): Level and
     amino acid acceptor activity of mouse brain tRNA
     during neural development. J. Neurochem. 20: 405-
     414.

20. Kaplan, B.B. and Sirlin, J.L.   (1975): Macromole-
     cules and behavior. II. Training induced alter-
     ation in leucine transfer RNA of goldfish brain.
     Brain Res. 83: 451-468.

21. Klee, H.J., DiPietro, D., Fournier, J.J. and
     Fischer, M.S. (1978): Characterization of
     transfer RNA from liver of the developing amphibian,
     Rana catesbeiana. J. Biol. Chem. 253: 8074-8080.

22. Lowry, O.H., Rosebrough, N.J., Farr, A.L. and
     Randall, R.J. (1961): Protein measurement with
     Folin phenol reagent. J. Biol. Chem. 193: 265-
     275.

23. Ofengand, J.   (1977): tRNA and aminoacyl-tRNA
     synthetases. Molecular mechanisms of protein bio-
     synthesis. (H. Weissbach and S. Pestka, eds.)
     Academic Press, New York, pp. 7-79.

24. Ortwerth, B.J. and Der, O.C. (1974): Studies on
     the specialized transfer RNA population of the
     lens. Exp. Eye Res. 19: 521-532.

25. Ortwerth, B.J., Yonuschot, G.R., Heidlege, J.F.,
     Chu-Der, O.M.Y., Juarez, D. and Hedgcoth, C.
     (1975): Induction of new species of phenylalanine
     transfer RNA during lens cell differentiation.
     Exp. Eye Res. 20:   417-426.

26. Ramberg, E.S., Ishaq, M., Rulf, S., Moeller, B.
     and Horowitz, J. (1978): Inhibition of transfer
     RNA function by replacement of uridine and uridine-
     derived nucleosides with 5-fluorouridine. Bio-
     chemistry 17:   3878-3885.

27. Roy, K.L., Bloom, A. and Soll, D. (1971): tRNA
     separations using benzoylated DEAE-cellulose.
     Procedures in Nucleic Acid Research (G.L. Cantoni
     and D.R. Davis, eds.) Harper and Row, New York,
     pp. 524-541.

28. Salas, C.E., Cummins, C.J. and Sellinger, O.Z.
     (1976): The developmental pattern of homologous
     and heterologous tRNA methylation in rat brain:
     differential effects of spermidine. Neurochem.
     Res. 1:   369-384.

29. Salas, C.E., Ohlsson, W.G. and Sellinger, O.Z.
     (1977): The stimulation of cerebral N^2-methyl
     and N_2^2-dimethyl guanine-specific tRNA methyl-
     transferases by methionine sulfoximine: an in vivo
     study. Biochem Biophys. Res. Commun. 76:   1107-
     1115.

30. Schimmel, P.R. (1979): Recent results on how
     aminoacyl transfer RNA synthetases recognize
     specific transfer RNAs. Molecular & Cell. Biochem.
     25:   3-14.

31. Sellinger, O.Z., and Salas, C.E. (1979): Transfer
     RNAs in brain. Biochemistry of Brain (S. Kumar,
     ed.) Pergamon Press, pp. 279-297.

32. Wittig, B., Reuter, S. and Gottschling, H. (1977):
     Comparative characterization of four purified
     lysine-specific transfer ribonucleic acids from
     chicken embryos. J. Biochem. 81:   1705-1713.

33. Yang, W.K. and Novelli, G.D. (1971): Analysis of
     isoaccepting tRNAs in mammalian tissues and cells.
     Methods in Enzymology, vol. XX, part C (K. Moldave
     and L. Grossman, eds.) Academic Press, New York,
     pp. 44-55.

GENERAL AND REGIONAL TURNOVER OF RIBOSOMAL RIBONUCLEIC

ACIDS IN THE BRAIN OF MALE AND FEMALE RATS DURING

POSTNATAL DEVELOPMENT

S. L. Petrović, M. B. Novaković, L. M. Rakić,
J. J. Ivanuš, R. I. Tepavac and A. I. Berner

Department of Molecular Biology
Boris Kidrich Institute

Department of Biochemistry
School of Medicine
Belgrade, Yugoslavia

ABSTRACT

The turnover rates of brain rRNAs were measured in male and female Wistar rats during the postnatal (7-31 days) and "adolescent" (38-58 days) periods in either the whole hemispheres or in the various substructures. These rates were considerably accelerated during the postnatal development, with T/2 values being reduced by 50 percent or more in the adult rats (26) when compared to the younger group. A difference in rRNA replacement rates was observed throughout the postnatal life between males and females. It was moderately significant in the younger group and highly significant in adolescents and young adults. These differences were also demonstrable in most of the studied cerebral substructures. The growth-correlated and sex-linked differences found are discussed in terms of RNA conservation as related to growth rate, the evolving male and female sexual steroid patterns, and the possible sexual dimorphism unrelated to sex hormones.

INTRODUCTION

In the confluent interphase cells and integrated
"resting" cells, ribosomal RNAs are subjected to an
orderly replacement (21, 31) at a rate that could largely
reflect the overall metabolic rate and protein turnover
(5). The similarity of the protein and rRNA or ribosome
turnover rates has been established in several organs,
including liver, kidney, lung, submaxillary glands, and
the brain (13, 16, 22, 30). This correlation allows the
evaluation of the general protein replacement rates on
the basis of rRNA turnover parameters, which can be readily determined to a satisfactory degree of accuracy since
protein replacement rates require a fairly complex schedul
(22). It should also be possible to discern the local
(e.g., brain regional) variations in protein replacement
rates, and the age-related or sex-linked differences in
these rates, by following turnover of ribosomes or rRNAs.
Determination of the rRNA replenishment rates could also
be helpful in characterization of hypertrophic and hyperplastic changes in mammalian cells and organs (4, 30).

Few studies of ribosomal turnover in the brain have
been made (3, 16, 23, 26) and its relationship to the
above factors is entirely lacking. We have therefore
measured brain rRNA replacement rates as related to postnatal development and sex and found consistent changes
and differences which are the basis for this report. A
preliminary survey of regional rRNA turnover rates was
also done.

MATERIALS AND METHODS

The sex of Wistar albino rats was ascertained at
birth and the animals were separated into groups of six
males and six females after weaning. Orotic acid 5-^{3}H
(NEN, Boston, Mass.) was injected subcutaneously on the
7th day after birth (70 µCi/animal), or intraperitoneally
on the 38th day of age (100 µCi/animal). Nucleic acid
isolation and characterization was performed as described
previously (26). The radioactivity level was determined
using Packard Prias liquid scintillation spectrometer.
Other details are specifid in the text.

RESULTS
Brain and Body Weight Changes in the Course of Regression
Measurements

As could be expected during the phase of rapid postnatal brain growth in the rat (8, 12), there was a con-

siderable increase in brain weight (Fig. 1, curve 1, right) in young animals, and some increase was also present during the "adolescent" period (38-58 day). Between the 12-31 postnatal day, the weight of whole brain increased at an average linear rate of 25.7 mg/day in males, vs. 25.4 mg/day in females. The weight of whole brain was at every point of measurement somewhat higher in males, but the overall regression difference was quite insignificant (F_B only 3.3×10^{-4} in Snedecor's (34) F regression test). Some of the increase in total brain size was due to a rapid growth of cerebellum between the 10th and 30th day, as seen from Fig. 1, curve 2, right (also compare refs. 8, 12, 14). The weight of the whole brain and the hemisphere was slightly but consistently greater in the male from 10 to 60 days (Fig. 1, curve 3, right).

On the other hand, the body weights (Fig. 1, left), starting from an almost identical level at the age of 12 days, increased almost linearly up to 58 days, but there-after at quite a different rate in males (5.5 g/day) and females (4.6 g/day), as indicated by the increasing dif-ference in the real weight and in the "weight-to-age ratio" (Fig. 1, inset, left). The data for body and brain weight generally corresponded with those thus far published for Wistar rats (7, 8, 12, 26).

Concentration and Amount of Brain Nucleic Acids in the Course of Postnatal Development

The relative recoveries of rat brain nucleic acids using the standard purification and separation procedures (26, 29), rather than the "absolute" values, are presented in Fig. 2. While phenol-detergent purification employed could be made nearly quantitative by including a reextrac-tion of phenolic phases at 60° (26, 29), this was imprac-tical for the more than 300 samples that had to be pro-cessed in the course of this work. To get values fairly close to actual nucleic acid concentration from data in Fig. 2 and Table 1, a correction for 70 percent recovery should be included. The values for DNA then correspond well with recent data for brain regions (32) and the same should be true for rRNAs, although the existing data are rather scant (26).

The concentration of rRNA decreases in a steep expo-nential fashion with an inflexion around the 30th day and reaches a plateau around the 50th day of life (Fig. 2). This pattern was less apparent in the whole hemisphere; indeed, there was a nearly linear drop in rRNA concentra-

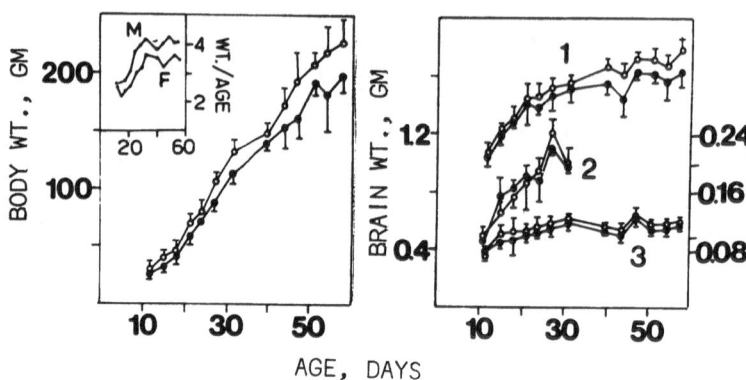

AGE, DAYS

Figure 1.   Left: Body weight of male (o) and female (•)
Wistar rats from 12 to 58 days of age.   Three
animals constituted each experimental group;
the standard deviations are denoted by bracketed vertical lines.   The inset shows the
ratio of body weight in grams to age in days
for the same set of experimental groups (age
in days on abscissa).
Right: Weight of whole brain (curve 1), of
left brain hemisphere (curve 3) and of cerebellum (curve 2) for animals described above..
Note that scale of weight is different for
cerebelli (right ordinate) and hemispheres
or total brain (left ordinate).
Equations describing overall change in weight:
Brain weight in males = 0.15179 + 0.40291 ln x
Brain weight in females = 0.17535 + 0.31510 ln x
Hemisphere wt. in males = 0.21791 + 0.09966 ln x
Hemisphere wt. in females = 0.17535 + 0.Lo649
ln x (x = age of animal in days)

tion throughout the observation period, from the 3rd to
the 58th day, with an average rate of 0.022 mg/g per day
in males and 0.025 mg/g per day in females.   In the cerebral cortex, the decrease averaged 0.017 mg per day in
either sex; the corresponding figures in the caudate
were 0.015 and 0.012 mg per day, and in hippocampus 0.01
and 0.012.   At 100 days of age the level of rRNA could be
calculated, on the basis of fits listed in Table 1, as

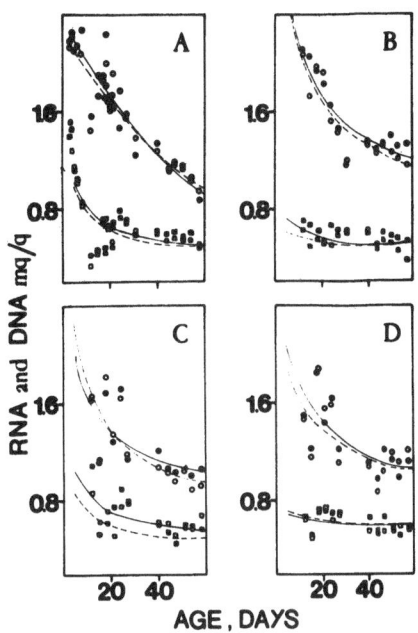

Figure 2.   Concentration of nucleic acids in rat brain
            hemispheres (A), cerebral cortex (B), the
            caudate area (C) and hippocampus (D) in
            animals aged 3-58 days.
            ■ Females, DNA; □ Males, DNA; ● Females,
            RNA; o Males, RNA. Results are expressed in
            mg nucleic acid per g of given structure.
            For further details see Materials and Methods,
            Table 1 and legend to Fig. 1.

1.34 vs. 1.25 mg/g in the whole hemisphere, as 1.24 vs.
1.32 in the cortex, as 1.01 and 1.16 mg/g in the caudate,
and as 1.23 vs. 1.18 mg per g tissue in hippocampus in
males and females, respectively. The net decrease of
rRNA between 10-60 days of age was thus 1.3-1.4 mg/g in
cortex, 0.9-1.0 mg/g in the whole hemisphere, 0.8-1.1 mg/g
in the caudate, and 0.6-0.7 mg/g in the hippocampus. As
seen in Table 1, all RNA regressions showed high correlation coefficients, indicating a very high probability of
such a change (as t test values were above 2.5 for all
regressions, at 10-21 degrees of freedom).

    Based on the averaged regression of the hemispheric
weight (Fig. 1) and that of rRNA concentration (Table 1),
it is found that the amount of rRNA per hemisphere decreases somewhat throughout the postnatal period. The

Table 1.   Regression of rRNA Concentration in Male and
            Female Rat Brain during Postnatal Development

Power regression estimates		Whole hemisphere	Cortex	Caudate	Hippocampus
a	Males	3.4117	5.0460	4.0845	2.7698
	Females	3.7879	4.8086	3.0721	3.3022
b	Males	-0.2592	-0.3623	-0.3604	-0.2325
	Females	-0.2909	-0.3375	-0.2677	-0.2797
r	Males	0.868	0.910	0.813	0.597
	Females	0.876	0.859	0.735	0.750
y_{10}	Males	1.878	2.191	1.781	1.554
	Females	1.971	2.211	1.659	1.648
y_{60}	Males	1.181	1.145	0.934	1.078
	Females	1.151	1.208	1.027	1.061

The power fits according to general equation $y = ax^b$
were estimated by a program for Hewlett-Packard 29C
calculator from the mean nucleic acid concentrations
shown in Fig. 2.   Each point included material from
three animals, analyzed separately.   To arrive at
approximate "true average" contents of rRNAs, the above
estimates should be corrected upward for 30 percent of
the listed values.
y = mg RNA/g tissue at day of age x
r = correlation coefficient (for 2$\overline{1}$ degrees of freedom
    in whole hemisphere regression, 11 d.f. in cortex
    regression, and 10 d.f. in caudate and hippocampus
    regressions)

decrease between the 10th and 60th day of age could be
calculated as 14 percent and 16 percent in males and
females, respectively.   This decrease in rRNA content
may differ in the various hemispheric substructures, but
in the absence of knowledge of their actual weight in the
course of development, no precise estimates could be made.
For instance, if it is assumed that cortex weight increased
to the same degree as the whole hemispheric weight (an
assumption that had to be made in calculation of the rRNA
turnover parameters), then the decrease of cortex rRNA
content would average 37 percent in males and 31 percent
in females for the above period.

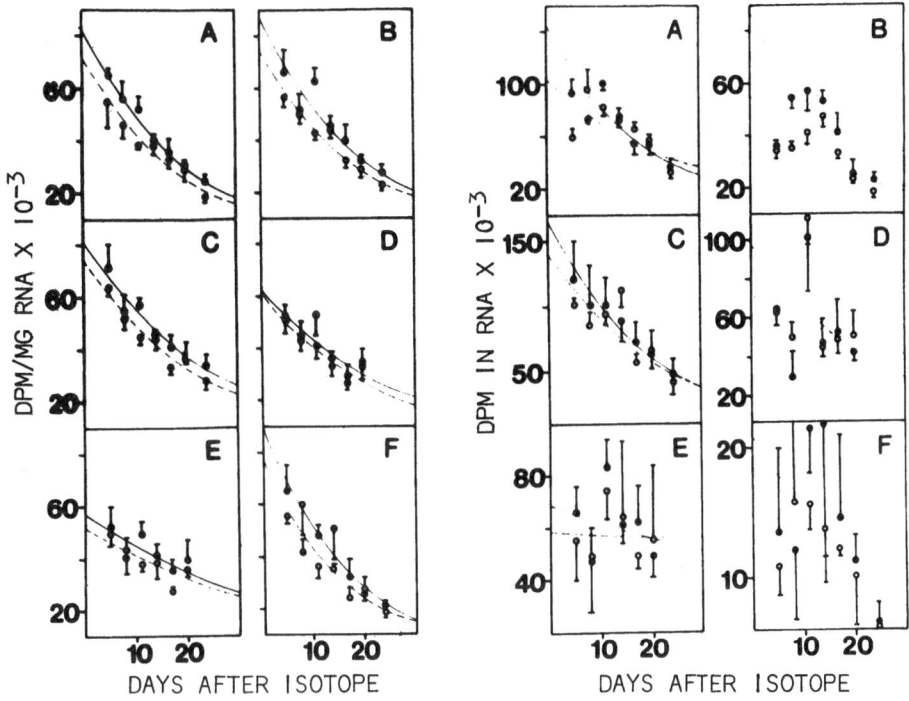

Figure 3.   Left: Specific radioactivities (disintegrations/min per mg RNA) in the brain hemispheres, their substructures and cerebelli of female (----, ●) and male (- - -, o) rats aged 12-31 days. (A) Total hemisphere; (B) "Cytoplasmic" hemisphere; (C) Cortex; (D) Nucleus caudatus; (E) Hippocampus; (F) Cerebellum.
Right: Total rRNA radioactivities (disintegrations/min per mg RNA/x/double left hemisphere weight/x/mg rRNA per g of the given substructure) of the brain tissue described above. Vertical bars correspond to standard deviations of the averages shown (see text).

Determination of rRNA Turnover Rates

The results for the specific rRNA radioactivities and the total radioactivities in the brain hemispheres and their substructures are shown in Figs. 3 and 4. To simplify the presentation, the measurements of three individual rRNAs were pooled together to obtain the averages shown although all individual values for the hemispheres are shown in Fig. 5.

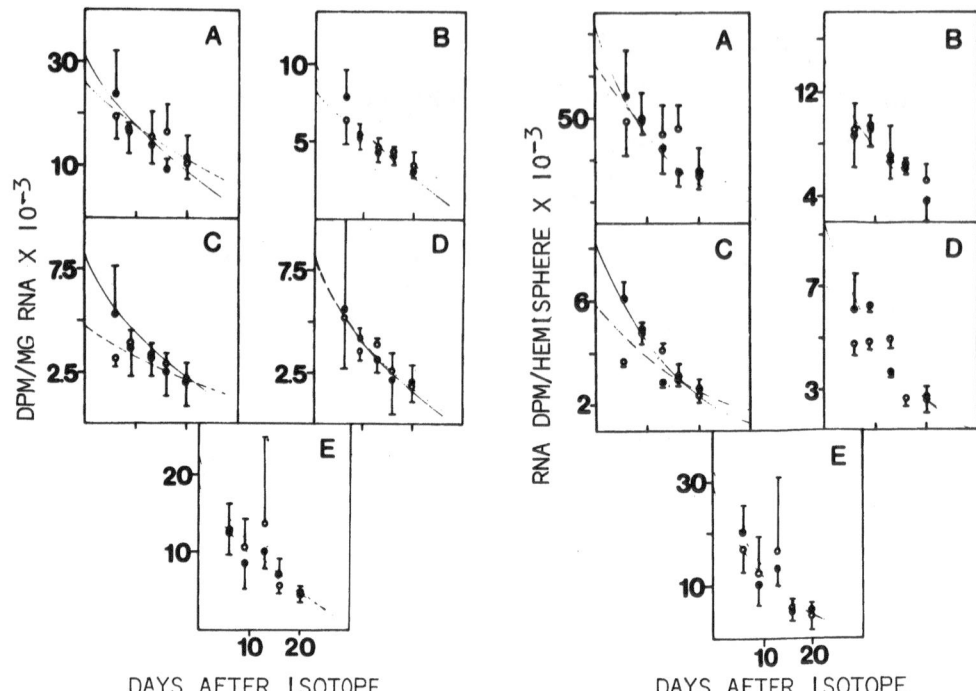

Figure 4.   Left: Specific radioactivities of rRNAs in
the brain hemispheres and in the individual
regions of female (——, ●) and male (- - -, o)
rats aged 44-58 days. The animals were intraperitoneally injected with 100 μCi each of
5-(³H)-orotic acid on the indicated days.
(A) Total hemisphere; (B) Cortex; (C) Nucleus
caudatus; (D) Hippocampus; (E) Thalamus.
Right: Total rRNA radioactivities for the
brain tissue described above. For further
details and explanations, see the legend to
Fig. 3 and Materials and Methods section.

In young animals, the specific rRNA radioactivities
were rather similar in all regions studied. The extrapolated initial labeling was almost the same for total
hemisphere material, "cytoplasmic" rRNA of the hemispheres
and for cortical rRNA, while that for the caudate and
hippocampal RNA was somewhat lower. On the other hand,
in adolescent animals there was a high degree of asymmetry in rRNA specific activities in the various cerebral
regions, thalamic structures being apparently more labeled
than the cortical elements. A similar asymmetry has been
reported in adult animals after intrathecal injection of
the labeled orotic acid (27).

The regression of specific activities in young animals displayed no sex differences (Table 2 and Fig. 3 left), while in adolescent animals the regression of specific activity was much slower for males than females in the whole hemisphere and in two of four substructures studied (Table 2 and Fig. 4 left). When the total rRNA radioactivities were compared, however, an obvious and consistent difference was found between male and female rRNAs in the young animals as well as in all the cortical structures and the whole hemisphere of the adolescent animals.

The turnover of predominantly cytoplasmic rRNAs obtained by cold phenol extraction of the sucrose homogenates obtained from cerebral hemisphere of young animals was also studied to determine whether the nuclear RNA turnover significantly alters the observed dynamics (3, 16). The specific radioactivities for this RNA were found to correspond well with the total hemisphere rRNAs, and the results were identical with the one seen for the whole cortical rRNAs (Fig. 3 left). The sex related differences were less obvious in the total radioactivities of rRNA for this fraction, but this could be the result of variable rRNA yield per unit of hemispheric weight.

The cerebellar rRNAs were difficult to extract reproducibly due to the interference of native deoxyribonucleoprotein, which was extremely viscous when released by detergent, and difficult to manage in the purification of nucleic acids from this brain region. This resulted in a very high scatter of the computed total radioactivities for this structure (graph F, Fig. 3 right). The cerebellar rRNAs were characterized by a low turnover rate in both young males and females. (Apparently low T/2 values obtained on the basis of specific rRNA radioactivities merely reflect high increase in cerebellar mass and rRNA content in the course of regression measurement -- see Fig. 1 right).

Hippocampal rRNAs in young animals apparently turned over more slowly than in any other brain region examined (Table 2 and Fig. 3). However, the regression data for total rRNA of this structure are only tentative since a great variation in rRNA concentration was found for this region (Fig. 2 D) and for the respective total rRNA radioactivities (Fig. 3 right, graph E). In adolescent animals, hippocampal rRNAs did not display such a slow turnover (Fig. 4 D left and right); it is possible that the slowness of rRNA replacement in hippocampi of young animals results from the high growth rate of this region during the observation period, as is the case in the cerebellum.

Table 2. Regression Parameters for rRNAs from the Various
         Brain Regions during the Postnatal Period of
         12-58 days.

Experimental Group and Brain Region	Sex	Specific rRNA Radio-activity Regression Parameters			Total rRNA Radio-activity Regression Parameters		
		a	b	T/2	a	b	T/2
Young Rats (7-31 days)							
Total hemisphere	Male	71665	0.0500	13.87	100318	0.0339	20.45
	Female	83108	0.0501	13.83	143437	0.0515	13.47
"Cytoplasmic" hemisphere	Male	74117	0.0465	14.92	54036	0.0353	19.63
	Female	87871	0.0478	14.51	71221	0.0408	17.01
Cortex	Male	74224	0.0394	17.61	139124	0.0425	16.30
	Female	81823	0.0379	18.30	164175	0.0484	14.33
Caudate area	Male	62048	0.0415	16.72	71682	0.0185	37.46
	Female	62966	0.0355	19.53	97284	0.0398	17.42
Hippocampus	Male	52123	0.0236	29.4	58143	0.0012	(100)
	Female	57179	0.0244	28.4	68291	0.0094	(74)
Cerebellum	Male	76672	0.0579	11.98	16980	0.0317	21.9
	Female	90735	0.0585	11.85	20753	0.0329	21.1
Adolescent Rats (38-58 days)							
Total hemisphere	Male	25409	0.0378	18.35	74180	0.0459	15.09
	Female	30737	0.0583	11.90	94873	0.0740	9.37
Cortex	Male	8056	0.0388	17.85	10387	0.0436	15.90
	Female	10387	0.0600	11.55	14249	0.0584	11.86
Caudate area	Male	4702	0.0331	20.9	5872	0.0392	17.7
	Female	6803	0.0668	10.4	8224	0.0615	11.3
Hippocampus	Male	7914	0.0691	10.0	7285	0.0450	14.4
	Female	8502	0.0764	9.1	9514	0.0617	11.2
Thalamus	Male	22251	0.0734	9.4	33285	0.0944	7.3
	Female	18648	0.0633	11.0	30213	0.0862	8.1

Natural logarithms of the specific rRNA radioactivities or the corresponding total rRNA radioactivities were fitted into a linear regression by a program for Hewlett-Packard 29C calculator, to obtain the above values.

a = Extrapolated zero time specific or total rRNA radioactivity, in dpm/mg or dpm per total hemisphere weight (or cerebellum weight).

b = The coefficient of regression in $\bar{y} = a + b\bar{x}$.

$\bar{T}/x$ = Calculated disappearance time for one-half of the a radioactivity parameter above.

The thalamic RNA, examined only in adolescent animals, appeared to turn over quite fast, with no clear
sex-linked difference (Fig. 4 E and Table 2). It should
be noted that data on rRNA metabolism in substructures
other than cortex in this experiment could be measured to
only 2.5 percent of the standard deviation of the corresponding count rate, and are therefore not precise. (The
counting of radioactive RNAs of total hemispheres and
cortex of adolescents, and of all materials from young
animals, was done to 0.5-1.0 percent of the standard
deviation.)

An analysis of regression data for total hemisphere
rRNAs of male and female rats in the 7-31 day and 38-58
day intervals is presented in Fig. 5. The regressions
for males and females showed a less significant difference (F_B 1.8-4, average 1.985) in young vs. adolescent
animals (F_B 1.6-11, average 5.35) in Snedecor's (34) F
regression test involving 15-21 regression points; nearly
all tested projections were significantly different between adolescent males and females in this test.

When rRNA half-lives for individual regressions,
including our recently published data on young adult
animals (26), were compared in Student's t test, a significant difference was found for young males compared to
young females, while those for the two higher age brackets
were highly significant (P < 0.01). It was also found
that adolescent males had significantly faster RNA turnover than young males, while rRNA turnover in adult males
was even faster; a parallel situation existed for T/2
values measured in females (Fig. 6 and Table 3).

DISCUSSION

Our results on the accretion of brain weight, an
important factor in the estimation of rRNA turnover rates,
differ somewhat from previous determinations reviewed by
Himwich (12). Most of the sex-linked differences in brain
weight were found in our experiments prior to the end of
the adolescent period (60 days) as defined by the saturation of plasma sexual steroid levels (19), while the older
data indicated a more slowly evolving and continuous
inter-sex divergence. The difference in brain weight
coupled with about the same average rRNA concentration
for males and females in either hemispheres or their substructures (Fig. 2) produced most of the sex-linked difference in brain RNA replacement found in young animals
(Fig. 3).

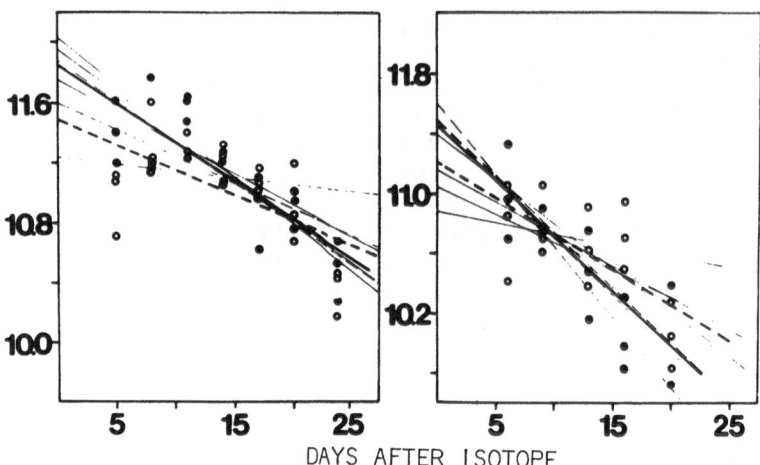

DAYS AFTER ISOTOPE

Figure 5.   Linear regression fits of the data for total
             hemisphere rRNA radioactivities of female (●)
             and male (o) rats aged 12-31 days (left) and
             44-58 days (right). The animals received
             ^{3}H-orotic acid at indicated intervals prior
             to killing. Conditions as in graphs A of
             Figs. 3 and 4 right, but the data for indi-
             vidual rRNA radioactivities are plotted sep-
             arately. The fits corresponding to whole
             series are rendered in bold lines (———,
             females; - - -, males). The fits for regres-
             sions derived by truncation, omitting either
             the first point or one to two last points,
             are represented by the corresponding thin
             lines. Regressions in the young animal group
             tended to accelerate with the increase in the
             sum of days of age for the animals included.

        The net loss of cerebral rRNAs and a considerable
decrease in cerebral rRNA concentration were found to
occur to a similar extent in males and females over the
postnatal period of 10-60 days. The decrease in rRNA
concentration is in accord with the known increase in
neuronal mass due to enlargement of the cell bodies (14).
The decrease, or at least stagnation, of brain rRNA content would be coupled to changes in protein synthetic
capacity, which perhaps follow the period of rapid growth
and brain enlargement prior to and immediately following
birth. A situation which might be interpreted in similar
terms was detected by Balázs and coworkers in developing
rat brain (1, 2). The decrease in rRNA content and concentration possibly results from a lowered demand for
synthesis of both exported (e.g., CSF) or intracellular

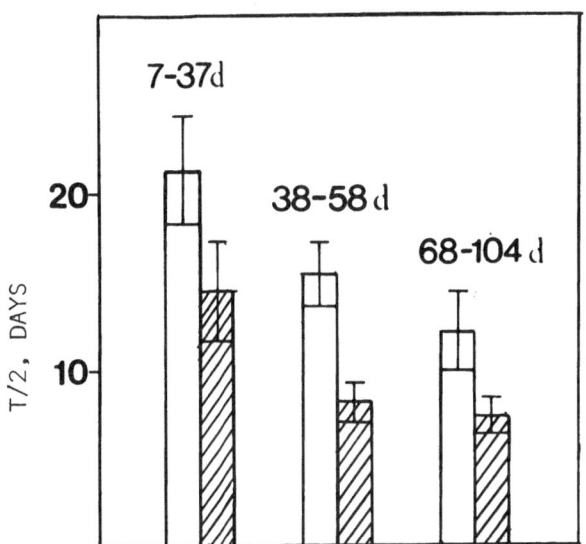

Figure 6. Half-lives of total brain hemisphere ribosomal RNAs in female (hatched bars) and male (blank bars) rats in three different age brackets (age in days indicated above bars). The standard deviations are represented by bracketed vertical lines. For detailed information concerning the corresponding regressions and their testing in Student's t test, see Tables 2 and 3, and ref. 26.

proteins after the onset of saturation in brain weight.

The decrease of DNA concentration in the rat cerebrum during maturation has been amply documented by earlier workers (15). It obviously results from an increase in cell volume, found especially in neurons (14, 15). As expected, there is no decrease in net DNA content (as could be deduced from Fig. 2, using equations given in the caption to Fig. 1).

The difference in turnover rates for males and females should not be considered as an adaptation of rRNA metabolic rates to different gene concentration (9), since rDNA in rat and other mammals appears to be located exclusively in acrocentric autosomes (24, 36). The females appeared to possess a higher overall rate of rRNA metabolism, including both a better utilization of rDNA transcripts (evidenced as higher labeling at the first point of measurement, 5 or 6 days after the isotope administration, as seen in Figs. 3 and 4), and a faster removal of ribosomes as end products; i.e., they had a higher turnover rate (5, 31).

Table 3. Comparison of rRNA Half-lifes in Animals of
         Various Ages

Age Bracket	Sex and Number of Regressions Analyzed		T/2, Days ± SD	t Value and Assessment of Significance
Young	Males	(3)	21.37 ± 3.24	2.23 (D.f. = 4)
(7-31 days)	Females	(3)	14.53 ± 2.89	(0.05 > P > 0.025)
Adolescent	Males	(3)	15.54 ± 1.78	2.84 (D.f. = 4)
(38-58 days)	Females	(3)	8.23 ± 1.05	(0.025 > P > 0.01)
Adult	Males	(4)	12.2 ± 2.25	3.18 (D.f. = 6)
(68-104 days)	Females	(4)	7.42 ± 1.32	(0.01 > P > 0.005)

t tests for males in various age brackets

Young and adolescent: t = 2.23 (d.f. = 4), 0.05 > P > 0.025
Young and adult: t = 3.73 (d.f. = 5), 0.01 > P > 0.005

t tests for females in various age brackets

Young and adolescent: t = 2.90 (d.f. = 4), 0.025 > P > 0.01
Young and adult: t = 3.68 (d.f. = 5), 0.01 > P > 0.005

For statistical procedures, consult ref. 34.

Our data concerning slowness of rRNA turnover in young
animals are in line with known characteristics of rRNA metabolism in growing and dividing cells (21, 31). Increased
conservation of ribosomes in the brain seems to coincide
with the rapid growth of its regions and structures, particularly in cerebellum and the subcortical regions in
young rats (Fig. 3, Table 2). Our data are not necessarily
at variance with reports on either better DNA transcription
(18) or better utilization and transport to cytoplasm of
rRNA in young animals (3). In the above essentially short-
term observations, much of the outcome will depend on the
selected age period and other conditions of the analysis.
The asymmetry of rRNA labeling in the cerebral regions
during (Fig. 4) and after adolescence (27) could partly
indicate a differential in regional ribosome replacement
rates, which can give rise to a variety of seemingly controversial short-term observatons. In view of the pronounced decline in rRNA concentration in the course of the
postnatal period from 10-60 days (Fig. 2) on one hand, and
the simultaneous acceleration of rRNA turnover (Fig. 6) on
the other, the net number of newly finished ribosomes entering the cytoplasm need not decrease, and could even increase

continuously after a certain point in brain development.
In favor of this concept are the data on young adult male
rats (16, 26) which exhibited a T/2 of about 12 days,
while in older adult males the T/2 decreased to 7 days
(23). The stability of ribosomes appears to decline
continuously throughout the postnatal life up to 12 months,
and possibly even beyond (as could be predicted from our
data in Table 1), until the onset of senescence which may
bring about changes in the opposite direction (23).

Decrease in RNA concentration and stability appears
to coincide largely with an increase in protein concentration in the brain, which is known to occur with an inflection at 30 days (12), and peaks within 50-60 days,
an inverse relationship to the RNA concentration change
(Fig. 2). As the fate of rRNA is ultimately determined
by subcellular ribonucleases (31), it is possible that a
major factor in the observed acceleration of ribosome
turnover might be a stochastic increase in the activity
of "soluble" subcellular ribonucleases, brought about by
both concentration and accretion of protein during brain
maturation.

The difference between males and females should,
however, be viewed also from the point of regulatory
influences of the evolving sexual steroid patterns, both
prenatally [when androgen levels already differ considerably in males and females (33)] and during maturation (19).
Much of the sex-bound difference in turnover rates is
greater in the adolescent period and adulthood than in
the very young animals (Table 3, Fig. 6).

Androgens are known to increase biopolymer conservation in a number of organismic systems and a variety of
metabolic settings (4, 11, 29, 30). Androgens could increase ribosome survival and also influence the preservation of rDNA transcripts during processing, by stimulating
synthesis of polyamines via induction of the corresponding
enzymes (10), and also by enhancing cathepsin organization
and removal from cells through increased oxocytosis, as
in the kidneys (20). Androgens also strongly influence
the development of the patterns of sexual behavior in the
rat, especially when given to very young animals (17).
If the considerable difference in rRNA turnover rates
between males and females, seen especially in adolescent
and mature animals, stems mainly from actions of androgens,
it should be eliminated by androgen treatment of females,
or even driven to a pattern slower than in males at large
enough doses, as in the case of mouse kidney (29).

The established differences in rRNA turnover should
be largely reflected in the turnover of mRNAs in an essentially "resting" system of complex parenchymal cells, such
as neurons and glial cells. Among the mRNAs that could
differ most in their average lifetime should be those for
sexually dimorphic enzymes, e.g., catecholamine metabolism
(6), lysosomal hydrolases (25, 28) and possibly other
androgen-inducible enzymes, such as arginase (35). Further work should establish the extent of actual difference
in lifetimes of these and other mRNAs in male and female
rats.

REFERENCES

1. Balazs, R., Kovacs, S., Cocks, W.A., Johnson, A.L.
    and Eayrs, J.T. (1971): Effect of thyroid hormone
    on the biochemical maturation of rat brain: postnatal
    cell formation. Brain Res. 25: 555-570.

2. Balazs, R. and Richter, D. (1973): Effects of hor-
    mones on the biochemical maturation of the brain.
    Chapter 7. In: Biochemistry of the Developing Brain,
    Vol. 1, W. Himwich (ed.), pp. 253-299, Marcel Dekker,
    Inc., New York.

3. Berthold, W. and Lim, L. (1976): Nucleo-cytoplasmic
    relationships of high-molecular-weight ribonucleic
    acid, including polyadenylated species, in the devel-
    oping rat brain. Biochem. J. 154: 529-539.

4. Borota, J.S. and Petrovic, S.L. (1979): Differences
    in turnover rates of ribosomal RNA in the kidney of
    female mice in hypertrophy induced by uninephrectomy
    and by androgen treatment. Iug. Phys. Pharm. Acta,
    in press.

5. Buchanan, D.L. (1961): Total carbon turnover mea-
    sured by feeding a uniformly labeled diet. Arch.
    biochem. biophys. 94: 500-511.

6. Crowley, W.R., O'Donohue, T.L. and Jacobowitz, D.M.
    (1978): Sex differences in catecholamine content in
    discrete brain nuclei of the rat: effects of neonatal
    castration or testosterone treatment. Acta endocrinol.
    89: 20-28.

7. Dellweg, H., Gerner, R. and Wacker, A. (1968):
    Quantitative and qualitative changes in ribonucleic
    acids of rat brain dependent on age and training
    experiments. J. Neurochem. 15: 1109-1119.

8. Donaldson, H.H. and Hatai, S. (1931): On the weight
    of the parts of the brain and on the percentage of
    water in them according to brain weight and to age,
    in albino and in wild Norway rats. J. Comp. Neurol.
    53: 263-307.

9. Fraser, R.S.S. and Nurse, P. (1979): Altered pat-
    terns of ribonucleic acid synthesis during the cell
    cycle: a mechanism compensating for variation in
    gene concentration. J. Cell Sci. 35: 25-40.

10. Henningson, S., Persson, L. and Rosengren, E. (1979):
    Polyamine metabolism in the kidneys of castrated and
    testosterone-treated mice after administration of
    methylglyoxal-bis/guanylhydraxone/. Biochim. biophys.
    acta 582: 448-457.

11. Hill, J.M. and Malamud, D. (1974): Decreased pro-
    tein catabolism during stimulated growth. FEBS Let-
    ters 46: 308-311.

12. Himwich, H.E. (1973): Early studies of the develop-
    ing brain. Chapter 1. In: Biochemistry of the
    Developing Brain, Vol. 1, W. Himwich (ed.), pp. 1-53,
    Marcel Dekker, Inc., New York.

13. Hirsch, C.A. and Hiatt, H.H. (1966): Turnover of
    liver ribosomes in fed and in fasted rats. J. Biol.
    Chem. 241: 5936-5940.

14. Howard, E. (1973): DNA content of rodent brains
    during maturation and aging, and autoradiography
    of postnatal DNA synthesis in monkey brain. Progr.
    Brain Res. 40: 91-114.

15. Howard, E. (1974): Hormonal effects on the growth
    and DNA content of the developing brain. Chapter 8.
    In: Biochemistry of the Developing Brain, Vol. 2,
    W. Himwich (ed.), pp. 1-68, Marcel Dekker, Inc.,
    New York.

16. Von Hungen, K., Mahler, H.R. and Moore, W.J. (1968):
    Turnover of protein and ribonucleic acid in synaptic
    subcellular structures from rat brain. J. Biol.
    Chem. 243: 1415-1423.

17. Jacobson, M. (1978): Developmental Neurobiology, 2nd edition. Plenum Press, New York.

18. Johnson, T.C. (1967): The effects of maturation on in vitro RNA synthesis by mouse brain cells. J. Neurochem. 14: 1075-1081.

19. De Jong, F.H. and Sharpe, R.M. (1977): The onset and establishment of spermatogenesis in rats in relation to gonadotrophin and testosterone levels. J. Endocrinol. 75: 197-207.

20. Koenig, G., Goldstone, A. and Hughes, C. (1978): Lysosomal enzymuria in the testosterone-treated mouse. A manifestation of cell defecation of residual bodies. Lab. Invest. 39: 329-341.

21. Kolodny, G.M. (1975): Turnover of ribosomal RNA in moust fibroblasts (3T3) in culture. Exp. Cell Res. 91: 101-106.

22. Lajtha, A., Latzkovits, L. and Toth, J. (1976): Comparison of turnover rates of proteins of the brain, liver and kidney in mouse in vivo following long term labeling. Biochim. biophys. acta 425: 511-520.

23. Menzies, R.A. and Gold, P.H. (1972): The apparent turnover of mitochondrial ribosomes and sRNA of the brain in young adult and aged rats. J. Neurochem. 19: 1671-1683.

24. Miller, D.A., Breg, W.R., Warburton, D., Dev, V.G. and Miller, O.J. (1978): Regulation of rRNA gene expression in a human familial 14p+ marker chromosome. Hum. Genet. 43: 289-297.

25. Morrow, A.G., Carroll, D.M. and Greenspan, E.M. (1951): A sex difference in the kidney glucuronidase activity of inbred mice. J. Natl. Cancer Inst. 11: 663-669.

26. Novaković, B.B., Petrović, S.L., Rakić, L.M. and Ivanus, J.J. (1979): Different turnover rates of brain ribosomal ribonucleic acids in male and female rats. J. Neurochem. 33: 661-667.

27. Novaković, M.B., Rakić, L.M. and Petrović, S.L. Manuscript in preparation.

28. Paigen, K. (1961): The genetic control of enzyme activity during differentiation. Proc. Nat. Acad. Sci. USA 47: 1641-1649.

29. Petrović, S.L., Novaković, M.B., Tepavac, R.I. and Wilson, C.W. (1977): Androgen-induced accretion of ribonucleic acids in kidney of female mouse (Mus musculus). Int. J. Biochem. 8: 193-198.

30. Petrović, S.L., Borota, J.S., Novaković, M.B. and Marinković, D.V. (1979): Coordinacy in the induction of acid glycosidases by testosterone in mice. Proc. Special FEBS Meeting on Enzymes, Dubrovnik, P. Mildner (ed.), Pergamon, London, in press.

31. Scott, J.F. (1977): Turnover of ribosomal RNA in cells in culture. Exp. Cell Res. 108: 207-219.

32. Seiler, N. and Schmidt-Glenewinkel, T. (1975): Regional distribution of putrescine, spermidine and spermine in relation to the distribution of RNA and DNA in the rat nervous system. J. Neurochem. 24: 791-795.

33. Slob, A.K., Ooms, M.P. and Vreeburg, J.T.M. (1978): Sex ratio in utero and the plasma concentration of testosterone in male and female rat foetuses. J. Endocrinol. 79: 395-396.

34. Snedecor, G. and Cochrane, W. (1968): Statistical Methods. Sixth edition. University of Iowa Press, Ames, Iowa.

35. Swank, R.T., Davey, R., Joyce, L., Reid, P. and Macey, M.R. (1977): Differential effects of hypophysectomy on the synthesis of beta-glucuronidase and other androgen-inducible enzymes in mouse kidney. Endocrinology 100: 473-480.

36. Tantravahi, R., Miller, D.A., D'Ancona, G., Croce, C.M. and Miller, O.J. (1979): Location of rRNA genes in three inbred strains of rat and suppression of rat rRNA activity in rat-human somatic cell hybrids. Exp. Cell Res. 119: 387-392.

CHANGES OF THE SOLUBLE AND INSOLUBLE GLYCOPROTEIN GLYCANS DURING RAT BRAIN DEVELOPMENT

C. Di Benedetta, P. Corsi, G. Gennarini,
F. Vitiello

Istituto di Fisiologia Umana
Università di Bari (Italy)

INTRODUCTION

The high complexity of the nervous tissue structures, which underlines the complexity of the brain function, requires that proper connections are made during the nervous system organizational stage. The formation of neural circuitry is, therefore, an important step in order to equip the brain with a correct and adequate substrate for its function.

The purpose of our research is to follow the evolution of a peculiar class of molecules during the development of the brain, i.e., the glycoproteins, since they are claimed to be involved in the structuring of the nervous tissue (5, 15) and to participate in the cell to cell recognition phenomena (1, 3, 12, 21). Previous studies have indicated that the cerebral glycoproteins are very heterogeneous in their polysaccharide moiety which can be grossly separated into two main classes having different molecular size (6). It was also shown that at least two classes of cerebral glycoproteins can be distinguished according to their solubility in different solvent systems (7) and also developmental pattern (8, 11), turnover rate (2, 18, 23) and composition (7): the so-called soluble glycoproteins (more easily solubilized) and the insoluble. On the basis of some previous observations the soluble glycoproteins have been related to the insoluble in a precursor-product association and recently it has been demonstrated, by means

of immunochemical methods, that some glycoproteins
present in a soluble form can be trailed in the membrane-
bound fraction (14, 24).

Since both classes of cerebral glycoproteins contain
the low and high molecular weight glycans, our aim was to
follow the developing pattern of the sugars present in
these heterosaccharide chains, in order to describe their
modification during rat brain maturation.

Information about compositional and metabolic changes
of cerebral glycoproteins as well as their glycans have
been already reported, even though the separation in
classes according to their solubility was not performed
and some results were contradictory (13, 16, 17, 19, 20).

MATERIALS AND METHODS

Sprague-Dawley rats were used, selected at birth in
the body weight range of 5.8-6.8 g and randomly distributed in litters of 9 animals. This particular litter
size was chosen after our previous experience on the
effect of different litter size on the development of rat
brain (9). The rats were properly marked at birth and
their body weight increment was recorded throughout the
experiment. The animals showing growth rate different
from that of the whole population were discarded. The
animals of two litters were killed by decapitation at
birth and at 5, 8, 11, 13, 15, 17, 21, 26, 31 days of
life and the whole brain was dissected out. The preparation of soluble and insoluble glycoproteins was performed
according to Di Benedetta et al. (7). The low molecular
weight (LMW) and high molecular weight (HMW) glycans
were obtained by the procedure already described (6). The
mucopolysaccharides, nucleic acids and other impurities
were separated from HMW glycopeptides by precipipitation
with cetylpiridinium chloride and from the LMW glycopeptides by gel filtration on a G-15 Sephadex column
(2.5 x 42 cm).

The monosaccharide determination was performed by
colorimetric (6) and gaschromatographic methods (25).

RESULTS AND DISCUSSION

The N-acetyl-glucosamine (GlcNAc) content of the
developing brain increases in the HMW chains of soluble
and insoluble glycoproteins, while no variation can be
detected for the LMW glycans in both fractions (Fig. 1-a).
It is worth noting that its increment is larger in the

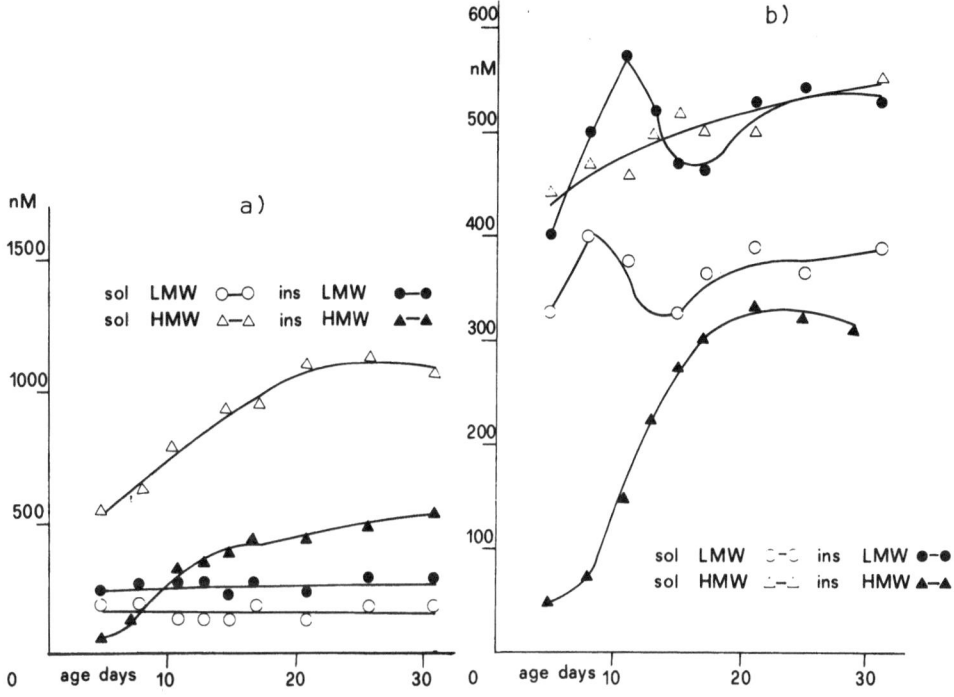

Fig. 1. a) GlcNAc changes (nM/g w.t.) during rat brain development in soluble (sol) and insoluble (ins) glycoproteins.

b) Mannose changes (nM/g w.t.) during rat brain development in soluble (sol) and insoluble (ins) glycoproteins.

insoluble than in the soluble fraction, even though the concentration in the latter is somewhat higher. It is known that the GlcNAc occupies different positions within the heterosaccharide chain including that of linking sugar between the saccharidic and protein moieties. Its amount, therefore, could indicate, with a good approximation, the number of heterosaccharide chains existing in this group of molecules. From the present data it can be, then, inferred that the formation of larger size glycans in the insoluble glycoproteins during brain development outnumbers appreciably that of the soluble.

The developmental pattern of Mannose is similar to that of the GlcNAc as far as the larger size glycans, since its concentration increases in the soluble and insoluble glycoproteins (Fig. 1-b). Also the Man concentration is higher in the soluble fraction with a larger increment in the insoluble glycoproteins. On the other hand, the pattern of the Man modifications in the LMW glycans is very complex and different from that of GlcNAc since its concentration presents a peak around 9-10 days of life, reaching adult values thereafter. A similar finding has been reported (26) during cerebellum development about some concanavalin A positive glycoproteins which appear around the 13th day of life, disappearing soon after.

The modifications of GalNAc (N-acetyl-D-galactos-amine) content, reported in Fig. 2-a, also show a different pattern. Its concentration increases in all glycopeptides, in the soluble and insoluble fractions, and, even if shifted in time, it occurs between day 10-20. The increment of this hexosamine is very large in the LMW of soluble glycoproteins and in the HMW of the insoluble.

The developmental pattern of galactose (Gal, Fig. 2-b), indicates that, unlike those of the other monosaccharides, some variations are occurring only in the HMW glycans of the insoluble glycoproteins (four-fold increase) and in the LMW of the soluble glycoproteins (less than two-fold increment).

The modifications reported in these sugars, usually present in the core of the polysaccharidic chain, show no univocal change. The variations, however, when present, are characterized by an increment, occurring between day 10 and 20, larger for the HMW than for the LMW glycopeptides.

In contrast the modifications of N-acetyl-neuraminic acid (NeuNac) and fucose (Fuc), monosaccharides more externally located, are more similar to each other. The Fuc concentration in the LMW glycans (Fig. 3-a) increases very slightly in both soluble and insoluble fractions, while there is a very appreciable increment in the HMW of the two fractions (four-fold in the soluble and almost eight-fold in the insoluble glycoproteins). The variations of NeuNAc are similar to those of Fuc. No increment for the smaller size glycans, a slight increase for the HMW chains of the soluble glycoproteins is detectable. For this latter group of molecules a period

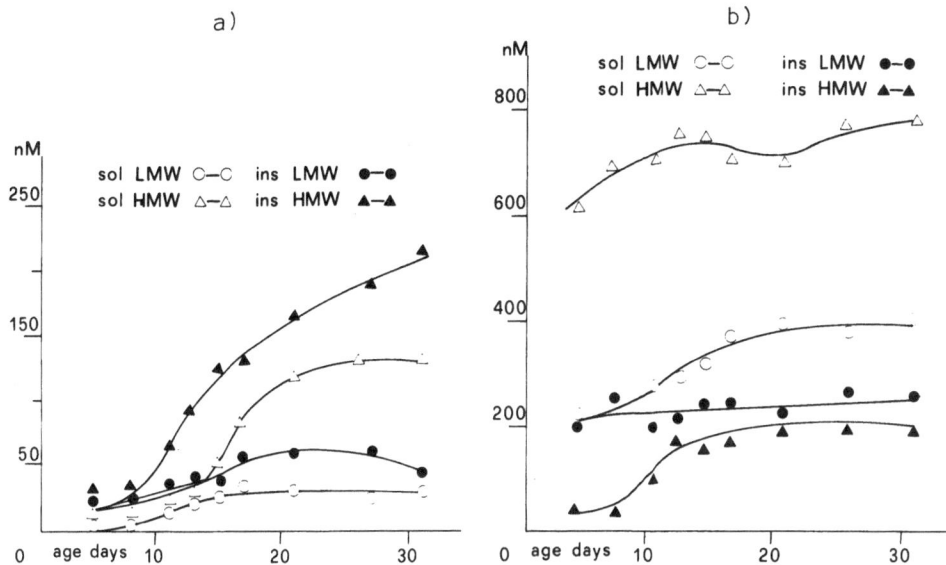

Fig. 2.   a) GalNAc changes (nM/g w.t.) during rat brain
          development in soluble (sol) and insoluble
          (ins) glycoproteins.

          b) Gal changes (nM/g w.t.) during rat brain
          development in soluble (sol) and insoluble
          (ins) glycoproteins.

of high rate of accumulation of NeuNAc is evident between day 11 and 16, suggesting a critical period for
this sugar in this specific class of cerebral glycoproteins (Fig. 3-b).

     The modifications of monosaccharides in the HMW of
both fractions compared to the protein content of the
developing nervous tissue (Table 1), show a common trend
of increment for all sugars with the exception of Mannose
in the soluble glycoproteins and of Gal in both fractions.
The sugar concentration of the soluble glycoprotein LMW,
referred to the protein content, does not change appreciably during development. The pattern of the insoluble, instead, shows a decrease for all sugars
(Table 2). These data suggest that during the development most probably insoluble glycoproteins with a higher

Table 1. Variations of monosaccharide concentration of HMW glycans during rat brain development (expressed as nM/mg prot).

AGE (days)	GlcNAc		Mannose		GalNAc		Gal		NeuNAc		Fuc	
	SOL	INS	SOL	INS	SOL	INS	SOL	INS	SOL	INS	SOL	INS
5	6.76	20.61	12.36	34.13	0.14	1.55	8.13	16.34	14.54	3.65	0.98	4.04
8	7.36	14.01	14.45	26.35	0.37	1.07	8.24	15.22	15.08	3.42	0.67	2.74
11	4.81	15.41	13.18	30.52	0.48	1.84	9.38	10.52	13.59	2.18	0.63	3.88
13	5.33	9.28	11.33	17.55	0.67	1.34	10.96	7.39	16.01	1.39	1.30	2.38
15	4.24	7.45	11.23	15.93	0.93	1.28	10.83	8.13	15.70	1.52	1.37	1.92
17	5.68	7.16	13.22	12.91	1.12	1.54	12.87	6.56	16.00	1.37	1.67	2.06
21	4.47	6.34	13.67	15.40	1.05	1.78	12.92	6.12	14.81	1.45	1.22	1.62
26	5.31	7.13	10.46	13.82	0.75	1.35	11.05	6.66	11.76	1.41	1.16	1.48
31	4.49	6.72	11.48	12.81	0.90	1.03	11.78	6.13	14.17	1.41	1.47	1.63

Table 2. Variations of monosaccharide concentration of LMW glycans during rat brain development (expressed as nM/mg prot).

AGE (days)	GlcNAc		Mannose		GalNAc		Gal		NeuNAc		Fuc	
	SOL	INS	SOL	INS	SOL	INS	SOL	INS	SOL	INS	SOL	INS
5	6.76	20.61	12.36	34.13	0.14	1.55	8.13	16.34	14.54	3.65	0.98	4.04
8	7.36	14.01	14.45	26.35	0.37	1.07	8.24	15.22	15.08	3.42	0.67	2.74
11	4.81	15.41	13.18	30.52	0.48	1.84	9.38	10.52	13.59	2.18	0.63	3.88
13	5.33	9.28	11.33	17.55	0.67	1.34	10.96	7.39	16.01	1.39	1.30	2.38
15	4.24	7.45	11.23	15.93	0.93	1.28	10.83	8.13	15.70	1.52	1.37	1.92
17	5.68	7.16	13.22	12.91	1.12	1.54	12.87	6.56	16.00	1.37	1.67	2.06
21	4.47	6.34	13.67	15.40	1.05	1.78	12.92	6.12	14.81	1.45	1.22	1.62
26	5.31	7.13	10.46	13.82	0.75	1.35	11.05	6.66	11.76	1.41	1.16	1.48
31	4.49	6.72	11.48	12.81	0.90	1.03	11.78	6.13	14.17	1.41	1.47	1.63

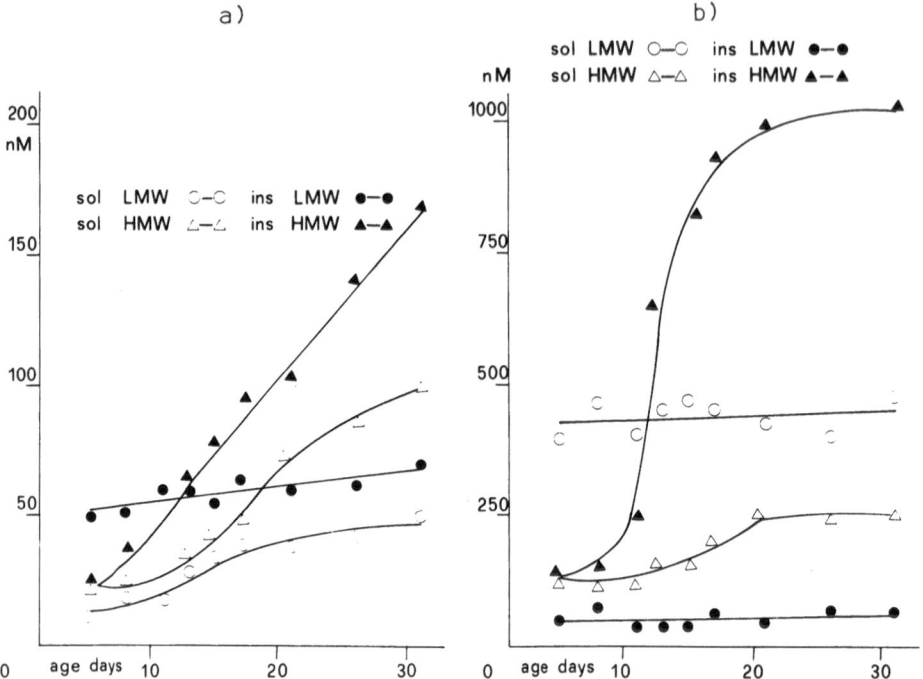

Fig. 3.   a) Fuc changes (n/Mg w.t.) during rat brain
          development in soluble (sol) and insoluble
          (ins) glycoproteins.

          b) NeuNAc changes (nM/g w.t.) during rat brain
          development in soluble (sol) and insoluble (ins)
          glycoproteins.

amount of larger size glycans and with a smaller number
of LMW chains are synthesized, while soluble glycoproteins richer in HMW glycans are formed during brain
maturation.

       CONCLUSIONS

       From the present data it can be inferred that many
variations are taking place in the polysaccharide moiety
of cerebral glycoproteins during the rat brain development and that the larger size glycans are mainly involved.
Their modifications are not comparable, however, even

though all the trends show an increment in their sugar concentration. This observation would suggest that, during brain development, an enormous variety of hetero-saccharide chains is formed, giving rise, therefore, to glycoproteins of different composition. These findings give further support to the hypothesis that these macromolecules could be considered mediators of cell adhesion or repulsion.

These results, moreover, confirm previous reports, which demonstrated that appreciable modifications in the composition (10, 17, 19, 20) and in the metabolism (4, 13, 16, 22) of brain glycoproteins are occurring during a well defined stage of brain development, which coin-cides with the formation of the complex neuronal net. The data of the literature and our present findings indi-cate, on the other hand, that the study of a mixture of glycoproteins cannot achieve any definite result, while the analysis of the pattern of glycoprotein(s), marker of a specific structure, could be more rewarding.

At present no definite conclusions can be drawn on the possible relationship between the soluble and in-soluble glycoproteins. The coincidence of the time of sugar accumulation in both fractions, mainly in the HMW glycans, and the demonstration of a critical period in the turnover of soluble glycoproteins (4) around the 13th day of life might support this hypothesis.

ACKNOWLEDGMENTS

The authors are most grateful to Prof. L. A. Cioffi for helpful discussions and stimulating criticism. The authors wish to thank Mrs. Grazia Salomone for the accurate preparation of the manuscript.

Partially supported by Ministero P. I. and CNR (n° 76.01388.04/78-2081) grants. G. Gennarini is a CNR fellow.

REFERENCES

1. Barondes, S.H. and Rosen, S.D. (1976): Cell surface carbohydrate-binding proteins: role in cell recogni-tion. In: Neuronal Recognition. S.H. Barondes (ed), pp. 331-356, Chapman and Hall, London.

2. Bosmann, H.B., Hagopian, A. and Eylar, E.H. (1969):
    Cellular membranes: the biosynthesis of glycoprotein
    and glycolipid in Hela cell membranes. Arch. Biochem.
    Biophys. 130: 573-583.

3. Brunngraber, E.G. (1969): The possible role of
    glycoproteins in neural function. Perspect. Biol.
    and Med. 12(3): 467-470.

4. Corsi, P., Gennarini, G., Vitiello, F. and
    Di Benedetta, C. (1979): Apparent half life time
    of soluble fucosylglycoproteins during brain
    development. In: A Multidisciplinary Approach to
    Brain Development. C. Di Benedetta, R. Balazs,
    G. Gombos, G. Porcellati (eds.), Elsevier, North
    Holland, Amsterdam, in press.

5. Di Benedetta, C. (1974): Structural and functional
    significance of brain glycoproteins. In: Central
    Nervous System, Studies on Metabolic Regulation
    and Function. E. Genazzini and H. Herken, (eds.).
    pp. 33-40, Springer-Verlag, Berlin, New York.

6. Di Benedetta, C., Brunngraber, E.G., Whitney, G.,
    Brown, D.B. and Aro, A. (1969): Compositional
    patterns of sialofucohexosaminoglycans derived from
    rat brain glycoproteins. Arch. Biochem. Biophys.
    131: 404-413.

7. Di Benedetta, C., Chang, I. and Brunngraber, E.G.
    (1971): Electrophoretic separation and properties
    of soluble glycoproteins extracted from whole rat
    brain tissue. Ital. J. Biochem. 20: 49-65.

8. Di Benedetta, C. and Cioffi, L.A. (1972): Glyco-
    proteins during the development of the rat brain.
    Adv. Exp. Med. Biol. 25: 115-124.

9. Di Benedetta, C. and Cioffi, L.A. (1972): Early
    malnutrition, brain glycoproteins and behavior in
    rats. Bibliotheca Nutritio et Dieta, 17: 69-82.

10. Di Benedetta, C., Corsi, P., Gennarini, G. and
    Vitiello, F. (1979): Developmental pattern of brain
    glycoproteins. In: A Multidisciplinary Approach to
    Brain Development. C. Di Benedetta, R. Balazs,
    G. Gombos and G. Porcellati (eds.), Elsevier North
    Holland, Amsterdam, in press.

11. Di Benedetta, C., De Luca, B. and Cioffi, L.A. (1970): Rat brain proteins and glycoproteins during development. Protides Biol. Fluids, 18: 181-184.

12. Dische, Z. (1966): The informational potentials of conjugated proteins. Protides Biol. Fluids 13: 1-20.

13. Dutton, G.R. and Barondes, S.H. (1970): Glycoprotein metabolism in developing mouse brain. J. Neurochem. 17: 913-920.

14. Gennarini, G., Iannelli, D., Corsi, P. and Di Benedetta, C. (1978): Immunochemical relationship between cytosol and synaptosomal glycoproteins of rat brain. Proc. ESN, 1: 607.

15. Gombos, G., Vincendon, G., Reeber, A., Ghandour, M.S. and Zanetta, J.P. (1978): Membrane glycoproteins in synaptogenesis. Proc. ESN, 1: 174-188.

16. Holian, O., Dill, D. and Brunngraber, E.G. (1971): Incorporation of radioactivity of D-glucosamine-I-^{14}C into heteropolysaccaride chains of glycoproteins in adult and developing rat brain. Arch. Biochem. Biophys. 142: 111-121.

17. Krusius, T., Finne, J., Kärkkäinen, J. and Järnefelt, J. (1974): Neutral and acidic glycopeptides in adult and developing rat brain. Biochim. Biophys. Acta 365: 80-92.

18. Margolis, R.K. and Gomez, Z. (1973): Rapid turnover of fucose in the water-soluble glycoproteins of brain. Biochim. Biophys. Acta 313: 226-228.

19. Margolis, R.K. and Gomez, Z. (1974): Structural changes in brain glycoproteins during development. Brain Res. 74: 370-372.

20. Margolis, R.K., Preti, C. and Margolis, R.U. (1976): Developmental changes in brain glycoproteins. Brain Res. 112: 363-369.

21. Moscona, A.A. (1976): Cell recognition in embryonic morphogenesis and the problem of neuronal specificities. In: Neuronal Recognition. S.H. Barondes (ed.) pp. 205-226, Chapman and Hall, London.

22.   Quarles, R.H. and Brady, O. (1971): Synthesis of
      glycoproteins and gangliosides in developing rat
      brain. J. Neurochem. 18: 1809-1820.

23.   Truding, R., Shelanski, M.L., Daniels, M.P. and
      Morell, P. (1974): Comparison of surface membranes
      isolated from cultured murine neuroblastoma cells
      in the differentiated or undifferentiated state.
      J. Biol. Chem. 249: 3973-3982.

24.   Van Nieuw Amerongen, A. and Roukema, P.A. (1975):
      The appearance of the soluble and the membrane-
      bound fractions of the nervous tissue-specific
      sialoglycoprotein GP-350 in the developing rat
      brain.   Brain Res. 89: 358-362.

25.   Zanetta, J.P., Breckenridge, W.C. and Vincendon,
      G. (1972): Analysis of monosaccharides by gas-
      liquid chromatography of the O-methyl-glycosides
      as trifluoroacetate derivatives.   Application to
      glycoproteins and glycolipids.   J. Chromatogr.
      69: 291-304.

26.   Zanetta, J.P., Roussel, G., Ghandour, M.S.,
      Vincendon, G. and Gombos, G. (1978): Postnatal
      development of rat cerebellum: massive and trans-
      ient accumulation of concanavalin A binding glyco-
      proteins in parallel fiber axolemma.   Brain Res.
      142: 301-319.

ACID MUCOPOLYSACCHARIDES AND MYELIN DEVELOPMENT

M. Rusić, M. Levental and Lj. Rakić

Department of Neurochemistry
Institute for Biological Research
Belgrade, Yugoslavia

Several biochemical (3, 5, 6, 24) and histochemical (1, 26) studies have demonstrated the presence of acid mucopolysaccharides (AMPS) in the central and peripheral nervous tissue. Many authors have investigated the presence of individual AMPS in various species. Although there is some disagreement about the percentage of individual AMPS in the nervous tissue, there is a general agreement about the occurrence of hyaluronic acid (HA), chondroitin sulphate A and C, dermatan sulphate, heparin and heparitin sulphate (HS) in the brain (4, 6, 7, 9, 19, 20). The need for AMPS in the maintenance of normal brain function is suggested by Young (25), who has demonstrated reversible neurological and electroencephalographic changes in cat after chronic intraventricular administration of hyaluronidase. Autoradiographic (10), histochemical (11) and biochemical studies (8, 12, 14) show that AMPS are present in neurons as well as in glial cells, but the presence of these macromolecules in myelin has not been proved until recently. The findings of Szabo and Roboz-Einstein (24) and of Wolman and Hestrin-Lerner (27) both showed that the white matter of the brain consists mainly of HA. The presence of "acid polysaccharides" observed during degeneration of myelin sheath led these authors to infer the participation of HA in the construction of myelin sheath. The results reported by Singh and Bachhawat (18) on AMPS content in rat brain before and during the peak of myelination support the previous suggestion (3) concerning the possible participation of AMPS in the process of the development of the myelin sheath. Dorfman and Ho (8)

showed that clonal glial cells from the rat glial tumor
and from mouse neuroblastoma (22, 23) are capable of
synthesizing AMPS. These authors emphasized the role
which these compounds might also have in the development
of normal myelin in the brain.

We were able to demonstrate biochemically the presence of AMPS in purified myelin fraction of rat brain
and their changes during development and myelination.

MATERIAL AND METHODS

White-Hood male rats, 14, 21, 28, 34, 42, 75 and
114 days old, were used. After decapitation, the brains
were rapidly removed and a 5% homogenate was prepared
from the whole brain. Myelin was isolated and purified
after triple exposure to osmotic shocks according to the
Norton-Poduslo technique (15). From the lipid free and
dry myelin, AMPS were extracted after proteolysis with
activated papain suspension and macrochromatography on
Ecteola-Cellulose column (21). Following the elution
with 3M NaCl and dialysis of the eluate for 48 hours,
AMPS were quantitated via uronic acid by the orcinol
method in the aliquot of this dialysate (21). The
values for total AMPS found in the analyzed sample were
obtained by multiplying the values for uronic acid content by 2.5, assuming 40% content of hexuronic acids in
the total AMPS. Following lyophilization, individual
AMPS were separated by microelectrophoresis on cellulose-
polyacetate strips in buffered $CuSO_4$ solution (pH - 3.4).
Individual AMPS were identified after staining with
Alcian blue and subsequent densitometry by comparing
them with appropriate standards. On the basis of
densitometric scanning (Fig. 1), the percentage of
individual AMPS in total AMPS was calculated, and their
amounts were determined from those percentages. For the
studies of the brain development and AMPS, whole brains
from rats of different age groups were used. The brains
were dried in acetone and after sequential delipidation
in chloroform-methanol (2:1) and (1:2), AMPS were investigated by the method of Stefanovic and Gore (21). The
results are expressed as µg AMPS per gram of dry weight.

In the second part of the experiment, 75-day-old
rats were submitted to electroconvulsive shock using an
electric stimulator. An electric stimulus of 0.2 sec, at
a frequency of 100 cycles per second and an 80 V amplitude, was applied using clips for both ears. The convulsion treatment was repeated (6 convulsions per day
with 1 hour intervals between shocks). The animals were

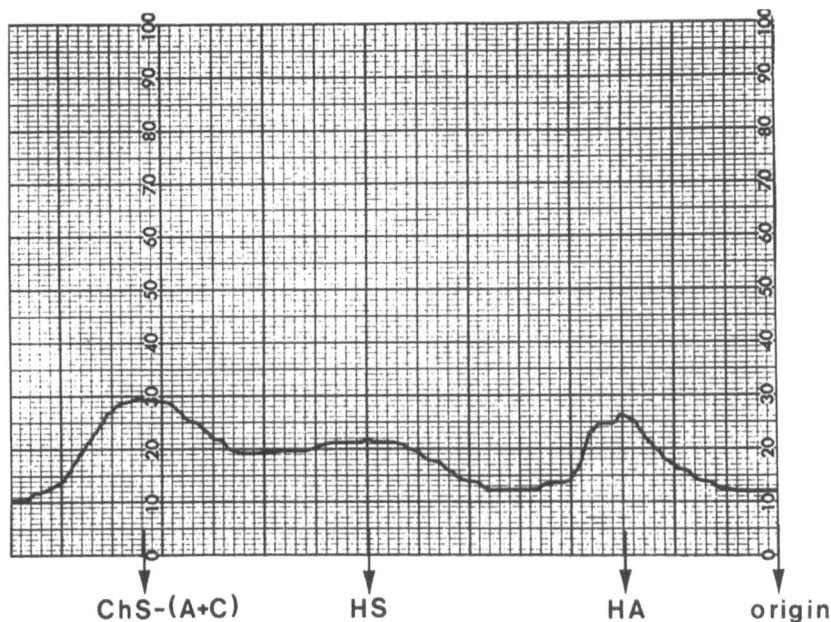

ChS-(A+C)   HS   HA   origin

Fig. 1.   Densitometer scans of individual AMPS (from
          28-day-old rat myelin) resolved electrophore-
          tically and identified by comparing electro-
          phoretic mobility, after staining by Alcian
          blue, of clearly marked fractions from the
          analyzed samples. Migration of individual AMPS
          from the standard solutions consisting of highly
          pure AMPS was submitted to electrophoresis
          simultaneously.

sacrificed either immediately or at 4, 24, 48 or 72 hours
after the last electroconvulsive shock. Extraction of
myelin from these brains and the changes in AMPS content
were determined as described above.

    RESULTS

    Our results (Table 1) concerning the yields of
myelin at various stages of postnatal development show
a high degree of correlation with the findings of Norton
and Poduslo (15).

    AMPS amount in the purified myelin fraction mark-
edly changes in the course of the brain development and

Table 1.   Myelin yield - % of brain dry weight.

Age in Days	Rustić, Levental and Rakić	Norton and Poduslo
14	1.75 - 1.80	1.70 ; 1.70
21	4.63 - 5.68	4.60 ; 5.80
28	7.70 - 7.80	
		7.90 ; 7.70
34	8.20 - 8.60	
42	9.20 - 9.70	
		9.50 ; 10.8
75	12.50 - 12.90	
114	13.80 - 14.10	

myelination (Table 2). On the 14th postnatal day, myelin contains high quantity of AMPS (3.7 mg per gram of dry myelin). In the following week, the content of these heteromacromolecules becomes 3.5 times lower, and on the 21st day is 1.14 mg per gram of dry myelin. At 28, 34 and 42 days postnatally, the AMPS level, which continues to decrease drastically, is over 11 times lower, ranging from 0.38 to 0.32 mg per gram of dry myelin as compared to the 14 day level. In adult rats, a slight increase (0.42 mg) is observed on the 75th day which subsequently reaches a low value of AMPS (0.34 mg per gram of dry myelin). On the 114th day, changes in total AMPS content of the whole rat brain and AMPS within myelin show similar trends, since both compounds decrease in the course of postnatal development. It should be noted, however, that from the 14th day the AMPS decrease in total brain homogenate is proportionally smaller than that of myelin (Table 2). On the 114th day, AMPS concentration of the total brain homogenate becomes 3 times lower, as compared to the myelin, which contains 10 - 11 times less than on the 14th postnatal day.

Table 2.   Content of total acid mucopolysaccharides
           during development.

Age In Days	Myelin		Whole Brain	
	µg AMPS/ g. dry wt.	+ S.E.M.	µg AMPS/g. dry wt.	+ S.E.M.
14	3725	+ 82	7972	+ 190
21	1143	+ 42	6337	+ 324
28	387	+ 8.7	5951	+ 280
34	378	+ 6.2	4254	+ 118
42	325	+ 10.2	3654	+ 233
75	420	+ 17.2	3820	+ 239
114	340	+ 11.7	2722	+ 85

The values for Myelin (5 experiments) and Whole Brain
(10 experiments) are represented as means + S.E.M.

Total AMPS are calculated via uronic acid multiplied by
2.5 (assuming there was 40% hexuronic acid in the total
AMPS).

    The separation of individual AMPS from total AMPS
by electrophoresis on cellulose-polyacetate strips and
their densitometric evaluation (Fig. 1) show the presence of HA, HS and chondroitin sulphate A and C (ChS-
A+C) in myelin, i.e., the same components as in the
total brain homogenate. The values of individual AMPS
(Table 3) within myelin are proportionally smaller as
compared to the values of the whole brain, while the
trend of their changes during postnatal development together with the trend of changes of total AMPS in myelin
and in total brain homogenate appear to be fairly
similar. In contrast, the participation of individual
AMPS in total AMPS (Table 4) is different from that in
the myelin and in the whole brain. HA is higher in

Table 3.   Individual acid mucopolysaccharide content
           during development.

Age In Days	Myelin HA	HS	ChS-(A+C)	Whole Brain HA	HS	ChS-(A+C)
14	468	1887	1370	1036	2236	4704
21	226	317	600	380	1585	4372
28	81	116	189	476	1545	3927
34	102	106	170	425	1276	2552
42	75	96	153	438	730	2484
75	92	155	185	573	993	2254
114	85	92	163	408	789	1524

On the basis of densitometric scanning, the percentage
of individual AMPS was calculated, and their amounts
(in μg AMPS) determined from these percentages.

myelin (12-27%) as compared to the whole brain tissue
(6-15%), and the trend of its changes is somewhat different in the course of postnatal development. HS in the
total brain homogenate is slightly lower and shows no
changes in the whole brain, while within myelin its
values are high on the 14th day but decrease markedly
on the 21st day and then stabilize during adulthood.
On the other hand, the percentage of ChS-(A+C) in the
whole brain is higher compared with the amount in myelin.
The trend of their changes is, however, alike.

    DISCUSSION

    Autoradiographic (10) and biochemical studies over
the last few years (12, 13) point to an even distribution
of total AMPS in neuronal and neuroglial cells. By
applying the Norton-Poduslo method for the isolation of
purified myelin from the rat brain together with the very
sensitive method for the separation of AMPS (21), we
have demonstrated biochemically the presence of AMPS

Table 4. Percentage distribution of individual acid mucopolysaccharides from total mucopolysaccharides

Age In Days	MYELIN			WHOLE BRAIN		
	HA	HS	ChS-(A+C)	HA	HS	ChS-(A+C)
14	12	51	37	13	28	59
21	19	28	53	6	25	69
28	21	30	49	8	26	66
34	27	28	45	10	30	60
42	23	30	47	12	20	68
75	22	32	44	15	26	59
114	25	27	48	15	29	56

HA - Hyaluronic acid

HS - Heparitin sulphate

ChS-(A+C) - Chondroitin sulphate

content in myelin fraction of the rat brain, which undergoes changes during myelination and biochemical maturation of myelin. At an early stage of myelination, the AMPS levels are highest (3.7 mg per gram dry myelin), while during rapid myelination (from 21st to 42nd day), AMPS quantity markedly decreases. At the stage of complete brain maturation, comprising biochemically mature myelin (on the 75th day), a certain increase in AMPS content is observed, while in the adult animals (on the 114th day) the values of these macromolecules are as low as on the 42nd day.

The changes in individual AMPS content identified in myelin (HA, HS and ChS-A+C) partially overlap with the profile of total amps. From the 14th to 28th day, HA amount becomes 5 times lower, and then stabilizes during the following period. HS content at the same time becomes 17 times lower, while ChS-(A+C) decreases by about 7 times. The level of both HS and ChS-(A+C)

Table 5.   The effects of electroconvulsions (ECS) on
           myelin AMPS isolated from the rat brain.

	µg AMPS/g.dry wt.	± S.E.M.
CONTROL	420	+ 17.6
ECS	158	+˙15.2
4[hr] after ECS	206	+ 12.0
24[hr] after ECS	238	+ 10.0
48[hrs] after ECS	290	+ 14.0
72[hrs] after ECS	380	+ 18.0

The results are expressed as the means + S.E.M. for 4
ECS experiments.

stays the same from 28 to 114th day, except for the 75th
day, when a certain increase in these sulphate MPS is
observed.   It should be noted that the proportion of HA
is higher in myelin than in the whole brain, which is in
good agreement with the results reported by Szabo and
Roboz-Einstein (24) who described larger amounts of HA
in the white matter compared with the whole brain.

     The results presented here, like our previous
studies (16) with ^{35}S-sulphate, indicate that sulphate
was incorporated into AMPS not only from the total brain
homogenate and myelin, but also from myelin.   At the same
time we were able to demonstrate the characteristic dynamic changes in the turnover of myelin AMPS.   This undoubtedly proves that AMPS are an integral part of myelin,
and not a by-product of contamination, as has been suggested by some authors (13).

     The profile of AMPS distribution within the total
brain during development (18), the high level in pre-
myelination stage and the rapid fall during myelination,
accompanied by a similar AMPS -sulphotransferase fluctu-
ation (2, 17), support the hypothesis that there is a
close relationship between brain maturation and AMPS
synthesis.   The fact that myelin in the central nervous

system is morphologically an extension of the oligoden-
droglial cells, and the findings that AMPS form the surface layer of cell membranes and are at the same time
constituents of myelin lead us to suppose that acid
mucopolysaccharides play a role in the process of
myelination.

The experiments of the last few years implicated
the role of myelin in many important brain functions,
apart from its known role as an insulator. To prove
this hypothesis, and further to see whether or not
there is any correlation between altered function and
the changes of these macromolecules as components of the
myelin, we have studied the behavior of myelin AMPS in
the course of electroconvulsive seizure in rat brain
(Table 5). Indeed, repetitive electroconvulsive shocks
immediately alter the amount of AMPS in the myelin,
which recovers only at 72 hours after the last electroconvulsive seizure. However, further studies are necessary in order to elucidate the significance on these
changes in relationship to the brain function.

REFERENCES

1. Abood, L.G. and Abul-Haj, S.K. (1956): Histochem-
    istry and characterization of hyaluronic acid in
    axons of peripheral nerve. J. Neurochem. 1: 119-
    125.

2. Balasubramanian, A.S. and Bachhawat, B.K. (1970):
    Sulphate methabolism in brain. Brain Res. 20: 341-
    360.

3. Brante, G. (1959): Mucopolysaccharides and mucoids
    of the nervous system. In: Biochemistry of CNS,
    F. Brucke (ed.) pp. 291-300, Pergamon, N. Y.

4. Chandrasekaran, E.V., Mukherjee, K.L. and
    Bachhawat, B.K. (1971): Isolation and character-
    ization of glycosaminoglycans from brain of child-
    ren with protein-calorie malnutrition. J.
    Neurochem. 18: 1913-1920.

5. Clausen, J. and Hansen, A. (1963): Acid mucopoly-
    saccharides of human brain: Identification by
    means of infra-red analysis. J. Neurochm. 10:
    165-168.

6. Cunningham, W.L. and Goldberg, J.M. (1968): The
    determination of glycosaminoglycans present in
    various mammalian brains. Biochem. J. 110: 35.

7. Custod, J.T. and Young, I.J. (1968): Cat brain
    mucopolysaccharides and their in vivo hyaluronidase
    digestion. J. Neurochem. 15: 809-813.

8. Dorfman, A. and Ho, P.L. (1970): Synthesis of acid
    mucopolysaccharides by glial tumor cells in tissue
    culture. Proc. Nat. Acad. Sci. 66: 495.

9. George, E. and Bachhawat, B.K. (1970): Brain
    glycosaminoglycans sulphotransferase in Sanfilippo
    syndrome. Clin. Chim. Acta 30: 317-324.

10. Hirosawa, K. and Young, W.R. (1971): Autoradio-
    graphic analysis of sulfate metabolism in the
    cerebellum of the mouse. Brain. Res. 30: 295-309.

11. Lampert, I.A. and Lewis, P.D. (1974): Demonstration
    of acidic polyanions in certain glial cells during
    postnatal rat brain development. Brain Res. 73:
    356-361.

12. Levental, M. and Rakic, Lj. (1973): Quantitative
    distribution of acid mucopolysaccharides in neuronal
    and glial cells isolated from rat cerebral crotex.
    Abstr. Commun., Fourth Meet. Int. Soc. Neurochem.,
    Tokyo, 119.

13. Margolis, R.U. (1967): Acid mucopolysaccharides
    and proteins of bovine whole brain, white matter
    and myelin. Biochim. Biophys. Acta 141: 91-102.

14. Margolis, R.U. and Margolis, R.K. (1974): Distri-
    bution and metabolism of mucopolysaccharides and
    glycoproteins in neuronal perikarya, astrocytes
    and oligodendroglia. Biochemistry 13: 2849-2852.

15. Norton, W.T. and Poduslo, S.E. (1973): Myelination
    in rat brain: method of myelin isolation. J.
    Neurochem. 21: 749-757.

16. Rusić, M., Levental, M. and Rakić, Lj. (1978):
    35S-sulphate incorporation into myelin sulphated
    mucopolysaccharides during rat brain development.
    Experientia 34: 696-697.

17. Saxena, S., George, E., Kokrady, S. and Bachhawat,
     B.K. (1971): Sulphate metabolism in developing rat
     brain: a study with subcellular fractions.   Indian
     J. Biochem. Biophys. 8: 1-8.

18. Singh, M. and Bachhawat, B.K. (1965): The distri-
     bution and variation with age of different uronic
     acid-containing mucopolysaccharides in brain.
     J. Neurochem. 12: 519-529.

19. Singh, M. and Bachhawat, B.K. (1968): Isolation and
     characterization of glycosaminoglycans of human
     brain of different age groups.   J. Neurochem. 15:
     249-258.

20. Singh, M., Chandrasekaran, E.V., Cherian, R. and
     Bachhawat, B.K. (1969): Isolation and character-
     ization of glycosaminoglycans in brain of different
     species. J. Neurochem. 16: 1157-1162.

21. Stefanovic, V. and Gore, I. (1967): A micromethod
     for determination of acid mucopolysaccharides in
     vascular tissue. J. Chromat. 31: 473-478.

22. Stoolmiller, A.C. (1972): Biosynthesis of muco-
     polysaccharides by neuroblastoma cells in tissue
     culture. Fed. Proc. 31: 910.

23. Stoolmiller, A.C., Dawson, G. and Dorfman, A.
     (1973): In: Tissue Culture of the Nervous System.
     G. Sato (ed), p. 247, Plenum Press, New York, N.Y.

24. Szabo, M.M. and Roboz-Einstein, E. (1962): Acidic
     polysaccharides in the central nervous system.
     Arch. Biochem. Biophys. 98: 406-412.

25. Young, I.J. (1963): Reversible seizures produced
     by neuronal hyaluronic acid depletion. Exp.
     Neurol. 8: 195-202.

26. Young, I.J. and Abood, L.G. (1960): Histological
     demonstration of hyaluronic acid in the central
     nervous sytem. J. Neurochem. 6: 89-94.

27. Wolman, M. and Hestrin-Lerner, S. (1960): A histo-
     chemical contribution to the study of the molec-
     ular morphology of myelin sheath. J. Neurochem.
     5: 114.

NEURONAL-GLIAL INTERACTIONS DURING

NEURAL GROWTH IN CULTURE

Antonia Vernadakis

Departments of Psychiatry and Pharmacology
University of Colorado School of Medicine
Denver, Colorado 80260

INTRODUCTION

Although there exists a vast body of information concerning brain maturation, we know relatively very little about neuronal-glial interrelationships during neural growth and differentiation. In view of the fact that glial cells in the central nervous system (CNS) out-number neurons by a factor of ten, their role in CNS function continues to be the subject of much investigation. Glial cells have been proposed to be involved in myelina-tion (3) to act as spatial ionic buffers (10), to be electrically coupled to neurons (23), and to proliferate with increased neuronal activity (4, 11). More recently we (22), and others (6), have proposed that glial cells may also modulate neuronal activity by regulating the amount of neurotransmitter substances in the microenviron-ment. In this paper we will review and discuss responses of glial cells to hormones and glial cell-hormone-neuro-transmitter interrelationship.

NEURAL CULTURE SYSTEMS

In recent years, neural culture systems have pro-vided in vitro models with which to study various aspects of neural growth (19). We have used the following neural culture systems in our research: organ culture using neural tissue explants in which neural cells maintain their original organization for over 24 hours; organotypic

culture in which the original organization of the tissue
may be lost, but the constituent cells emerge into the
zone of outgrowth; dissociated brain cell cultures consisting of neurons and glial cells at different ratios
depending on the age of the neural tissue at the time of
cultivation and the days in culture; and C-6 glial cells,
a rat astrocytoma cell line. The characteristics of these
neural systems have recently been reviewed by Vernadakis
and Culver (19).

GLIAL CELLS AND NEUROTRANSMITTER ACCUMULATION

     In early in vivo studies using the chick embryo as
an animal model, we investigated the uptake of ^{3}H-nore-
pinephrine (^{3}H-NE) in cerebral hemispheres and cerebellum
slices of chicks from early embryonic age up to 3 years
after hatching (9, 16). We found that in both the
cerebral hemispheres and cerebellum of chicks ^{3}H-NE
accumulation progressively increased during brain maturation. In the cerebral hemispheres, ^{3}H-BE uptake increased
up to 3 months after hatching, leveled off up to 1 year
and markedly declined during aging; in the cerebellum,
^{3}H-NE accumulation had already reached maximum levels at
20 days of embryonic age and did not significantly change
thereafter. The increases in ^{3}H-NE accumulation during
brain development are attributed to maturation of the
uptake and storage processes which appear to develop
earlier in the cerebellum than in the cerebral hemispheres.
The continuous increase in the accumulation of ^{3}H-NE after
6 weeks, when neuronal maturation is completed, is interpreted to represent accumulation of ^{3}H-NE in glial cells
actively proliferating during this time (5, 17). Henn
and Hamberger (6) were the first to report that NE accumulates in brain fractions enriched in glial cells.

     Using C-6 glial cells, we have found that NE, 10^{-6}M,
accumulates in these cells (20) (Fig. 1). Moreover, in
cells treated with dibutyryl cyclic AMP (DBcAMP), 10^{-3}M,
for 2 days, NE accumulation is markedly lower than in
controls at similar cell densities. DBcAMP-treated cells
show astrocytic-like morphology. Thus, morphologically
differentiated glial cells accumulate less NE. Moreover,
during glioblast growth, NE acts preferentially on glial
cells. As glial cells mature, the shift of NE is to
neuronal growth and differentiation, including neurotransmission.

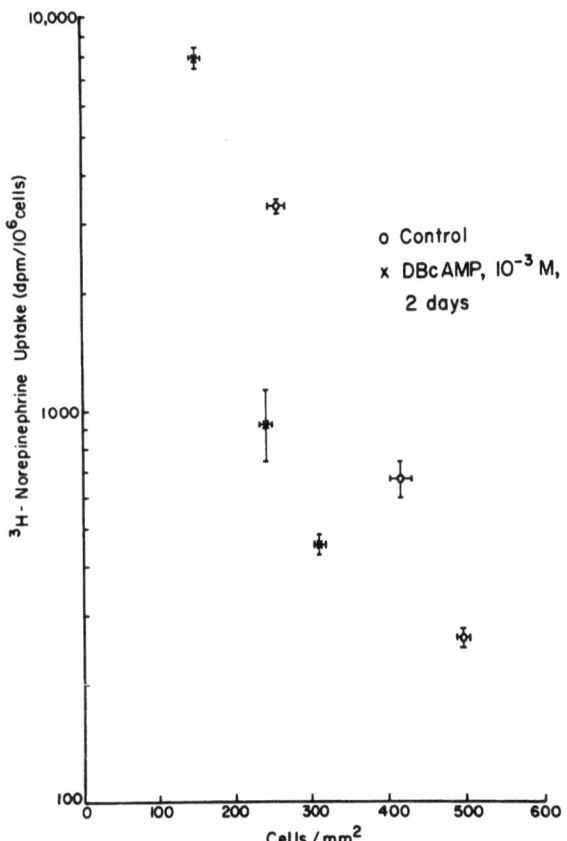

Fig. 1.   Effect of dibutyryl cyclic AMP (DBcAMP) on
          ^{3}H-norepinephrine (^{3}H-NE) uptake.   C-6 glial
          cells were cultured for 2 days in DBcAMP,
          10^{-3}M, and uptake of ^{3}H-NE was measured for
          15 minutes.   Points with bracketed lines
          represent Mean ± SE [From Vernadakis and
          Nidess (20)].

GLIAL CELLS AND HORMONES

For several years, we have been interested in the
role of glial cells in neural growth and particularly
in their response to hormones. We have found, using
the organ and organotypic neural culture systems, that
some steroid hormones enhance the proliferation of glial

cells (15). Cerebellar explants were removed from 16-day-old chick embryo and were maintained in organ culture for 24 hours. DNA content was higher in cerebellar explants treated with cortisol (2.76×10^{-5}M) for 24 hours (Table 1). This increase in DNA after hormone treatment is interpreted to represent a further increase in the proliferation of glial cells induced by the hormones, since neurogenesis has been established by this age period. This effect of steroid hormones, to enhance proliferation of glial cells, is further evidenced by the results obtained from another study using the organotypic neural culture system (Table 2), in which we found that the migration rate of glial cells is higher than in control cultures.

UPTAKE OF HORMONES IN GLIAL CELLS

The mechanism by which hormones accelerate proliferation of glial cells is not understood. Recent evidence from our laboratory using C-6 glial cells and dissociated brain cell cultures has shown that steroid hormones accumulate intracellularly in both neurons and glial cells (20). Dissociated brain cell cultures were prepared from 8-day-old chick embryo cerebral hemispheres. The method for dissociation was a modidication of that described by Sensenbrenner and associates (2, 14). Cell cultures were prepared from cerebral hemispheres from 6- or 10-day-old chick embryos E6 or E10. At various times (days) in culture the medium was discarded, serum-free medium was added, and cultures were incubated for one hour to equilibrate; ^{3}H-corticosterone, 1.87×10^{-9}M (0.27 µC; S.A. 54 Ci/mmole) final concentration, was added for one hour. Corticosterone was extracted with dichloromethane according to the method of McEwen (13). Protein content was determined by the Lowry method (12).

At 9-10 days in culture (c9, c10), when there is an abundant neuronal population in dissociated cultures from either 6- or 10-day-old chick embryo cerebral hemispheres, the accumulation of ^{3}H-corticosterone was higher in the cultures from 10-day-old chick embryos than from 6-day-old embryos. Since at this stage of culture neurons are predominant, the higher hormone accumulation in cultures from older chick embryos may reflect the maturation of additional neural hormone receptors. In 30-day cultures, accumulation of ^{3}H-corticosterone (0.9×10^{-9}M) was significantly higher than in the 9- to 10-day cultures. Since the 30-day cultures consist predominantly of glial cells, the high accumulation of corticosterone reflects uptake of this hormone in glial cells.

Table 1. Effects of Cortisol, Corticosterone and
          Estradiol on DNA Concentration of Cerebellar
          Explants Removed from 16-day-old Chick Embryos
          and Maintained as Organ Cultures for 24 Hours

Culture Medium[1]	DNA Concentration (μg/mg wet tissue)
Noncultured	5.04 ± 0.23[2]
Basal	6.61 ± 0.27 (< 0.001)[3]
Cortisol	8.07 ± 0.49 (< 0.02)[4]
Corticosterone	6.30 ± 0.40
Estradiol	7.87 ± 0.66 ($0.05 - 0.1$)

[1]The medium was Eagle's basal medium with Earle's salts.
Hormones, when added, were in the following concentra-
tions: cortisol 2.76×10^{-5}M, corticosterone 2.89×10^{-5}M, estradiol dipropionate 2.65×10^{-5}M.

[2]Each value represents the mean $\pm$ SE of 12-16 explants.

[3]Numbers in parentheses are values for comparison to
noncultured control group.

[4]Numbers in parentheses are P values for comparison to
basal medium group.

Table from Vernadakis (15)

     When cultures were preincubated for one hour with
10^{-9}, 10^{-7}, or 10^{-5}M unlabeled corticosterone, the retention of ^{3}H-corticosterone was exponentially decreased in
both young (9-10 days) and old (30-34 days) cultures.
These findings suggest that the retention mechanism is
saturable. The specificity of the hormone retention
mechanism is illustrated in Figs. 2 and 3. Whereas preincubation of E10cll cultures with unlabeled corticosterone (1×10^{-5}M) decreased the retention of ^{3}H-corticost-
erone (1×10^{-9}M) by more than 50% preincubation with
11-dehydrocorticosterone, a metabolite of corticosterone
in neural tissue (13) decreased the retention of labeled
corticosterone by only about 11%, and preincubation with
progesterone had no competitive effect on the retention

Table 2.   Effects of Hormones on the Migration Rate
           of Cells in Cultured Cerebellar Explants[1]

Culture Medium[2]	Number of Explants	Migration Rate[3] (mean ratio ± SE)
Control for cortisol	67	4.07 ± 0.12[4]
Cortisol (2.76 x 10^{-8}M)	50	5.06 ± 0.23 (< 0.001)
Control for corticosterone	16	3.55 ± 0.16
Corticosterone (2.89 x 10^{-8}M)	21	4.56 ± 0.28 (< 0.01)[5]
Control for estradiol	21	4.36 ± 0.24
Estradiol dipropionate (2.65 x 10^{-8}M)	24	5.14 ± 0.25 (< 0.05)
Control for progesterone	36	3.78 ± 0.15
Progesterone (3.18 x 10^{-8}M)	45	4.41 ± 0.21 (< 0.02)
Control for testosterone	21	3.88 ± 0.26
Testosterone (3.47 x 10^{-8}M)	19	3.76 ± 0.21

[1]Cerebellar explants were removed from 15-day-old chick embryos and maintained in culture (Maximow double coverslip assembly) for 5 days.

[2]Medium consisted of 45% Gey's BSS, 5% ascitic fluid, 5% chick embryo (9-day-old) extract and glucose.

[3]Migration rate was calculated during the growth period between the first and fifth days in culture. Migration rate was calculated as follows: Outlines of the explants were traced at 1 and 5 days by focusing the image of the explant on translucent paper placed on the translucent back of a camera attached to the micro-scope. The images were cut from the paper and the papers were weighed to determine relative surface areas. The ratio between the surface of 1-day and that of 5-day culture was calculated as "migration" rate.

[4]Mean ± SE.

[5]Numbers in parentheses are P values for comparison to appropriate control group.

Table from Vernadakis (15)

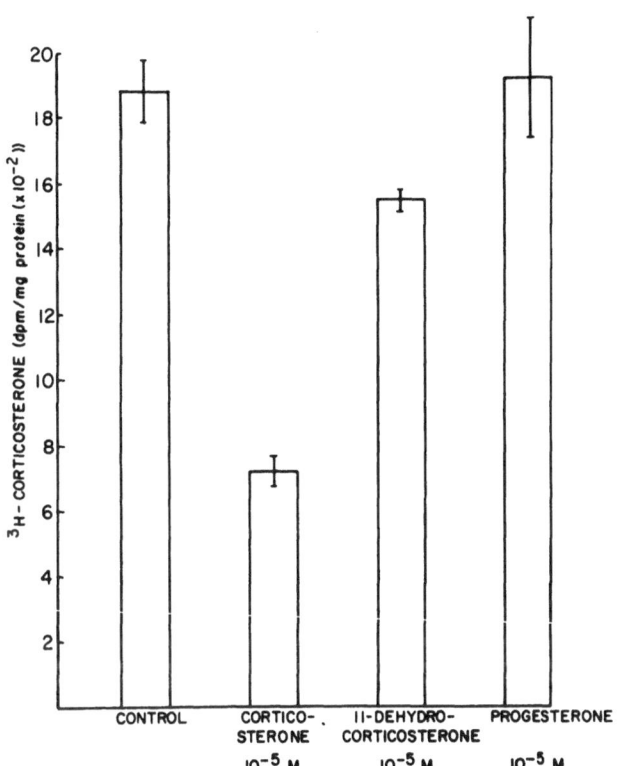

Fig. 2.   Uptake of 1,2-^{3}H-corticosterone (dpm/mg)
          protein in 12-day cultures of dissociated
          cerebral hemispheres from 10-day-old chick
          embryos after preincubation of 1 hour with
          unlabeled hormones (10^{-5}M). Bars represent
          means ± SE [From Vernadakis et al. (21)].

of labeled corticosterone (Fig. 2). Since cultures of
this age consist predominantly of neurons, the retention
of corticosterone appears to be specific in maturing
neurons. However, in E10c31 cultures, consisting primarily of glial cells, unlabeled corticosterone decreased
the retention of ^{3}H-corticosterone by more than 50%,
unlabeled progesterone decreased the retention by approximately 30%, unlabeled 11-dehydrocorticosterone had only
a slight decreasing effect, and testosterone had no

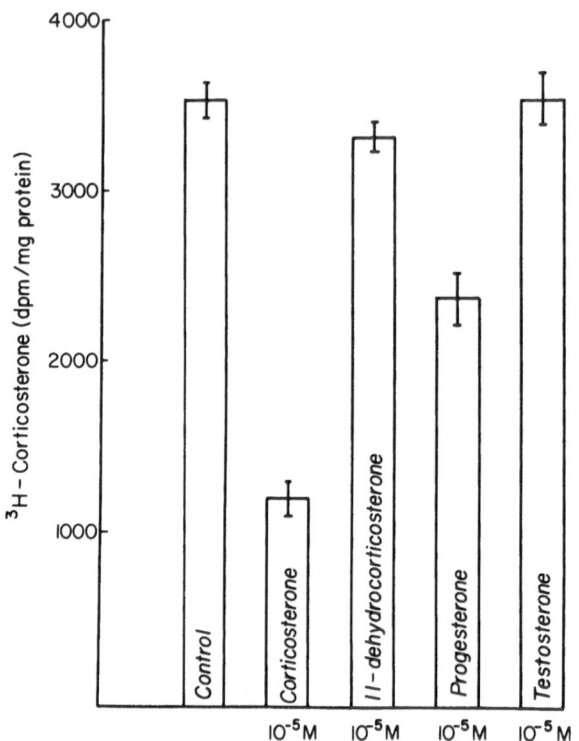

Fig. 3.   As in Fig. 2 except that cultures were 31 days.

effect on the retention of ^{3}H-corticosterone (Fig. 3).
Thus, the retention of labeled corticosterone is non-
specific in cultures of predominantly glial cells.

The foregoing findings are of importance because
they demonstrate differences in corticosterone retention
between neurons and glial cells. That the retention of
corticosterone in glial cells is nonspecific is further
demonstrated in our studies using C-6 glial cells as the
glial cell model. Retention of ^{3}H-corticosterone was
reduced by unlabeled corticosterone as well as by cortisol, 11-dehydrocorticosterone, progesterone, and testosterone (fig. 4).

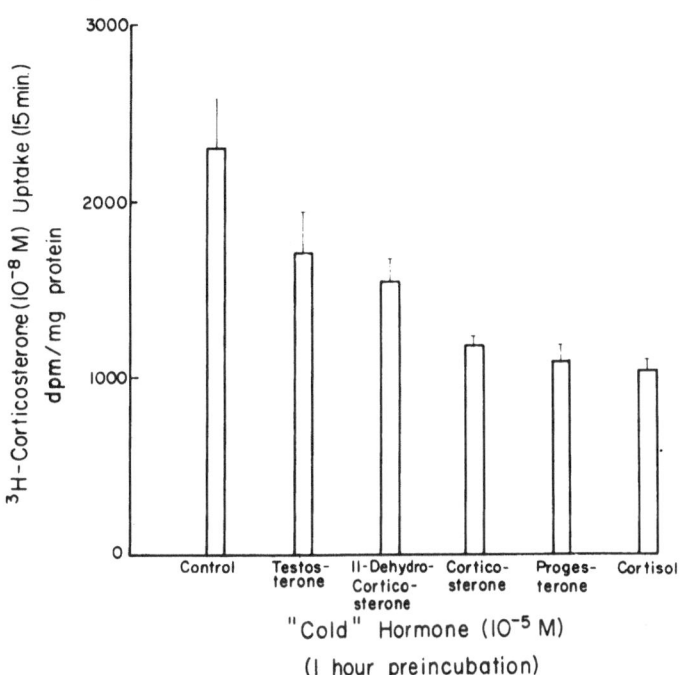

Fig. 4.   As in Fig. 2 except that cultures were C-6
glial cells (2B-clone).

GLIAL CELL-HORMONE-NEUROTRANSMITTER INTERRELATIONSHIP

That there may be an interaction of hormones and
neurotransmitters on glial cells has derived from a
recent study using cerebellar explants in organ culture
(18)(Table 3). Cerebellar explants were removed from
16-day-old embryos, a time when glial cells begin to
actively proliferate, and were cultured for 4 hours in
the presence or absence of cortisol. After 4 hours,
explants were incubated for 20 minutes in ^{3}H-NE at either
10^{-6}M or at 10^{-7}M. Accumulation of ^{3}H-NE at 10^{-6}M but
not at 10^{-7}M was lower in explants cultured in the
presence of 2.76 x 10^{-5}M cortisol. Iversen and Salt (8)
have reported that high doses of cortisol inhibit the
extraneuronal uptake, Uptake$_2$ (7), which is the low

Table 3. ^{3}H-NE Accumulation (Tissue/Medium Ratios for
            20 Min.) in Cerebellar Explants Removed from
            16-day Chick Embryos and Cultured in the
            Presence of Cortisol for 4 Hours

Treatment	^{3}H-NE Concentration[1]	
	10^{-7}M	10^{-6}M
Control	2.17 ± 0.24	2.27 ± 0.34
Cortisol, 1 x 10^{-5}M	---	2.52 ± 0.41
Cortisol, 2.76 x 10^{-5}M	2.25 ± 0.30	1.61 ± 0.11 (< 0.02)[2]

[1]Each value represents the mean ± SE of 6-8 samples
  from two experiments.

[2]Represents P value for comparison to control

Table from Vernadakis (18).

affinity uptake system for NE in the peripheral nervous
system (7). We propose that, since the inhibitory effect
of cortisol was observed with a high concentration of
NE (10^{-6}M), this may be interpreted to mean that cortisol
inhibits the uptake of NE in glial cells.

    The significance of NE accumulation in glial cells
must at present be speculative. A schematic representation of the fate of NE in the central nervous system is
represented in Fig. 5. It has been proposed that glial
cells function as a "safety valve" in neurotransmission
by limiting the extracellular accumulation of neurotransmitter substances. Inhibition of neurotransmitter
accumulation in glial cells by hormones or other substances may therefore alter extracellular-neuronal
transmitter balance and affect cellular homeostasis.
During neural growth the role of glial cells and their
response to hormones may have a different function. We
have reported that NE levels are high during early neural
growth (17), and that neuronal uptake may precede synthesis of NE (1). Therefore, inhibition of NE uptake in
glial cells by an intrinsic substance such as a hormone
may result in increasing the availability of the neuro-
humor substance for facilitating the growth of neuronal
mechanisms. In conclusion, we propose that glial cells

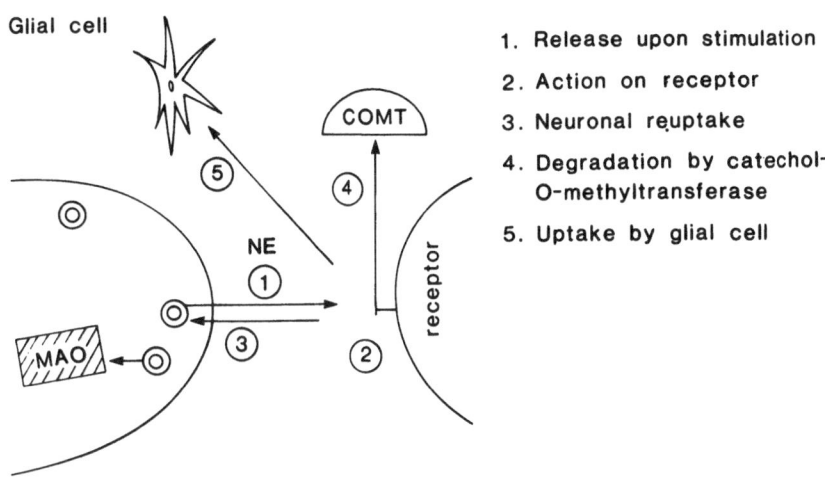

Glial cell

COMT

NE

receptor

MAO

1. Release upon stimulation

2. Action on receptor

3. Neuronal reuptake

4. Degradation by catechol-
   O-methyltransferase

5. Uptake by glial cell

Presynaptic nerve ending   Postsynaptic neuron

Fig. 5.   Diagrammatic representation of the fate of
          norepinephrine in the central nervous system.

modulate neuronal growth by regulating the extraneuronal
microenvironment via their response to intrinsic substances such as hormones and neurohumors.

REFERENCES

1.   Arnold, E.B. and Vernadakis, A. (1979): Development
     of tyrosine hydroxylase activity in dossociated
     cerebral cell cultures. Develop. Neurosci. 2: 46-50.

2.   Booher, J. and Sensenbrenner, M. (1972): Growth and
     cultivation of dissociated neurons and glial cells
     from embryonic chick, rat and human brain in flask
     cultures. Neurobiology 2: 97-105.

3.   Bunge, R.P. (1968): Glial cells and the central
     myelin sheath. Physiol. Rev. 48: 197-251.

4.   Diamond, M.C., Law, F., Rhodes, H., Lidner, B.,
     Rosenzeig, M.R., Krech, D. and Bennett, E.L. (1966):
     Increases in cortical depth and glial numbers in
     rats subjected to enriched environment. J. Comp.
     Neurol. 128: 117-126.

5.   Hanaway, J. (1967): Formation and differentiation
     of external layer of the chick cerebellum. J. Comp.
     Neurol. 131: 1-14.

6.   Henn, F.A. and Hamberger, A. (1971): Glial cell
     function uptake of transmitter substances. Proc.
     Nat. Acad. Sci., USA 68: 2686-2690.

7.   Iversen, L.L. (1971): Role of transmitter uptake
     mechanisms in synaptic neurotransmission. Brit. J.
     Pharmacol. 4: 571-591.

8.   Iversen, L.L., and Salt, P.J. (1970): Inhibition
     of catecholamine uptake$_2$ by steroids in the isolated
     heart. Brit. J. Pharmacol. 4: 571-591.

9.   Kellogg, C., Vernadakis, A. and Rutledge, C.O. (1971):
     Uptake and metabolism of ^{3}H-norepinephrine in the
     cerebral hemispheres of chick embryos. J. Neuro-
     chem. 18: 1931-1938.

10. Kuffler, S.W. (1967): Neuroglial cells: physiologi-
     cal properties and a potassium mediated effect of
     neuronal activity on the glial membrane potential.
     Proc. Roy. Soc. (London) B. 168: 1-21.

11. Kuffler, S.W. and Nicholls, J.G. (1966): The
     physiology of neuroglial cells. Ergeb. Physiol.
     57: 1-90.

12. Lowry, O.H., Rosebrough, N.J., Farr, A.L. and
     Randall, R.J. (1951): Protein measurement with the
     tolin phenol reagent. J. Biol. Chem. 193: 265-275.

13. McEwen, B.S., Weiss, J.M. and Schwartz, L.S. (1969):
     Uptake of corticosterone by rat brain and its con-
     centration by certain limbic structures. Brain Res.
     16: 227-241.

14. Sensenbrenner, M., Booher, J. and Mandel, P. (1971):
     Cultivation and growth of dissociated neurons from
     chick embryo cerebral cortex in the presence of
     different substrates. Z. Zellforsch. 117: 559-569.

15. Vernadakis, A. (1971): Hormonal factors in the proliferation of glial cells in culture. In: Influence of Hormones on the Nervous System, D.H. Ford (ed.), Proc. Int. Soc. of Psychoneuro-endocrinology, Brooklyn, New York, 1970, S. Karger, Basel.

16. Vernadakis, A. (1973): Uptake of ^{3}H-norepinephrine in the cerebral hemispheres and cerebellum of the chicken throughout the lifespan. Mech. Age. Dev. 2: 371-379.

17. Vernadakis, A. (1973): Changes in nucleic acid content and butyrylcholinesterase activity in CNS structures during the life span of the chicken. J. Gerontol. 28: 281-286.

18. Vernadakis, A. (1974): Neurotransmission: A proposed mechanism of steroid hormones in the regulation of brain function. In: Proceedings of the Mie Conference of the International Society for Psychoneuroendocrinology, N. Hatotani (ed.), S. Karger AG, Basel.

19. Vernadakis, A. and Culver, B. (1979): Neural tissue culture: A biochemical tool. In: The Biochemistry of Brain, S. Kumar (ed.), Pergamon Press (in press).

20. Vernadakis, A. and Nidess, R. (1976): Biochemical characteristics of C-6 glial cells. Neurochem. Res. 1: 385-402.

21. Vernadakis, A., Culver, B. and Nidess, R. (1978): Actions of steroid hormones on neural growth in culture: Role of glial cells. In: Proceedings of the 7th International Congress of the International Society for Psychoneuroendocrinology, Psychoneuro-endocrinology 3: 47-64.

22. Walker, F.D. and Hild, W.J. (1969): Neuroglia electrically coupled to neurons. Science 165: 602-603.

EFFECTS OF CONVULSIONS ON RABBIT HIPPOCAMPAL

GANGLIOSIDES DURING HIPPOCAMPUS DEVELOPMENT

D. Kostić and A. Vranešević

Laboratory for Neurochemistry
Institute of Biochemistry
Faculty of Medicine
University of Belgrade
Yugoslavia

Brain tissue of animal species with lower as well as higher degree of biological organization contains large amounts of gangliosides. The distribution and the characteristics of these complex substances were studied in the whole brain of rats and other animals as well as in specific human brain regions during brain development (1, 4, 10, 22). Investigations concerned with the sub-cellular localization of sialic acid-containing compounds in the brain indicated that the synaptosomes and in particular the synaptosomal plasma membranes were rich in gangliosides (2, 34). High concentrations of tri-tetra- and pentasialogangliosides were observed in the central nervous regions which are important in electrogenesis of the electric fish (T. ocellate) (12, 16). The content of sialic acid was found to be changed during the excitation of the brain (9). It was shown also that the exposure to light caused an increase in synaptosomal and a decrease in mitochondrial ganglio-sides in the rat brain (17). However, Dreyfus observed that the biosynthesis of ganglioside was greater during the adaptation to light than to darkness in the chick retina (5). Thus, these substances may be involved in the transmission process.

These observations, as well as the implication of the hippocampal genesis in the epileptic seizures (6, 7, 37), led us to study the relationship of the seizure

activity to the gangliosides contained in developing
hippocampus.

MATERIAL AND METHODS

Male chincilla rabbits bred and raised in our
Institute were used at 15, 20 and 90 days of age for
these investigations. Epileptic seizures were induced
by intraperitoneal injection of 65 mg/kg body weight of
pentylenetetrazol (Knoll Pharmaceutical Co., New Jersey,
USA) in 0.9% naCl. The animals were decapitated 20-25
minutes after the onset of convulsions and the heads were
dropped into the liquid nitrogen. The brains were removed
and the hippocampus dissected out in a cold chamber (4°C).
Pooled tissue samples were used for all the assays. The
lipids were extracted in chloroform-methanol according
to Suzuki (26). Purification, determination and thin
layer chromatography of gangliosides have been described
in detail elsewhere (15). The separated gangliosides
from samples and standards were detected by iodine vapors
(Koch-Light Laboratories, England).

RESULTS

The contents of total gangliosides of rabbit hippocampus during ontogenesis are presented in Table 1. An
increase in the total lipid-bound N-acetylneuraminic
acid (NANA) content of hippocampus occurred from the 15th
to the 90th day of postnatal life. The rate of ganglioside accretion in this structure was low between the 20th
and 90th day and the amount of gangliosides on the 20th
day was close to the their maximum content at the 90th
day after birth.

Convulsions induced by pentylenetetrazol led to a
decrease of gangliosides NANA in early stages of hippocampal development. The concentrations of the hippocampus gangliosides were not affected when the seizures
were induced at the age of 90 days.

The relative concentrations of the individual
gangliosides were variable (Table 2). The mean monosialoganglioside G_{M1} values were lower on the 15th day
than on the 20th and 90th day of hippocampal development, being 12.91%, 15.60% and 15.72% of total gangliosides, respectively. The ganglioside G_{D1a} content was
very high in all examined periods of the postnatal
development of this structure. The relative G_{D1a} content was 36.60, 39.64 and 31.38% of the total lipid-

Table 1.   Total ganglioside n-acetylneuraminic acid
           in developing rabbit hippocampus.

Age (days)	Control	Pentylenetetrazol
15	0.400 ± 0.012	0.344 ± 0.016[+]
20	0.470 ± 0.024	0.396 ± 0.010[+]
90	0.494 ± 0.018	0.488 ± 0.023

Values expressed as mg per g fresh weight of tissue.

Mean ± SD from four separate experiments.

[+]Statistically significant difference ($p < 0.05$).

Table 2. Ganglioside distribution in developing rabbit hippocampus and after convulsions induced by pentylenetetrazol.

Ganglioside	15 day		20 day		90 day	
	Control	Pentylenetetrazol	Control	Pentylenetetrazol	Control	Pentylenetetrazol
G_{M3}	2.76 ± 0.54	1.77 ± 0.69	1.78 ± 0.50	2.80 ± 1.01	2.72 ± 1.67	2.19 ± 0.34
G_{M2}	1.99 ± 0.18	1.92 ± 0.67	1.54 ± 0.65	2.65 ± 0.86	2.76 ± 1.55	2.42 ± 1.88
G_{M1}	12.91 ± 1.17	18.93 ± 0.86[+]	15.60 ± 1.02	20.52 ± 1.41[+]	15.72 ± 4.11	16.39 ± 1.37
G_{D1a}	36.60 ± 3.49	42.54 ± 3.28	39.04 ± 1.04	44.40 ± 1.25[+]	31.38 ± 3.12	31.86 ± 2.73
G_{D2}	4.98 ± 0.36	5.06 ± 1.28	3.34 ± 1.11	3.29 ± 0.65	3.72 ± 1.76	9.28 ± 3.06
G_{D1b}	13.92 ± 2.72	9.66 ± 3.01	16.65 ± 0.90	9.69 ± 1.27[+]	16.82 ± 1.48	15.27 ± 4.12
G_{T1}	23.46 ± 1.27	16.97 ± 2.21[+]	19.57 ± 0.80	13.05 ± 1.75[+]	22.99 ± 2.32	19.49 ± 1.04
G_{Q1}	3.38 ± 0.49	3.15 ± 0.97	2.48 ± 0.35	3.60 ± 0.98	3.89 ± 2.28	3.10 ± 2.79

Ganglioside distribution is expressed as the percentage of the total N-acetylneuraminic acid.

Means ± SD from four separate experiments.

The Svennerholm numbering is used.

[+]Significance determined by t test ($p < 0.05$).

bound NANA on the 15th, 20th and 90th day, respectively. Ganglioside G_{D1b} and G_{T1} were also found in considerable amounts. Similar G_{D1b} values (about 16%) were observed on the 20th and 90th day, while somewhat lower (13.92%) values were present on the 15th postnatal day. The trisialoganglioside G_{T1} developmental features differed from those of the disialoganglioside G_{D1b} on the 15th and 20th day. This fraction had the highest values on 15th day (23%). On the other hand, G_{T1} content was found to be slightly decreased on the 20th day when compared to the value observed on the 15th and 90th days of the postnatal life. Among the minor fractions, the greatest values were those of disialoganglioside G_{D2} and tetrasialoganglioside G_{Q1}.

The convulsion induced by pentylenetetrazol led to a significant increase in the concentrations of monogangliosides. The G_{M1} fraction was affected on the 15th and 20th days while G_{DLA} was affected on the 20th day only. The other less polar gangliosides, G_{D1b} and G_{T1}, except G_{D1b1}, decreased in the hippocampus on the 15th and 20th days. However, the content of major and minor ganglioside fractions did not change when the seizures were induced on the 90th day of age (Table 2).

DISCUSSION

So far the chemical composition of the lipids, especially that for the gangliosides, has not been explored in the hippocampus, even though significant progress was made in the morphological study of this structure in experimental animals.

Our results clearly indicate that the total ganglioside content of the hippocampus increases to a maximal concentration on the 20th day of postnatal dvelopment. These observations are consistent with the behavior of the total gangliosides studied in the whole brain of rats and other animals (18, 25, 26, 31). The highest values of G_{DLa} and G_{m1} in the hippocampus coincide with the period of intense ganglioside synthesis. This fact tends to support the importance of these gangliosides in the biosynthesis of other gangliosides as described previously (13, 20, 27). Moreover, the concentration pattern of G_{DLa} during development of the hippocampus is in agreement with the concept that the G_{DLA} levels can be considered as a marker for the degree of dendritic arborization and neuronal interconnections in the brain tissue (10).

The reduction in the level of certain amino acids and the activity of their enzymes, the rapid depletion of energy reserves as well as the morphological changes in several models of epilepsy have been well documented (21, 28). Previously, we had demonstrated an altered composition of polar and less polar gangliosides in the temporal cortex, caudate nucleus and thalamus of the rabbit during pentylenetetrazol induced convulsions (16). The results presented here also showed clear differences in content of the total gangliosides in the hippocampus of the 15- and 20-day-old rabbits following the administration of this convulsant. On the other hand, the values for the glycoplipid NANA on the 90th day were not altered by the induction of seizures at this age. The contrast in the ganglioside response could be explained by the inadequate development in earlier stages of the inhibitory system in other structures and the absence of the cortico-hippocampal interaction.

The differences in the regional effects of pentyl-enetetrazol remain unexplained. (8), especially in the hippocampus. The available data on the participation of the gangliosides in the excessive cell discharge during chemically-induced seizures, while limited, are controversial. Whisler et al. (35) have found that the total gangliosides brain decreased in dog after pentyl-enetetrazol administration. In contrast, Winter and Bernheimer (33) did not observe any change in the ganglio-side composition of the cortex and caudate nucleus. However, the findings presented here indicate that hippo-campal gangliosides changed when the convulsions were induced during the development of this structure, parti-cularly in the earlier periods. The alteration in the hippocampal ganglioside levels induced by epilepsy might result from the removal of the sialosyl group from the ganglioside molecules (with several sialic acids) local-ized in the synaptic plasma membranes (22, 29). The massive discharges could conceivably disturb the enzyme system that releases NANA from gangliosides and therefore increase the content of monosialoganglioside G_{M1} with a single sialic acid. Neuraminidase is known to exist in synaptic membrane (30) and its function may be altered, leading to changes of ganglioside molecules. Although the particulate neuraminidase in the rat brain at birth is fully active and constantly maintained during the first month of the postnatal period, the level of soluble sialidase doubles in the fourth week as compared to that in the second week (3). The differentiation between the effects of these two neuraminidases during convulsions in the rabbit hippocampus is difficult, especially when the complex sialic compounds of nerve endings are degraded. It appears that the activation of the mechanisms responsible for the uncontrolled firing of neurons in the CNS, including the hippocampus, are probably due to a membrane disturbance.

REFERENCES

1. Avrova, N.F. (1971): Brain ganglioside patterns of vertebretes. J. Neurochem. 18: 667-674.

2. Avrova, N.F., Chenykaeva, E.Y. and Obhukova, E.L. (1973): Ganglioside composition of rat subcellular fractions. J. Neurochem. 20: 997-1004.

3. Carubelli, R. and Tulsiani, D.R.P. (1971): Neuraminidase activity in brain and liver of rats during development. Biochim. Biophys. Acta 237: 78-87.

4. Dalal, K.B. and Einstein, E.R. (1969): Biochemical maturation of the central nervous system. Lipid changes. Brain Res. 16: 441-451.

5. Dreyfus, H., Harth, S., Urban, P. and Mandel, P. (1976): Stimulation of chick retinal ganglioside synthesis by light. Vision Res. 16: 1365-1369.

6. Gambarian, L.S. and Koval, J.N. (1973): In: The Hippocampus-Physiology and Morphology. Erevan, pp. 73-77.

7. Gusel', V.A. and Grigor'eva, O.N. (1975): Age peculiarities of trace discharges of after-effect and the activity of epileptogenic zone in the hippocampus of rabbits. Zh. Evoc. Biokhim. Fiziol. 11: 410-418.

8. Hahn, F. (1960): Analeptics. Pharmacol. Rev. 12: 447-530.

9. James, F. and Fotherby, K. (1963): Distribution in brain of lipid-bound sialic acid and factors affecting its concentration. J. Neurochem. 10: 587-592.

10. Jusuf,H.K.M. and Dickerson, J.W.T. (1977): Effect of development on the gangliosides of human brain. J. Neurochem. 28: 1299-1304.

11.   Jusuf, H.K.M. and Dickerson, J.W.T. (1978): Disia-
       loganglioside G_{D1a} of rat brain subcellular parti-
       cles during development. Biochem. J. 174: 655-
       657.

12.   Kostić, D. (1974): Ph.D. Thesis, Medical Faculty.
       Univeriity of Belgrade.

13.   Kostić, D. (1978): Characteristics of the brain
       tissue ganglioside G_{2a}. In: Enzymes of Lipid
       Metabolism. S. Gatt, L. Freysz and P. Mandel,
       (eds.), pp. 667-678, Plenum Press, New York.

14.   Kostić, D., Nussbaum, J.L. and Mandel, P. (1969):
       A study of brain gangliosides in "Jimpy" mutant
       mice. Life Sciences 8: 1135-1143.

15.   Kostić, D. and Vranešević, A. (1979): Regional
       rabbit brain ganglioside composition after con-
       vulsions induced by pentylenetetrazol. In:
       Abstracts of XI Internat. Congress of Biochemistry,
       Toronto, p. 547.

16.   Kostić, D., Vranešević, A., Rakić, Lj. and
       Vrbaski, S. (1975): Gangliosides in various brain
       structures of the electric fish. T. ocellata.
       Acta Med. Iug. 29: 289-295.

17.   Maccioni, A., Gimenez, M. and Caputto, R. (1971):
       The labelling of the ganglioside fraction from brains
       of rats exposed to different levels of stimulation
       after injection of $(6-^3H)$-glucosamine. J. Neurochem.
       18: 2363-2370.

18.   Merat, A. and Dickerson, J.W.T. (1973): The effect
       of development on the gangliosides of rat and pig
       brain. J. Neurochem. 20: 873-880.

19.   Paldino, A.M. and Purpura, D.P. (1979): Quantita-
       tive analysis of the spatial distribution of axonal
       and dendritic terminals of hippocampal pyramidal
       neurons in immature human brain. Exp. Neurol. 64:
       604-620.

20.   Rosenberg, A. (1979): Biosynthesis and metabolism
       of gangliosides. In: Complex Carbohydrates of
       Nervous System. R.U. Margolis and R.K. Margolis
       (eds.), pp. 25-39, Plenum Press, New York.

21. Saradžišvili, P.M. and Geladze, M. (1977): In:
     Epilepsy, pp. 65-77, Ed. Medicina, Moskva.

22. Smith, A.P. and Loh, H.H. (1979): Architecture of
     the nerve ending membrane. Life Sciences 24: 1-20.

23. Spence, M.W. and Wolfe, L.S. (1967): Gangliosides
     in developing rat brain. Canad. J. Biochem. 45:
     671-688.

24. Stanfield, B.R. and Cowan, W.M. (1979): The develop-
     ment of the hippocampus and dentate gyrus in normal
     and reeler mice. J. Comp. Neur. 185: 423-460.

25. Suzuki, K. (1965): The patterns of mammalian brain
     gangliosides. II. J. Neurochem. 12: 621-638.

26. Suzuki, K. (1965): The patterns of mammalian brain
     gangliosides. Regional and developmental differences.
     J. Neurochem. 12: 969-979.

27. Suzuki, K., Poduslo, S. and Norton, W.T. (1967):
     Gangliosides in the myelin fraction of developing
     rats. Biochim. Biophys. Acta 144: 375-381.

28. Svennerholm, L. (1979): Gangliosides and neuronal
     transmission. In: Abstracts of CNRS Internat.
     Symposium "Structure and function of gangliosides".
     Le Bischenberg, France, pp. 23-27.

29. Swanson, P. (1975): Convulsive disorders. In:
     Biochemistry of Neuronal Disease. M. Cohen (ed.),
     pp. 19-31, Harper and Row, Maryland.

30. Tettamanti, G., Pretti, A., Lombardo, A., Bonali,
     F. and Zambotti, V. (1973): Parallelism of sub-
     cellular location of major particulate neuraminidase
     and gangliosides of rabbit brain cortex. Biochim.
     Biophys. Acta 306: 466-477.

31. Vanier, M.T., Holm, M. Öhman, R. and Svennerholm,
     L. (1971): Developmental profiles of gangliosides
     in human and rat brain. J. Neurochem. 18: 581-592.

32. Vavilova, N.M. (1972): O strukturno-funkcionalnom
     sozrevanij gippokampa v ontogeneze. In: Evolucija
     funkcij v onto geneze. E.M. Kreps (ed.),
     pp. 152-165, "Nauk", Leningrad.

33. Whisler, K.E., Tews, J.K. and Stone, W.E. (1968):
    Cerebral aminoacids and lipids in drug-induced
    status epilepticus. J. Neurochem. 15: 215-220.

34. Wiegandt, H. (1967): The subcellular localization of
    gangliosides in the brain. J. Neurochem. 14: 671-674.

35. Winter, E. and Bernheimer, H. (1975): Brain ganglio-
    sides in drug induced status epilepticus. J.
    Neurochem. 24: 591-592.

36. Yu, R.K. and Yen, S.I. (1975): Gangliosides in
    developing mouse brain myelin. J. Neurochem. 25:
    229-232.

37. Zuckermann, E.C. (1968): Hippocampal epileptic
    activity induced by localized ventricular per-
    fusion with high-potassium cerebrospinal fluid.
    Exp. Neurol. 20: 87-110.

BRAIN DEVELOPMENT AND LEARNING

M. E. Bitterman

Békésy Laboratory of Neurobiology
University of Hawaii
Honolulu, Hawaii 96822

Learning is generally conceived to be an associative process, as is evident from the structure of learning experiments (2). In classical conditioning, two stimuli are paired in anticipation that the first will acquire certain properties of the second, as when a tone that is followed repeatedly by the presentation of food to a dog comes to elicit salivation. Instrumental conditioning differs from classical in that the pairing of stimuli is contingent upon the behavior of the animal. For example, a tone is followed by food only if the animal presses a lever in response to the tone, and the contingency serves also, of course, to pair the response with the food. As to the nature of the associations formed in such experiments and the conditions necessary for their formation, there has been considerably less agreement (5, 16). Are sensory events associated, or are there associations between sensory and motor events, or both? Is temporal contiguity the only condition for the formation of associations, or do motivational processes play a critical role? A related question of considerable importance is whether learning has undergone any fundamental change in the course of evolution. If it has, then no general answers to questions about associative mechanisms are to be expected; work with different animals may provide quite different answers.

Vertebrate and Invertebrate Learning

That learning may not have undergone any fundamental evolutionary change has been suggested by comparative experiments which show striking similarities in the

performance of taxonomically diverse animals -- not only
vertebrates of different classes, but invertebrates as
well.   The results for invertebrates are especially interesting, since the brains of those invertebrates that are
capable of associative learning and the brains of vertebrates seem to have evolved quite independently.   While
the experiments for the most part provide little more
than demonstrations of classical or instrumental conditioning, there are some invertebrates -- notably, octopuses
and honey bees -- in whose performance more detailed
similarities to the performance of vertebrates have been
discovered (9).   I choose some illustrations from work
with honey bees now underway in my own laboratory.

A phenomenon first described by Pavlov (18) and
subsequently found in widely divergent vertebrate species
(5, 16) is that one component of a compound conditioned
stimulus may "overshadow" -- that is, interfere with the
conditioning of -- another component.   In one of Pavlov's
experiments, for example, a dog trained with a tactile-
thermal compound and then tested with the separate components responded to the tactile component as to the
compound, but did not respond to the thermal component
although the thermal conponent alone could be readily
conditioned.   A demonstration of overshadowing in honey
bees is provided by an experiment in which the animals
were trained to feed from a gray target marked either
with an odor (jasmine), or with a color (an orange disk),
or with both.   Two groups were trained with the compound,
then one was extinguished- that is, given an unrewarded
test- with the odor alone and the second was extinguished
with the color alone. A third group was trained and extinguished with the odor, while a fourth was trained and
extinguished with the color.   The training consisted of
three visits on each of which there was a large drop of
40% sucrose solution on the target and the animal was
permitted to drink to repletion.   In extinction, the
sucrose solution was replaced with tap water, and the
number of contacts with the target in a 5-min period was
measured.   The results are shown in Figure 1.   Response
to the odor extinguished no less rapidly after training
with the compound (curve 3) than after training with odor
alone (curve 4), but there was much less response to
color after training with the compound (curve 1) than
after training with color alone (curve 2); that is, color
was overshadowed by odor in training with the compound.

Another ubiquitous phenomenon of vertebrate learning
is the "partial reinforcement effect" (5, 16).   In the
prototypical experiment, two groups of animals are trained

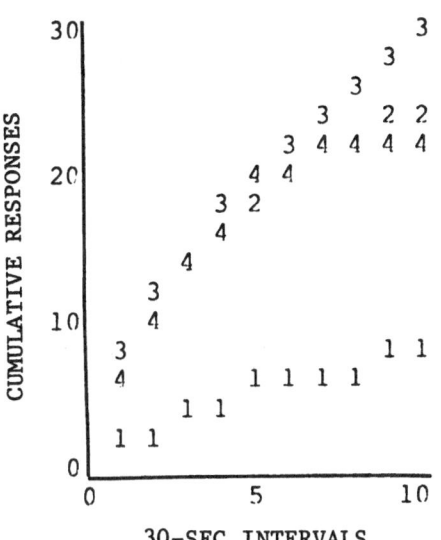

Fig. 1. Overshadowing in honey bees. 1, trained with
         orange-jasmine, extinguished with orange; 2,
         trained with orange, extinguished with orange;
         3, trained with orange-jasmine, extinguished
         with jasmine; 4, trained with jasmine, extin-
         guished with jasmine.

to make some simple instrumental response, a "continuous"
group rewarded on each trial, and a "partial" group rewarded only on a random half of the trials; when both
groups are subsequently extinguished, the partial group
shows greater resistance to extinction -- more persistence -- than the continuous group. In an analogous experiment with honey bees, there were 12 visits to a colored
target containing a large drop of 40% sucrose solution
from which the animals drank to repletion. The sucrose
was available immediately on every visit for a continuous
group, but six of the 12 visits for a partial group began
with a 2-4 min period during which there was no reward --
the animal first found a target containing tap water,
which was replaced after the scheduled interval by a
target containing sucrose. In extinction, each animal
was tested with a target containing tap water for a period
of 10 min during which each contact with the target was
recorded. Here again the results for bees resemble the
results for vertebrates; the extinction curves (very much
like those illustrated in Figure 1) show substantially
more responding by the partial group than by the continuous.

What can these resemblances mean? One possibility
is that they reflect the operation of common associative
mechanisms inherited from a common ancestor. Our attempts
to understand the mechanisms of association would, of
course, be enormously facilitated if they actually were
the same in these relatively simple animals as in vertebrates, and other students of learning in bees have, in
fact, already concluded on the basis of behavioral resemblances that the "neural strategies" of vertebrate learning
can be discovered in work with bees (17). If association
is a synaptic event, the similar processes of conduction
and synaptic transmission in vertebrate and invertebrate
nervous systems may support that conclusion, but there are
also impressive differences -- in cell structure, dendritic elaboration, and neural organization -- whose functional significance remains to be assessed. It is possible
too, that the behavioral resemblances merely reflect some
common functional properties of fundamentally different
associative mechanisms which have evolved independently.
In this view, overshadowing in dogs and overshadowing in
honey bees are behavioral analogues rather than homologues,
and further analysis may be expected to turn up clear
functional differences on the assumption that convergence
to the point of identity is unlikely in richly polygenic
systems (21). The overshadowing and partial reinforcement
effects found in bees may not even be associative phenomena
at all in the sense of being critically dependent on the
nature of the associative mechanisms involved. The first
may be due to a process of stimulus-selection that ante-
dates association, and the second to the fact that the
repeated responding measured in extinction is elicited
and ultimately rewarded in the training of the partial
group whenever reward is delayed. The analysis of invertebrate learning still is in too early a stage to permit
any confident conclusion about the similarity of vertebrate and invertebrate mechanisms. What will be required,
of course, for a satisfactory characterization of invertebrate learning is a much more detailed analysis, not in
one or two, but in a variety of divergent groups.

An Evolutionary Divergence in Vertebrate Learning

Learning experiments with vertebrates of different
classes show some striking similarities in performance
that have long been held to reflect the operation of
common mechanisms (23), but they show, as well, some
interesting differences in performance that point to
divergence in mechanism. The differences appear in experiments on instrumental conditioning and are best considered
in relation to the theory of instrumental conditioning.

When a rat is rewarded with food in some situation, S, for making some simple instrumental response, R, the tendency for S to evoke R is strengthened. Why should that be so? One explanation is that the reward produces an association between S and R -- the famous "S-R reinforcement principle" (12, 23). In this view, contiguity of stimulus and response is a necessary but not a sufficient condition of association; the action of reward must be assumed to account for the fact that the training strengthens the tendency for S to evoke R rather than any of the other responses that may occur frequently at the outset. It is clear, however, that the S-R reinforcement interpretation is incomplete. Whatever other role reward may play in instrumental conditioning, experiments on "incentive contrast" demonstrate that reward is in some sense "learned about" (5, 16). Rats trained with large reward become disturbed when small reward is given instead, and they now perform less well for small reward than do control animals trained from the beginning with small reward. The results suggest that the experimental animals come to expect large reward and are disturbed by the discrepancy between the magnitude of reward expected on the basis of past experience in the situation and the magnitude of reward presently encountered. In an interesting variation of this procedure, one group is trained with large reward, a second group with small reward, and then both groups are extinguished. Although, as Figure 2 illustrates, large reward produces better performance in training (lower latency of responding) than does small reward, extinction is more precipitous after training with large reward where the discrepancy between expectation and reality is greater (11). It is important also to note that rats show not only "negative" contrast but "positive" contrast as well; that is, performance for large reward is better after training with small reward than in control animals trained from the outset with large reward.

With expectation of reward found in experiments on incentive contrast, the insufficiency of the S-R reinforcement principle becomes evident, but there seems to be no way to dispense with it entirely because the strengthened tendency for S to evoke R cannot be understood in terms of expectation alone. What can it mean to say that an animal expects food in S? The simplest interpretation is that S and food have been associated. To account for the strengthened tendency of S to evoke R, it seems necessary to say, not simply that the animal expects food in S, but that R is expected to produce food in S, and here again a simple associative interpretation

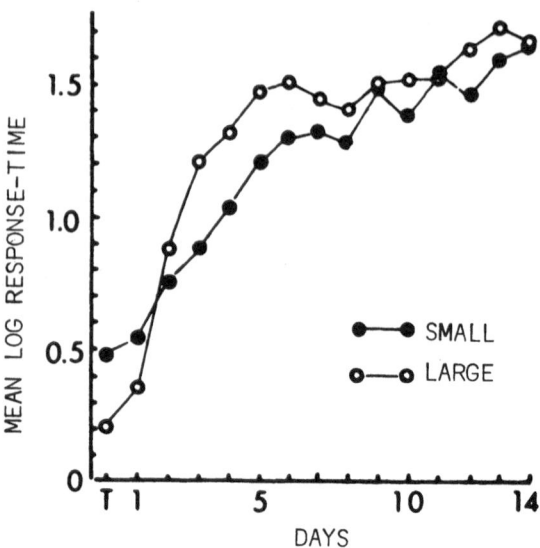

Fig. 2.   Extinction in rats as a function of magnitude
          of reward in training (11). The asymptotic
          levels of performance established in training
          are shown at point T.

may be offered: a compound stimulus consisting of S and
feedback from R is associated with food. Evidence that
not S alone, but the compound of S and feedback from R,
is associated with food in the course of instrumental
conditioning is provided by experiments in which instrumental and consummatory responses are measured concur-
rently. For example, a dog required to push a panel 33
times to earn each reward begins each series of panel-
pushes without salivating and only begins to salivate
when the series is almost complete (24). Where an audi-
tory stimulus is used to signal that the required number
of responses has been made and that food soon will be
presented, the dog begins to salivate only after the
signal is given (10). What remains to be explained,
however, is why the animal pushes the panel in the first
place. If salivation indicates the expectation of food,
these results suggest that it is not the expectation that
produces R, but R that produces the expectation. To deal
with all of the data of instrumental conditioning, it may
be necessary to postulate two different associative

mechanisms -- an S-R reinforcement mechanism that is
responsible for selection of the rewarded response, and
another mechanism that is responsible for learning about
reward.

An interesting implication of this theory is that,
if the two mechanisms did not evolve at the same time,
there may be animals in which only one of them is to be
found, and work with the goldfish suggests that it is
such an animal. Like rats, goldfish perform better for
large reward than for small, but shifts from large reward
to small reward or from small reward to large reward give
no indication of contrast (15). Furthermore, as Figure 3
illustrates, resistance to extinction in goldfish varies
directly with magnitude of reward in training, rather than
inversely as in rats (3). In general, there is nothing
in these results for goldfish that indicates learning
about reward, nor are the results for goldfish unique.
The relation between resistance to extinction and magnitude of reward is direct in painted turtles (19), as it is
in goldfish, although the relation is inverse in pigeons
(7), as it is in rats. If the phyletic picture sketched
in Figure 4 is correct, and if the results for the four
species studied can be generalized to their respective
classes, it may not be too far-fetched to suppose that the
ancestral vertebrate was an S-R reinforcement animal, and
that the capacity to learn about and to anticipate reward
developed only later in some common ancestor of birds and
mammals. The hypothesis suggests that analogous experiments with amphibians should give results like those for
goldfish and painted turtles, while crocodilians and mar-
supials should perform like pigeons and rats (4).

The idea that goldfish and painted turtles do not
learn about reward may seem to be contradicted by demonstrations of classical conditioning in both species. For
example, when illumination of a target is followed repeatedly by liquid food pumped through an orifice in its center,
goldfish begin to strike at the target as soon as it is
illuminated (6). Is this evidence that the animals expect
food? When illumination of an enclosure is followed
repeatedly by shock, goldfish begin to swim actively about
as soon as the light is turned on (25). Is this evidence
that the animals learn to expect shock? Not necessarily,
if expectation of a stimulus is defined as an afferent
representation of that stimulus based on a purely afferent
(S-S) linkage. Both instances of classical conditioning
may reflect the formation of sensory-motor linkages
(between light and the response evoked by the contiguous
food or shock), and there is some reason to believe that

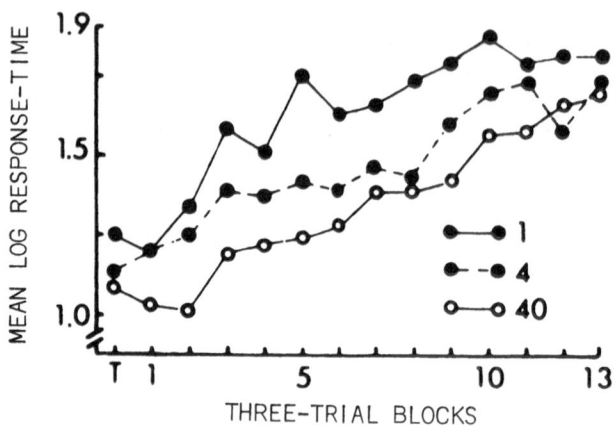

Fig. 3.   Extinction in goldfish as a function of magni-
          tude of reward (1, 4 or 40 Tubifex worms) in
          training (3). The asymptotic levels of per-
          formance established in training are shown at
          point T.

the S-R reinforcement mechanism may play a role, especially
in the appetitive case. Although food is scheduled independently of the animal's behavior, any generalized
response to the illuminated target that occurs is, in fact,
soon followed by food, and it may be the adventitious
contiguity of response and food in the presence of the
light rather than the scheduled contiguity of light and
food that is responsible for the strengthened tendency
to respond in the presence of the light (6). The possibility that the painfulness of shock is reduced by the
response to it also should be taken seriously, but it is
intuitively less likely that classical conditioning with
aversive unconditioned stimuli can be understood entirely
in terms of adventitious reward, and it may be, therefore,
that the evolutionary divergence suggested by the comparative data on incentive contrast is a divergence in the
mechanism of classical conditioning. To begin with,
perhaps, a conditioned stimulus merely acquired certain
motivational and motor properties of the unconditioned

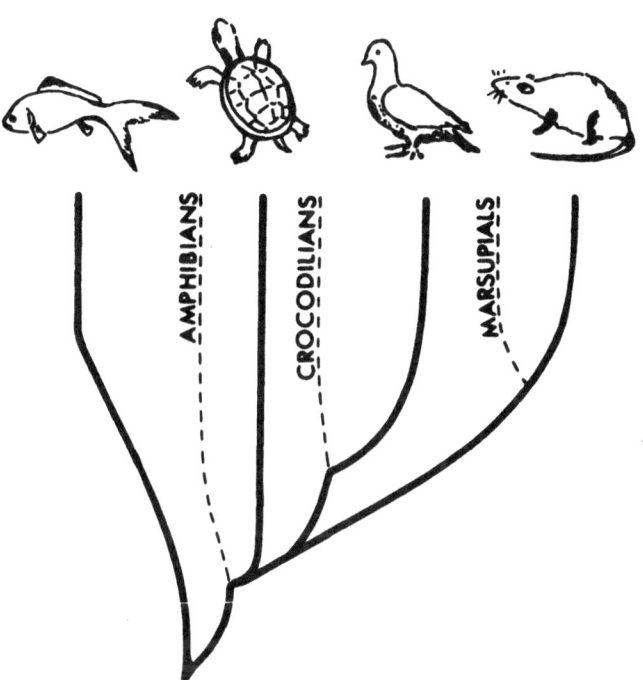

Fig. 4.   Rough sketch of the evolutionary relationships
          among the four species studied in experiments
          on incentive contrast. The dashed lines repre-
          sent lineages for which performance in anala-
          gous experiments is predicted.

stimulus, and only later did there develop the capacity
for sensory linkage that made it possible for the unconditioned stimulus to be remembered and anticipated in
the presence of the conditioned stimulus. Evidence that
classical conditioning in the more advanced vertebrate
lineages produces both S-S and S-R associations supports
this view (16), which represents, of course, a complication of the two-process conception derived from the
analysis of instrumental conditioning in mammals.

     Of the several different methods that have been
developed to look for S-S associations, there is place
here to consider only one. In experiments on "sensory

preconditioning," two neutral stimuli, A and B, are paired, after which the animal is trained either classically or instrumentally to make some new response to B, and then it is tested with A. If A has been associated with B in the first stage of training, A also should evoke the new response, with which it never before has been contiguous. No evidence of A-B association appeared in a recent experiment with goldfish (1) directly analogous to experiments that show sensory preconditioning in rats and rabbits (20). These results support the hypothesis that incentive contrast is not found in goldfish because learning about reward is the product of an S-S process which does not operate in goldfish.

Ontogeny and Phylogeny in Vertebrate Learning

Since function is an expression of structure, it is reasonable to suppose that close relationships between ontogeny and phylogeny, such as are evident in morphology, should be evident also in behavior, and the results of recent work on instrumental conditioning in neonatal rats by Amsel and co-workers do indeed reflect the divergence discovered in comparative work with adult animals of the several vertebrate classes. In a series of experiments with rats 11-16 days old, large reward was found to produce better performance than small reward, but shifts from large to small reward gave no indication of contrast (22). The relation between magnitude of reward and resistance to extinction in these young rats is direct, as it is in goldfish, rather than inverse, as it is in adult rats (14) Plotted in Figure 5 are the results of an experiment with 14-day-old rats in which a locomotor response was rewarded with the opportunity to suck either at a dry nipple (small reward) or at a nipple that provided milk (large reward). The curves show acquisition (increasing speed of response over rewarded trials) in both groups, with better performance in the large-reward group and greater subsequent resistance to extinction.

It is important to note that rapid conditioning, both classical and instrumental, is found in rats as early as the first post-natal day. One-day-old rats may be trained, for example, to probe at one paddle rather than another when responses to one but not the other are rewarded with food (13). In an experiment on conditioned aversion, one-day-old rats developed a clear tendency to avoid an odor paired with a lithium injection (8). It is evident, then, that work with neonatal rats not only supports the idea of a significant evolutionary divergence

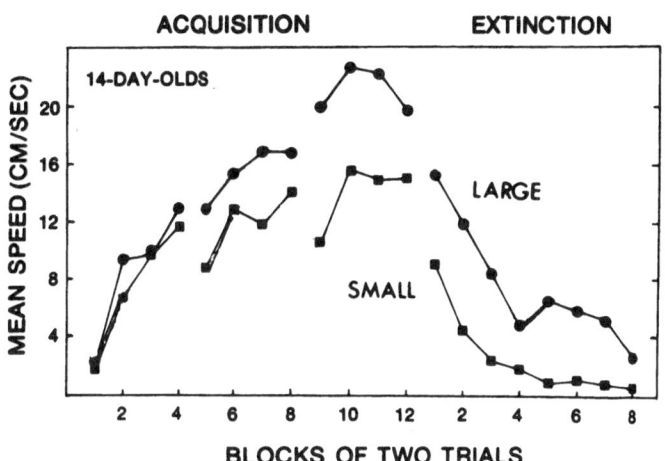

Fig. 5.   Acquisition and extinction in young rats as a
           function of magnitude of reward in training
           (14).   It should be noted, for purposes of com-
           parison with the curves of Figs. 2 and 3 which
           are plotted in terms of response-time, that
           these curves are plotted in terms of speed of
           response, the reciprocal of response-time.

in vertebrate learning, but provides an independent avenue
of inquiry into its structural basis.

REFERENCES

1.   Amire, T.W. and Bitterman, M.E. (in press): Second-
     order appetitive conditioning in goldfish.   J. Exp.
     Psychol., Anim. Behav. Proc.

2.   Bitterman, M.E. (1962): Techniques for the study of
     learning in animals: Analysis and classification.
     Psychol. Bull. 59: 81-93.

3.   Bitterman, M.E. (1969): Thorndike and the problem of
     animal intelligence. Amer. Psychol. 24: 444-453.

4. Bitterman, M.E. (1975): The comparative analysis of
    learning. Science, 188: 699-709.

5. Bitterman, M.E., LoLordo, V.M., Overmier, J.B. and
    Rashotte, M.E. (1979): Animal Learning: Survey and
    Analysis. Plenum, New York.

6. Brandon, S.E. and Bitterman, M.E. (1979): Analysis
    of autoshaping in goldfish. Anim. Learn. Behav. 7:
    57-62.

7. Brownlee, A. and Bitterman, M.E. (1968): Differential
    reward conditioning in the pigeon. Psychon. Sci. 12:
    345-346.

8. Cheatle, M.D. and Rudy, J.W. (1979): Ontogeny of
    second-order odor-aversion conditioning in neonatal
    rats. J. Exp. Psychol., Anim. Behav. Proc. 5: 142-151.

9. Corning, W.C., Dyal, J.A. and Willows, A.O.D.
    (1973, 1975): Invertebrate Learning, v. 1-3. Plenum,
    New York.

10. Ellison, G.D. and Konorski, J. (1964): Separation
    of the salivary and motor responses in instrumental
    conditioning. Science 146: 1071-1072.

11. Gonzalez, R.C. and Bitterman, M.E. (1969): Spaced-
    trials partial reinforcement effect as a function of
    contrast. J. Compar. Physiol. Psychol. 67: 94-103.

12. Hull, C.L. (1943): Principles of Behavior. D. Apple-
    ton-Century, New York.

13. Johanson, F.B. and Hall, W.G. (1979): Appetitive
    learning in 1-day-old rat pups. Science 205: 419-421.

14. Letz, R., Burdette, D.R., Gregg, B., Kittrell, E.M.W.
    and Amsel, A. (1978): Evidence for a transitional
    period for development of persistence in infant rats.
    J. Compar. Physiol. Psychol. 92: 865-866.

15. Lowes, G. and Bitterman, M.E. (1967): Reward and
    learning in the goldfish. Science 157: 455-457.

16. Mackintosh, N.J. (1974): The Psychology of Animal
    Learning. Academic Press, New York.

17. Menzel, R. and Erber, J. (1978): Learning and memory
    in bees. Scientif. Amer. 239: 102-110.

18.   Pavlov, I.P. (1927): Conditioned Reflexes. Oxford
      University Press, Oxford.

19.   Pert, A. and Bitterman, M.E. (1970): Reward and
      learning in the turtle. Learn. Motiv. 1: 121-218.

20.   Pfautz, P.L., Donegan, N.H. and Wagner, A.R. (1978):
      Sensory preconditioning versus habituation. J. Exp.
      Psychol.: Anim. Behav. Proc. 4: 286-295.

21.   Simpson, G.G. (1964): Organisms and molecules in
      evolution. Science 146: 1535-1538.

22.   Stanton, M. and Amsel, A. (1980): Adjustment to re-
      ward reduction (but no negative contrast) in rats
      11, 14, and 16 days of age. J. Compar. Physiol.
      Psychol. 94: 446-458.

23.   Thorndike, E.L. (1911): Animal Intelligence.
      Macmillan, New York.

24.   Williams, D.R. (1965): Classical conditioning and
      incentive motivation. In: Classical Conditioning,
      W.F. Prokasy (Ed.), Appleton-Century-Crofts, New York.

25.   Woodard, W.T. and Bitterman, M.E. (1973): Pavlovian
      analyses of avoidance conditioning in the goldfish
      (Carassius auratus). J. Compar. Physiol. Psychol.
      82: 123-129.

EARLY SEA URCHIN EMBRYO AS A MODEL FOR THE STUDY OF

PRE-NERVOUS FUNCTIONS OF NEUROTRANSMITTERS: NEW DATA

G. A. Buznikov, B. N. Manukhin, L. M. Rakić
and T. M. Turpaev

N. K. Kol 'tsov Institute of Developmental
Biology of the Academy of Sciences of USSR
Moscow, USSR, and
Department of Biochemistry
School of Medicine
Belgrade, Yugoslavia

The most extensively studied neurotransmitters,
such as acetylcholine, catecholamines and indolylalkylamines, are found already in the early, "pre-nerve"-
stage embryos of all examined classes of either non-
vertebrates or vertebrates (Fig. 1). A comprehensive
survey of the available data on the "pre-nerve"-stage
neurotransmitters permits the conclusion that the concentration of such compounds changes in a well-regulated
fashion at key stages of early embryogenesis. This
appears to be true especially at the time of the embryonic
cleavage. The common neurotransmitters appear to be
functional in "pre-nerve"-stage embryos, and prevention
of their function disrupts the course of embryogenesis.
This in turn seems to indicate a direct involvement of
"pre-nerve" neurotransmitters in the regulatory processes
of the early embryogenesis (1-6).

A sizable portion of the experimental material used
in the present study was obtained from the Mediterrannean
sea urchins in the International Laboratory for Brain
Research in Kotor, Yugoslavia. We will consider here
the data on serotonin-like substances in these embryos,
on the localization of the receptors (or the functional
analogues of the receptors) which determine the sensitivity

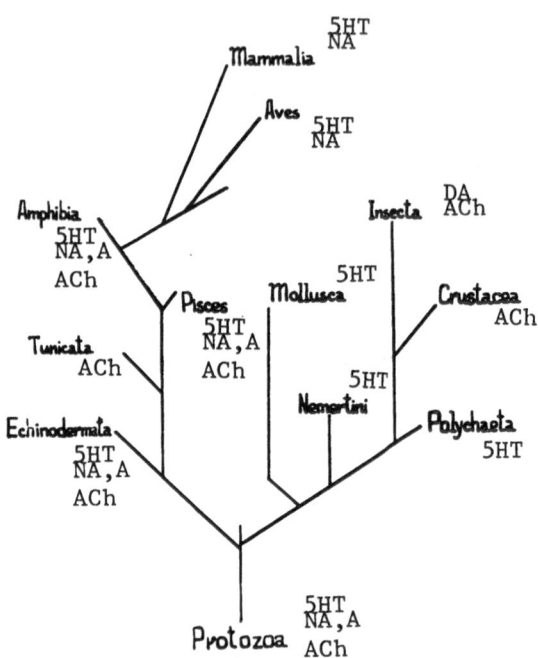

Fig. 1.   "Pre-nerve" synthesis of neurotransmitters in
various animals.   5-HT = serotonin; A = epinephrine; HA = norepinephrine; DA = dopamine;
AH = acetylcholine.

of the early embryos to "pre-nerve" neurotransmitters
and their antagonists, and also on the ability of early
embryos to accumulate structural analogues of acetylcholine, catecholamines, and indolylalkylamines.   These
data could probably illustrate in sufficient detail the
suitability of the early sea urchin embryo for studies
on "pre-nerve" neurotransmitters.

The biogenic amines of the early embryos.

Serotonin and adrenalin-like substances in the early
sea urchin embryos were characterized by us several years

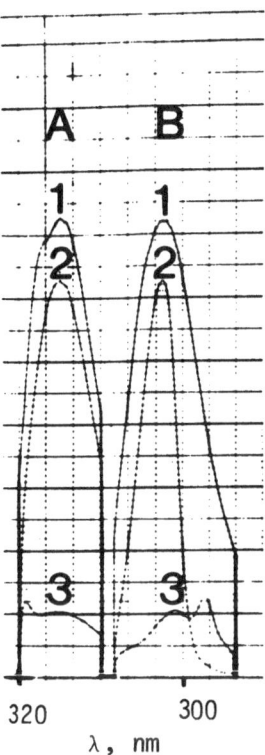

Fig. 2. Spectra of fluorescent emissions (A) and
excitation (B) of the extract of embryos of
sea urchin. Strongylocentrotus intermedius
at the stage of 4-8 blastomers (analyzed in
neutral medium). 1 = extract; 2 = 100 ng
tryptamine; 3 = control.

ago (1, 2). Using new micromethods for the quantitative
determination of the biogenic monoamines, these results
were confirmed and refined with the embryos of
Mediterrannean sea urchins. The early embryos of the
sea urchines Arbacia limula, Paracentrotus lividus and
Sphaerechinus granularis were used. Methods used to
prepare the materials for analysis, and the analytical
procedures were described in detail in a previous communication (2). Briefly, the materials were extracted
with 2 percent perchloric acid containing 0.5 percent
EDTA. The extracts were purified by two methods. One

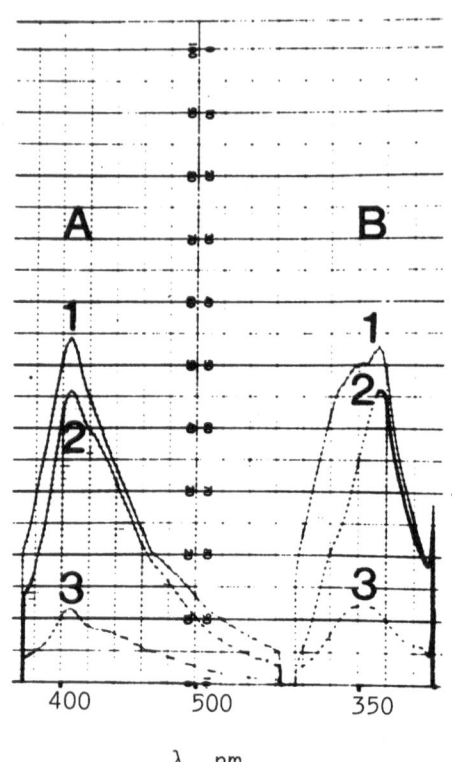

λ, nm

Fig. 3.   Spectra of fluorescent emission (A) and excit-
          ation (B) of an extract from the embryos of sea
          urchin Strongylocentrotus intermedius at the
          stage of 4-8 blastomers (determined after con-
          densation with o-phthalic aldehyde).   1 =
          extract; 2 = 100 ng serotonin; 3 = control.

was the extraction with butanol and the other consisted
of passing the extract through Dowex 50 resin.   The
indolylalkylamines were determined using a MPF-4 Hitachi
or an Aminco Bowman fluorescence spectrophotometer, or
analyzing the fluorescent products of their condensation
with o-phthalic aldehyde (OPA) or with ninhydrin.

     The major component in the determination of indolyl-
alkylamines in the embryonic extracts was a component
having the fluorescence excitation properties identical
to those of tryptamine (Fig. 2).   In some cases, a second

minor component was apparent, possessing a similar excitation and emission fluorescence maxima to that of serotonin (Fig. 3). These two components were identified in all stages of development where the nervous system was present, including that of the late larva and adults.

To identify these components, we have determined the fluorometric properties of a large number of indole derivatives which differ from serotonin and tryptamine in the position or the chemical nature of the substituent groups, or in the stereochemistry of the aminoethyl group substituents. The excitation and emission fluorescence maxima of the major tryptamine-like component of the embryos did not differ from that of tryptamine. It also did not differ from its derivatives possessing radial or lateral bonds at the amino nitrogen. This component clearly differed from other indole derivatives examined, especially from the compounds possessing one or more substituents bound to the carbon atoms of the indole ring. Therefore, on the basis of fluorescence analysis it was possible to identify this compound as tryptamine or its substituted aminoethyl derivative.

It is known that many biogenic monoamines and their nontoxic derivatives lower the sensitivity of the early sea urchin embryos toward the cytotoxic monoamine analogues. In experiments with Mediterrannean sea urchins, the best protection against cytotoxic derivatives of indole was afforded by tryptamine or its derivatives possessing a substituent at the amino nitrogen. Tryptamine derivatives with substitution at ring carbons did not protect. Such pharmacological tests permitted the identification of tryptamine-like substances of sea urchin embryos as tryptamine or an N-substituted derivative of tryptamine.

The less abundant serotonin-like component of the embryos by its excitation and emission fluorescence corresponded only with serotonin itself and differed from all examined serotonin analogues, including bufotenine. It is therefore highly likely that the minor serotonin-like component of the sea urchin embryonic extracts is serotonin. Our histochemical data also support the presence of serotonin in the early embryos.

Of the catecholamines analyzed in the early embryos of the Mediterranean sea urchins, only dopamine was identified. The dopamine concentration in the early embryos does vary and the changes apparently are related

to the cleavage. Such changes were also found for both
serotonin-like components.

Toneby has verified our previous results on sero-
tonin- and epinephrine-like substances in the early sea
urchin embryos (6). He also found that the chief catecholamine of the early embryos is dopamine, whereas the
most abundant indolylalkylamine was tryptamine. According to Toneby, serotonin appears only in the larva
having a nervous system, i.e., it cannot be considered
as a "pre-nerve" stage neurotransmitter. This discrepancy between our data and Toneby's results is possibly
due to a difference in the methods of monoamine determination.

Papers published in the last few years also confirmed
our old data on the presence of acetylcholine in the early
embryos. It is thus possible to conclude that the pre-
nerve stage of embryonic development is characterized by
presence of several neurotransmitters within the same
cell. Transition to the typical situation of "a single
neurotransmitter in any given cell" found in the multi-
cellular organisms, begins in sea urchins at the time
of cell differentiation corresponding to gastrulation.

The binding of neurochemicals to the embryos.

Many neuropharmacological agents of a structure
similar to acetylcholine or the biogenic monoamines
disturb or block sea urchin embryogenesis and inhibit
the macromolecular syntheses, especially protein biosynthesis (2, 3). These effects could be prevented,
modified or counteracted by the addition of the corresponding neurotransmitter to the medium (sea water).
It was found that penetration of a given agent into the
cells is necessary for its specific action. We therefore concentrated on the study involved in the binding
of neurochemicals to the cells of early embyros of
Arbacia lixula and Paracentrotus lividus. Measurement
of the binding was accomplished by two independent
methods, a spectrophotometric determination of the
amount of the given substance in the medium, and a
biological determination of the accumulation of the
agent in the embryonic cells. These methods, as a
rule, gave quite similar results.

We have found that that the structural analogues
of the neurotransmitters belonging to the class of
either primary, secondary or tertiary aromatic amines
are rather intensively bound to the early embryos.

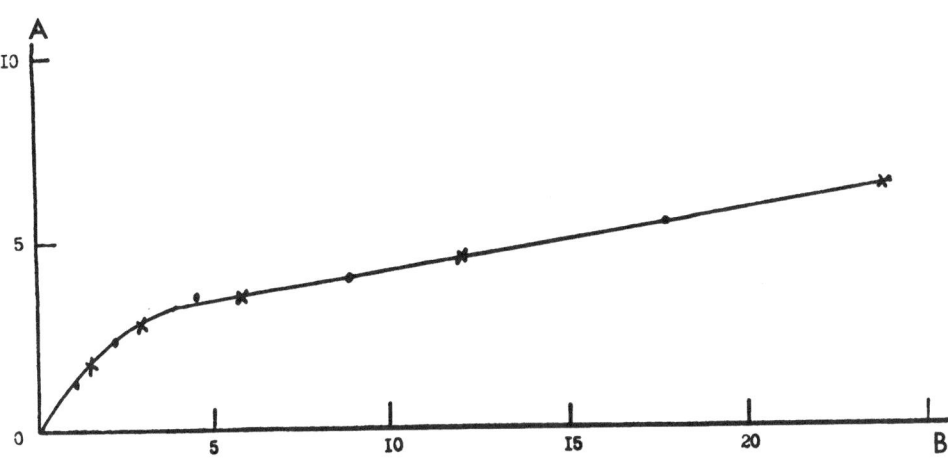

Fig. 4.   Binding of serotonin antagonist No. 407 to the
          early embryos of sea urchin <u>Arbacia lixula</u>.
          A, the binding (micrograms per 1000 embryos)
          after an exposition of 80 min. B, starting
          concentration of the substance (micrograms/
          1000 embryos).

This binding could be accomplished against a concentration
gradient. Work with cell homogenates showed that the
accumulation of the neurochemicals occurs chiefly in the
cytoplasm of the embryos. The transformation of tertiary amines to the quaternary ammonium bases resulted in
a large reduction of their binding by the embryonic
cells.

     The accumulation of neurochemicals by the embryos
depends on the length of exposure, the level of such
agents in the medium, and the concentration of embryos.
The binding exhibits a typical saturating kinetics, with
maximal binding occurring after 80-90 min of incubation.
The relationship between the extent of binding and the
concentration of the agent in the medium resembles the
empirical Langmuire adsorption isotherm (Fig. 4). The
analysis of the corresponding curves could be of value

in judging the biological specificity of a given system. The inflection portion of the curve is rather steep for the attained adsorption equilibrium. This can be explained by assuming the existence of a high and low affinity process. Since the amount of agent bound is not proportional to the extent of its cytotoxic effect, it appears that the structures responsible for the binding do not correspond with those determining sensitivity toward it. Moreover, a significant portion of the bound neurochemical appears to be incapable of acting upon the structures which mediate the toxic effect. The dissociation between the binding sites and the sensitivity-determining sites is reflected by the absence of a real correlation of the protective effect of exogenous neurotransmitters against the cytotoxic agents and the inability of neurotransmitters to influence the binding of these substances.

The ability of embryos to accumulate structural analogues of the monoamines is lowered by osmotic shocking or by a 10-min exposure to 70°C. At 55-65°C this ability is only partially abolished, while at 40°C (a temperature which completely arrests the embryo development) the accumulation is unaffected. Poisons like rotenone, vitamin A and mercuric chloride, at concentrations which completely block the embryonic development, are without influence on the binding of the neurochemicals. Highly effective inhibitors of the binding at non-toxic doses are the many polycyclic nitrogen-containing compounds, such as the tricyclic depressants and phenothiazine derivatives. These substances at the same time lower the sensitivity of the embryos to the corresponding neurochemicals. In this case, as different from experiments with exogenous neurotransmitters, there is a direct correlation between the extent of protective action and the effect of the given substance on the binding of a cytotoxic neurochemical.

To our knowledge, the binding of the various neurochemicals structurally akin to monoamines and acetylcholine has not been demonstrated in non-nerve cells in adult organisms. Thus, such a binding could be considered as a special property of the embryonic cells. It is possible that the intracellular structures responsible for this binding also have an important role in storage of the "pre-nerve" neurotransmitters. Examination of this possibility appears to be an important task in future investigation.

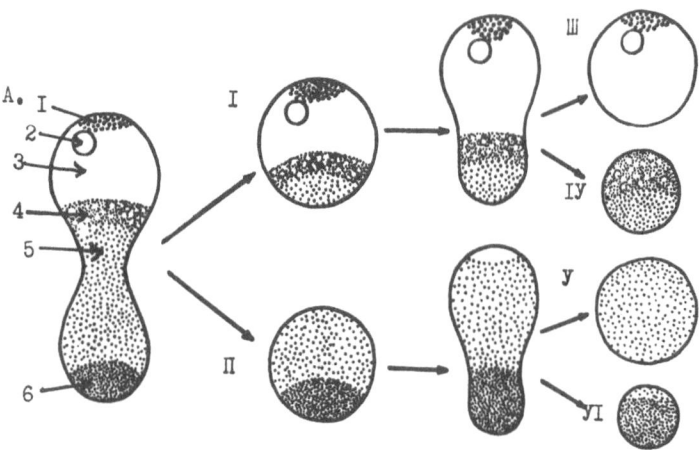

Fig. 5.   Stratification of the sea urchin egg (A) and
           egg fragments (B) for sea urchin Arbacia spp.
           [drawn according to Harvey (4)]. I - white
           half; II = the red half; III to VI = the
           transparent, granular, yolk and pigment quarter;
           1- lipid bodies; 2- pronucleus; 3- nongranular
           cytoplasm; 4- mitochondria; 5- yolk granules;
           6- pigment (granules ?).

Binding of "pre-nerve" neurotransmitters and their
antagonists to cell receptors.

We have already shown that both the cytotoxic effect
of neurochemicals and the protective action of neurotransmitters are exerted on an intracellular level.
According to this model, the corresponding receptors or
the functional analogues of the receptors are, as different from the synaptic receptors for neurotransmitters,
not localized on the cell surface but are found intracellularly. Yet another characteristic of "pre-nerve"
neurotransmitters receptors was uncovered during an
investigation on the sensitivity of the early embryos
toward various groups of CNS agents. The activity of

the substances on the early embryos sharply differed
from the analogous order from experiments using classical
models (2).

It is obvious that further concrete data on pre-
nerve receptors will be extremely useful for a more detailed elucidation of the system of "pre-nerve" neurotransmitters. The early embryos of Arbacia lixula were
found to be especially suitable for such investigations.
In these embryos the two specific types of sensitivity
to the antagonists of monoamines and acetylcholine were
defined by us as "usual sensitivity" and "supersensitivity". Many neurochemicals act upon embryos of Arbacia
lixula in the same way as upon the embryos of the other
species of sea urchins examined which exhibit normal
sensitivity. However, the Arbacia lixula embryos were
10-1000 times more sensitive than those of other sea urchi
species (supersensitivity) to each of the examined groups
of neurochemical substances. These two types of sensitivity apparently correspond with the existence of two
classes of intracellular neurotransmitter binding ligands
(3).

By ultracentrifugation of nonfertilized eggs of
Arbacia lixula, it is possible to obtain a fraction
which after fertilization could enter a more or less
normal embryonic development (Fig. 5). Some of these
embryo fragments (e.g., white halves and clear quarters)
are completely without supersensitivity, but preserve
the "usual sensitivity". Other types of fragments (red
halves, and granular, yolk and pigmented quarters) have
both types of sensitivity. A correlation of these data
with the data on ultrastructure of the corresponding
embryo fragments allows the formulation of a hypothesis
that either the receptors responsible for the "supersensitivity", or some other intracellular factors necessary for the enhanced sensitivity are bound to the yolk
granules.

It was previously proposed that the "usual sensitivit
receptors, and also a portion of the ligand responsible fc
supersensitivity, are linked to the elements of cyto-
skeleton (microfilaments and microtubules). The data
obtained from ultracentrifugation experiments do not
contradict this proposal. It was shown that the substances which cause the destruction of microtubules
(colchicine) and of microfilaments (cytochalasin B)
coincidentally lower the sensitivity of early embryos
toward antagonists of "pre-nerve" neurotransmitters.
Such a reduction in sensitivity was found for both types.

From these data, some of the functions of "pre-nerve"
neurotransmitters appear to be exerted at the level of
microtubules

CONCLUSION

The data presented in this work stress the functional
distinctions of pre-nerve neurotransmitters from the
synaptic neurotransmitters. The "pre-nerve" neurotransmitter system is characterized by the presence of several
neurotransmitter systems in the same cell, by an intracellular localization of receptors, by the synthesis,
transport and binding of neurotransmitters within the
same cell, and perhaps by the selective receptors for
each neurotransmitter. Investigation of these differences appears to be valuable for understanding the
evolution of neurotransmitter-dependent processes.

REFERENCES

1. Buznikov, G.A. (1967): The low molecular weight
    regulators of embryonal development. "Nauka,"
    Moscow.

2. Buznikov, G.A. (1973): 5-hydroxytryptamine, cate-
    cholamines, acetylcholine, and some related sub-
    stances in early embryogenesis. Comparative
    Pharamacology, Michaelson, M.J. (ed), Vol II,
    pp. 593-623, Pergamon Press, New York.

3. Buznikov, G.A., Rakić, L., Turpae, T.M.,
    Markova, L.N. (1974): Sensitivity of sea urchin
    embryos to antagonists of acetylcholine and mono-
    amines. Exp. Cell Res. 86: 2, 317-324.

4. Harvey, E.B. (1956): The american Arbacia and other
    sea urchins. Princeton University Press, Princeton,
    N.J.

5. Manukhin, B.N., Volina, E.V., Markova, L.N.,
    Rakić, L., Buznikov, G.A. (1980): The new data on
    the biogenic monoamines in developing embryos of
    sea urchins. Zh. Evol. Biokhim. Physiol. in press.

6. Toneby, M. (1977): Functional aspects of 5-hydroxy-
    tryptamine and dopamine in early embryogenesis
    of Echinoidea and Asteroidea. Stockholm.

EFFECT OF UNILATERAL DEAFFERENTATION ON THE DEVELOPMENT

OF THE LATERAL GENICULATE NUCLEUS OF THE DOG

Lidija Lazarević, Lj. M. Rakić and
N. N. Lyubimov

Brain Research Laboratory
Institute for Biological and Medical Research
of Montenegro
Kotor, Yugoslavia
                    and
Brain Institute
Academy of Medical Sciences
Moscow, USSR

INTRODUCTION

Study of the morphological changes in the lateral geniculate nucleus (LGN) following bilateral deafferentation has been the subject of numerous studies (2, 8, 14, 15, 16, 17, 22, 23, 27). Fewer papers have dealt with unilateral deafferentation of the LGN by sectioning the optic tract and keeping the opposite side intact as a control (3, 6, 7, 35). The changes in the LGN after deafferentation are not only determined by the deafferentation procedures used but also by the age and species of the animals. When one eyelid is sutured in a young kitten, some of the geniculate cells, deprived of their normal visual input, grow to a lesser degree than normal (37, 38). The transneuronal cell atrophy that occurs in the dorsal LGN after removal of one eye is more rapid in young than in adult cats (12, 13, 21, 37, 38), and atrophic changes continue well beyond the time at which geniculate cell growth has stopped (14). The role of a binocular or translaminar competition in the production of the transneuronal atrophy has been emphasized only in very young animals (7, 14). In contrast, the comparative studies of the optic tract transection and eye enucleation, by

Garey et al. (6, 7) suggested that even in adult cats there is some degree of cellular reaction present between the cells of the laminae. Total or partial deafferentation of the LGN in man (11) and monkeys (1, 31) produces severe degenerative changes with the disappearance of the nerve cells. In dogs and cats, the results obtained by various authors show considerable discrepancies. Klassovskij and Kosmarskaja (18) found in young dogs after a bilateral enucleation a disappearance of most cell elements in the LGN, while Figurina (4) performing the same operation found a slight loss of the cells, only in the proximal parts of that nucleus.

In our previous studies we stressed the "stability" of the LGN cells in dogs to the deafferentation caused by optic tract transection (35). Subsequently, we thought it worthwhile to examine the age-dependent effect of deafferentation of the LGN on the cytoarchitectural organization during the growth of individual regions of this structure.

MATERIALS AND METHODS

The 12 experimental dogs were bred in our laboratory and were of known age at the time of surgery. The optic tracts (unilateral deafferentation) were unilaterally transected either at 14 days, 45 days or one year (adults) of age (Fig. 1). Each group consisted of 3 dogs. They were sacrificed with Nembutal[R] 5-6 years after the operation. The brains were fixed by intraaortic perfusion with 10% neutralized formalin. The brains, embedded in paraffin, were serially sectioned (20 μm) and then stained with cresyl violet.

The size of the deafferented and undeafferented LGN of all the experimental (operated) and control (intact) animals was measured planimetrically according to Kononova (20). The outlines of each LGN section were magnified to three-fold for the projection. The size of cells was measured using the following formula:

$$V = \frac{h(K_1 + K_2 + \ldots \ldots K_n)}{225}$$

where K_1, K_2, K_n represent the size of the planimetrically measured LGN, h - thickness of the section, 225 - magnification of the projection squared.

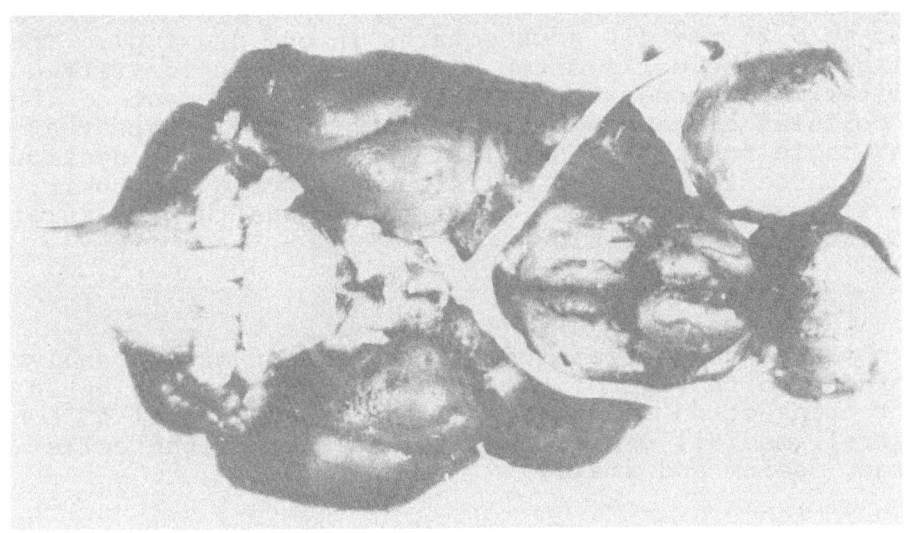

Figure 1.   The base of the brain with the transected
            left optic tract.

In addition, we also made a spatial reconstruction of
the external configuration of LGN by projecting the relatively enlarged outlines of the frontal sections.

For qualitative and quantitative analysis, three
levels of the LGN were selected: rostral, medial and
caudal. The reported ontogenic development of the LGN
emphasizes the importance of the rostral and caudal parts
in the understanding of the LGN development (24). The
rostral part of the LGN is connected with the phylogenetically newer peripheral visual projections, the caudal one
with the phylogenetically earlier differentiated (macular)
visual projections, and the medial part is a transitional
region evolving from the phylogenetically old to the phylogenetically new receptive functions of vision (10). Frontal
serial sections of the LGN were analyzed qualitatively. In
the animals examined, all the levels of the LGN were photo-
graphed and the cytoarchitectural photographs of the micro-
preparations, magnified 100 X, obtained from the experimental were compared with those of the control animals.

Quantitative analysis of the sections was done after
a previous reconstruction of the LGN cells with the aid
of a drawing apparatus (microscope AY-12, drawing apparatus
PA-4). The cell counting of the reconstructed frontal section included: (I) total number of cells in the whole section, (II) number of cells in 1 mm^2, and (III) number of

cells in 0.25 mm^2 (10 such squares in one drawing). The
outlines of the cells drawn with a camera lucida-fitted
Amplival microscope were used for the measurement of the
LGN cellular diameters representing (a) the greatest at a
right angle to (b) the smallest length. In each section
of the three levels, 250 cells were measured and their
mean size was determined; their volume was then calculated
according to the formula:

$$V = 10.4 \frac{ab \sqrt{ab}}{6 \pi} \ \mu m^3$$

The obtained data presented by the histogram were analyzed
according to the age of the animal at the time of optic
tract transection and to its effect on each level of LGN
(rostral, medial, caudal) and to the size of the cells
(large, medium and small).

RESULTS

Qualitative Cytoarchitectural Examinations of the LGN after Optic Tract Transection

Global cytoarchitectural examinations of the LGN
showed that deafferentation of that structure after optic
tract transection does not cause a deficit of nerve cells
or a disturbance in their laminar organization in either
control dogs (Fig. 2) or in experimental dogs after unilateral deafferentation (Figs. 3 and 4) of LGN.

The deafferented LGN was smaller than the non-deafferented LGN of the operated and control animals (Table 1).

In addition to the decrease in the general volume of
the deafferented LGN the individual volume of the single
cells also decreased along with an increase in the density
(distribution) of the cells in the LGN on the side of the
optic tract transection.

Quantitative Analysis of the LGN after Optic Tract Transection

Most of the classic investigations so far have emphasized a strict symmetry in the organization of the central
nervous system of the vertebrates, particularly in the
number of cell elements. In contrast, the findings of
Soviet authors (2, 5, 19, 32) point to certain deviations
in the cytoarchitecture of the left and right halves of
the cerebral cortex and some subcortical formations.

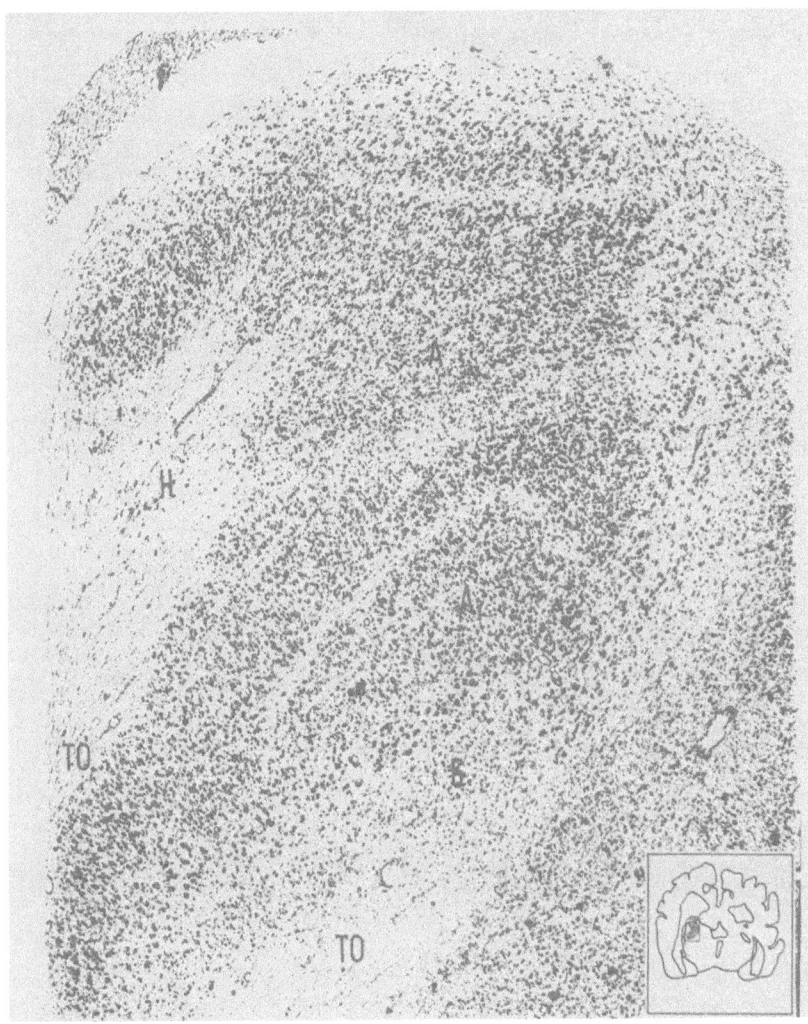

Figure 2.   Frontal section of the LGN of the control
            dog.   X 100
            A_1, A_2 and B - laminae; TO - optic tract;
            H - hilus

Therefore, the denervated LGN was compared not only with
the non-denervated LGN but also with both of the LGN from
control (non-operated) animals.

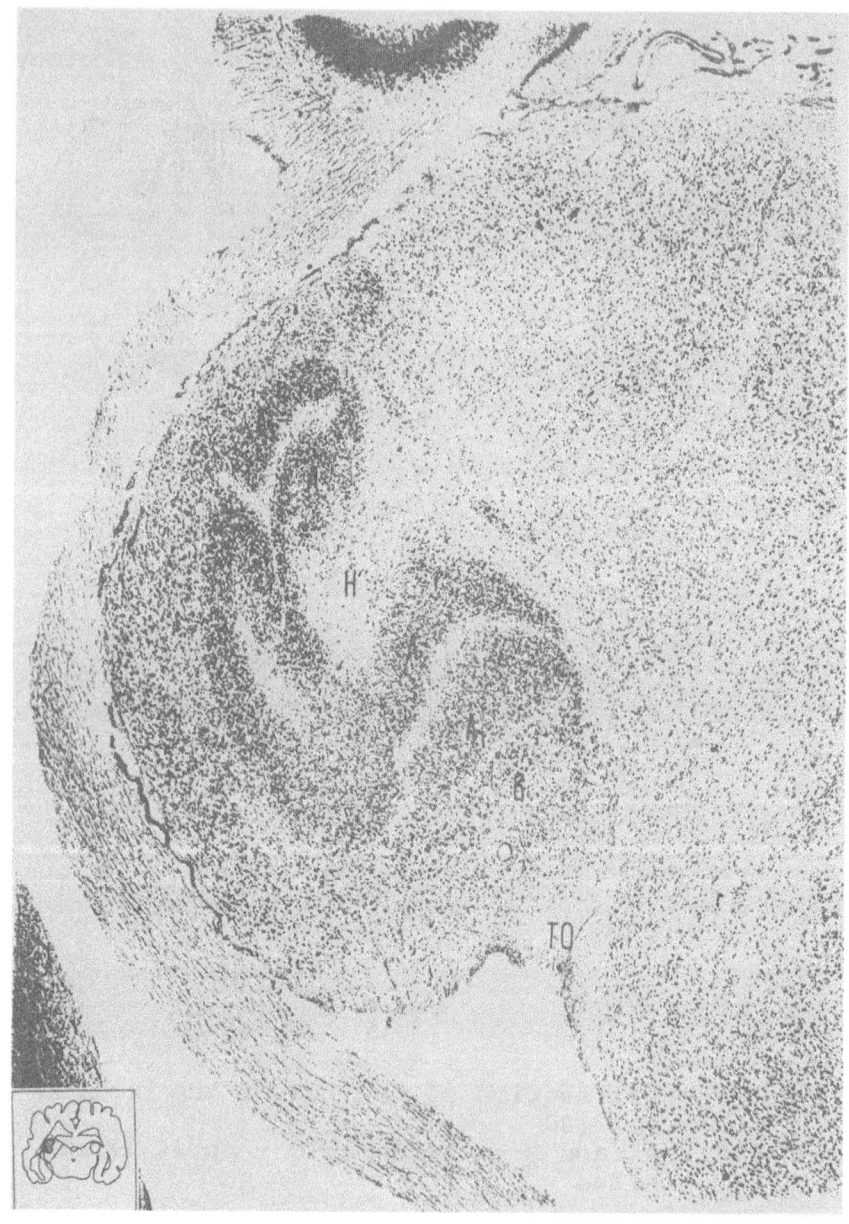

Figure 3.   Frontal section of the LGN ipsilateral to the
            optic tract transection.   Identification same
            as in Fig. 2.

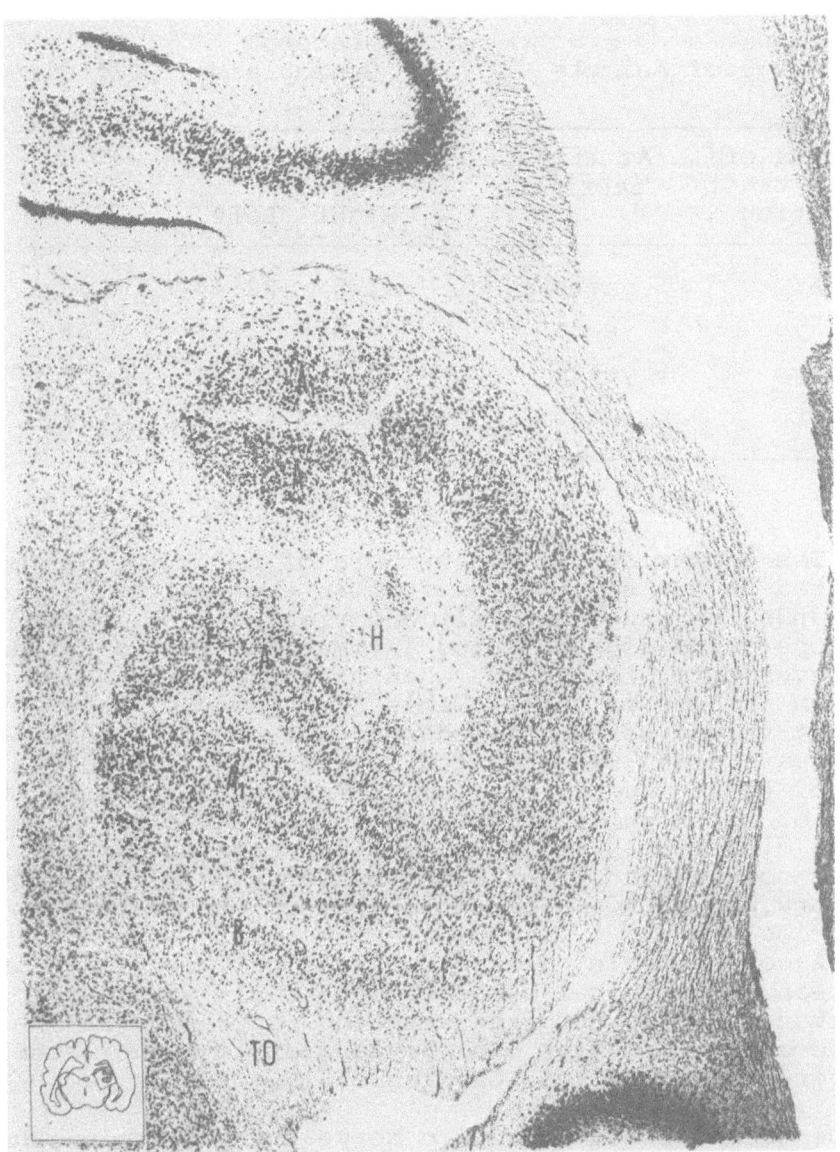

Figure 4.   Frontal section of the LGN of the side
            contralateral to the optic tract transection.
            Identification same as in Fig. 2.

Table 1. Volume of the Lateral Geniculate Body in mm^3

Age of Animals		Lateral Geniculate Body		Differences in Percent
At time of optic tract section	At time of sacrifice	Right	Left	
1 year	6 years	41.57	22.44	46.1
15 days	6 years	26.47	13.62	48.6
45 days	6 years	55.58	23.48	58.69
Intact	6 years	46.10	43.14	4.3

The comparison of the numbers of cells in 0.25 mm^2 (counted in ten squares) present in the rostral, medial and caudal parts of the left and right LGN of control (non-operated) dogs revealed asymmetries (Table 2) which were even more pronounced when expressed by minimum and maximum numbers of cells (Table 3). However, more consistent values were obtained when the cells of more than one square (ten) were counted for the comparative evaluation of the number of the cells present on the same LGN level of the right and left control LGN (Table 4). Although differences in the number of cells were observed in the counted individual LGN sections, a left to right symmetry was found in the total sum of the cells of LGN.

A considerable, statistically significant rise in the mean arithmetical number of the cells was seen in the denervated LGN as compared with the contralateral nondeafferented LGN after 5 or 6 years of total unilateral deafferentation of the LGN (Table 5).

Age Dependent Effect of Optic Nerve Transection on the Development of the LGN

Examination of the external configuration of the LGN, obtained by transferring the frontal sections on paper with the aid of a projector, showed that the intensity in the growth of this structure depended on the age of the animal at the time of the optic tract transection. The transection of the optic tract at the age of 15 days

Table 2. Number of Cells in 0.25 mm^2 at the Caudal (C), Medial (M) and Rostral (R) Level of Lateral Geniculate Body in a Control Dog

	C	Mean Value	M	Mean Value	R	Mean Value
	114		151		120	
	125		274		117	
	126		258		96	
	176		134		128	
Right	134	137	214	199	172	137
	149		280		134	
	123		227		140	
	140		195		160	
	122		201		167	
	166		156		140	
	172		188		186	
	153		190		167	
	165		143		93	
	133		149		181	
	166		193		168	
Left	180	157	205	185	180	162
	111		174		160	
	213		226		173	
	188		197		176	
	91		187		133	

Table 3. Minimum and Maximum Number of Cells in 0.25 mm^2 in the Lateral Geniculate Body of a Control Dog[*]

	Level	Maximum	Minimum
	C	176	114
Right	M	280	134
	R	172	96
	C	213	91
Left	M	226	143
	R	186	93

[*]Identification same as in Table 2.

Table 4. Values of the Comparison of Two Identical
Contralateral Levels of the Lateral Geniculate Body of a Control Dog[*]

	Level	Values for Identical Levels	Mean Values
	C	157	
Right	M	185	157
	R	162	
	C	137	
Left	M	199	167
	R	137	

[*]Identification same as in Table 2.

caused a marked retardation in LGN growth on the affected
side, especially in the rostral area. The growth in the
caudal and medial parts of LGN lagged behind on both sides
equally (Fig. 5A). The optic tract transection at 45 days
of age led to a significantly more advanced LGN development as compared with the preceding group. However, a
retarded growth was found in the caudal and medial parts
of the LGN. On the other hand, the rostral part continued
to grow on the side of the transected optic tract to a
length twice that of the LGN denervated on the 15th postnatal day (Fig. 5B). Optic tract transection at one year
induced noticeable changes on the side of optic tract transection but only in the medial part of the LGN (Fig. 5C).
In control dogs, the LGN growth was uniform on both sides
of the brain (Fig. 5D).

Age Dependent Effect of Optic Nerve Transection on LGN Cell Volume

Deafferentation of the LGN caused by optic tract
transection produced selective changes in the volume of
the cells depending on the animal's age at the time of
optic deafferentation. The optic tract transection performed at 15 and 45 postnatal days led to drastic cellular
changes, not only on the side of transection, but also on
the contralateral side. The number of large cells, and
to a somewhat lesser degree the number of medium-sized

Table 5. Number of Nerve Cells in 0.25 mm² in the Lateral Geniculate Body after Optic Tract Section

At time of optic tract section	At time of sacrifice		C	t	M	t	R	t
Intact	6 years	Right	137 ± 7.54	1.42	199 ± 26.80	0.49	147 ± 8.12	1.30
		Left	157 ± 11.74		185 ± 8.30		162 ± 8.18	
1 year	6 years	Right	127 ± 10.53	2.37*	138 ± 12.56	4.04*	182 ± 11.70	0.85
		Left	180 ± 19.26		215 ± 12.24		190 ± 7.21	
15 days	6 years	Right	169 ± 13.89	0.62	196 ± 15.26	1.62	185 ± 12.92	2.62*
		Left	284 ± 16.94		250 ± 29.17		257 ± 24.33	
45 days	6 years	Right	146 ± 7.61	0.36	142 ± 12.68	3.12*	126 ± 8.94	4.49*
		Left	140 ± 14.66		221 ± 21.93		188 ± 10.58	

* $p < 0.05$

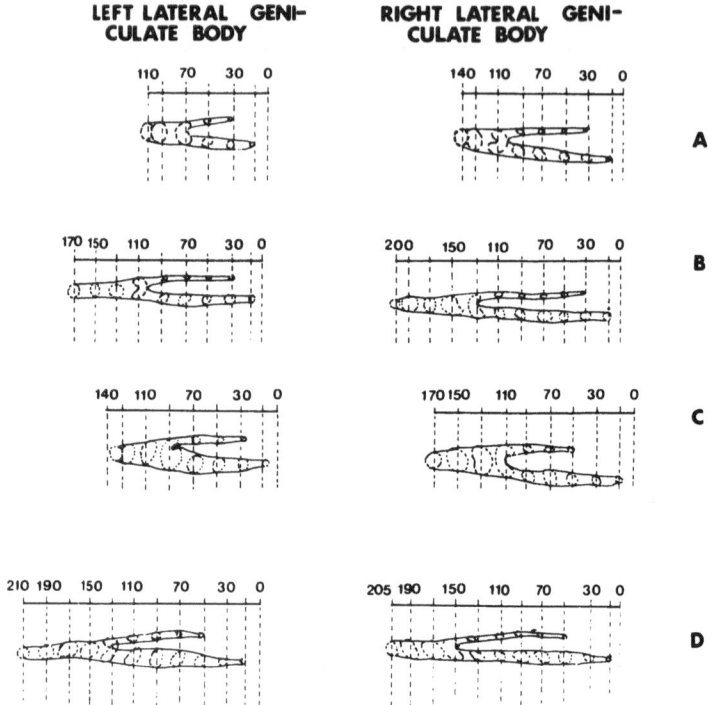

Figure 5.   Reconstitution of the volume of the lateral
             geniculate nucleus on the side of the optic
             transection (left) and the contralateral side
             (right).   The sections are represented by
             Arabic numbers.

cells, decreased particularly at the caudal level of LGN,
whereas the number of small cells decreased in all three
levels following the LGN denervation on the 15th postnatal day. A significant reduction in the number of large
cells in the caudal and medial levels in the medium-sized
cells of the medial and rostral, and in the small cells at
all levels was observed following optic tract transection
on the 45th postnatal day. In adult animals the changes
in cell volume in comparison with the preceding groups
were relatively small after LGN denervation. The large
cells decreased in the caudal and rostral, the medium-
sized cells in the caudal, and the small ones in the
rostral and caudal levels (Fig. 6).

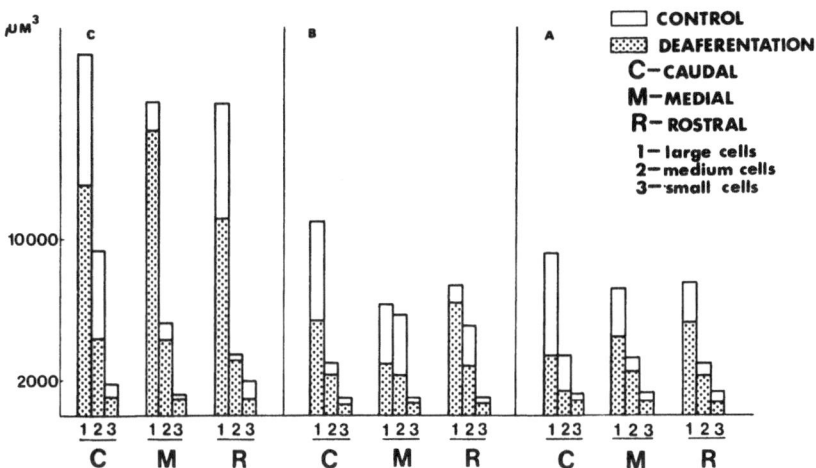

Figure 6.   Cell volume at the different levels of the
             lateral geniculate nucleus after deafferen-
             tation.

DISCUSSION

     The results of our investigations show that unila-
teral deafferentation of the LGN caused by optic tract
transection in dogs does not reduce the number of nerve
cells, which not only implies the existence of a compen-
satory phenomenon but also reflects a unique functional
organization of this structure. The number of cells on
the side of deafferentation increased, as shown by our
comparative quantitative-qualitative evaluations, because
of a decrease in the general volume of the structure and
in the individual volume of the cells themselves. Moreover, the unequal and regional relative increase in the
number of cells of the deafferented LGN in comparison
with the non-deafferented one seems to depend on the
stage of the LGN differentiation at the time of the denervation, as indicated by our data (Table 5).

     So far, the morphological investigations have shown
that, in contrast to the monkey, where deafferentation is
accompanied by massive atrophic changes in the LGN, visual

deprivation (21, 37) as well as deafferentation in the
dogs produces a decrease only in the volume of the nerve
cells without a marked reduction in their numbers in the
LGN (3, 7). Therefore, a certain degree of "stability" of
the LGN cells in relation to the peripheral deafferentation
exists in the dog. The degree and the duration period of
this "stability" are not indicated in the aforementioned
studies. Our results emphasize that this stability extends
over a long period, since severe degenerative changes were
absent in the LGN 5 and 6 years after deafferentation.
Postoperative changes stabilized after a few months and
did not progress; therefore, the decrease in cell volume
of this period represents the final functional-morphologica:
feature of the optic tract deafferentation. Otherwise a
cell loss would have been found, as previously described fo:
other species (3). This conclusion is prompted not only by
our own results but also by others who found, besides the
degenerated synapses, numerous normal ones in the LGN after
bilateral enucleations (28, 29, 30, 34, 36). Therefore,
these findings indicate that the deafferentation of LGN
induced by optic tract transection or enucleation causes
degeneration of the synapses bound to the axons contained
in the optic tract fibers. However, these results also
indicate that the LGN of the dogs receives other synapses
which maintain the functional-morphological integrity of
the nucleus. The decrease in LGN volume is based on the
decrease of the individual cellular volume and not on the
disappearance of their cell elements caused by the degeneration of the synapses arising from the optic tract fibers

The "stability" of the geniculate cells in peripheral
deafferentation observed in beasts (dogs and cats), which
does not exist in primates (1, 4, 12, 31) and man (2, 11)
is related to their specific visual receptor. The greater
number of synapses found in a single LGN neuron in beasts
(about forty) in comparison to that seen in primates (only
one in monkeys) (9) suggests the existence of a greater
number of afferent projections (particularly extravisual)
which may be responsible for the lack of cellular disappearance in the corpus geniculatum laterale after visual
deafferentation in this species.

The neurophysiological findings (26) described in dogs
and cats also suggest the existence of additional synaptic
connections with the LGN, since the evoked potentials
specifically induced by light stimuli and electrical stimulation of the contralateral optic tract do not disappear in
the side of the denervated nucleus. The LGN receives impulses from the opposite side through the commisure colli-
culi superior (25). Besides, the direct projections

received by the LGN from the reticular formation of the midbrain and brain stem are responsible for maintaining the tonic activity not only of their normal neurons but deafferented neurons as well (33).

REFERENCES

1. Clark , W.E. LeGros and Penman, C. G. (1934):   The projection of the retina in lateral geniculate body. Proc. Roy. Soc. 114: 291-314.

2. Clark , W.E. LeGros (1941):   The laminar organization and cell content of lateral geniculate body in monkey. J. Anat. 75: 419-433.

3. Cragg, B.G. (1971):   The fate of axon terminals in visual cortex during trans-synaptic atrophy of the lateral geniculate nucleus. Brain Res. 34: 53-60.

4. Figurina, I.I. (1965): Degeneracija naružnih kolen-častih tel posle enukleacii u šenkov i detenišej obezjen. AN SSSR Dokl. 161: 244-248.

5. Filimonov, J.N. (1933): Ueber die Variabilität der Grosshirnrindenstruktur. Mitt. III. Regio Occipitalis bei höheren und niederen Affen. J. f. Psych. und Neurol. 45: 69-137.

6. Garey, L.J., Fisken, R.A. and Powell, T.P.S. (1973): Effects of experimental deafferentation on cells in the lateral geniculate nucleus of the cat. Brain Res. 52: 363-369.

7. Garey, L.J., Fisken, R.A. and Powell, T.P.S. (1976): Cellular changes in the lateral geniculate nucleus of the cat and monkey after section of the optic tract. J. Anat. 121: 15-27.

8. Glees, P. (1961): Terminal degeneration and trans-synaptic atrophy in the lateral geniculate body of the monkey. In: Neurophysiologie and Psychopath. des visuellen Systems. Simposion Freiburg, 104-110.

9. Glees, P. and Clark, W.E. LeGros (1941): The termination of optic fibers in the lateral geniculate body of the monkey. J. Anat. 75: 295-309.

10. Glees, P., Hollerman, W. and Naeve, H. (1964):   Die
     Repräsentation Retinaler Sehstoren im Corpus Genicu-
     latum Laterale des Affen.   Albrecht Grafes Arch.
     Ophth. 167: 367-376.

11. Goldby, F.A. (1957):   A note on transneuronal atrophy
     on the human lateral geniculate body.   J. Neurol.
     Neuros. Psychiat. 20: 202-207.

12. Guillery, R.W. (1971):   Survival of large cells in
     the dorsal lateral geniculate after interruption of
     retinogeniculate afferents.   Brain Res. 28: 541-545.

13. Guillery, R.W. (1972):   Experiments to determine
     whether retinogeniculate axons can form translaminar
     collateral sprouts in the dorsal lateral geniculate
     neurons of cat.   J. Comp. Neurol. 146: 407-419.

14. Guillery, R.W. (1973):   Quantitative studies of trans-
     neuronal atrophy in the lateral geniculate nucleus of
     cats and kittens.   J. Comp. Neurol. 149: 423-438.

15. Hickey, T.L., Spear, P.D. and Kratz, K.E. (1977):
     Quantitative studies of cell size in the cat's dor-
     sal lateral geniculate nucleus following visual
     deprivation.   J. Comp. Neurol. 173: 265-282.

16. Hoffman, K.P. and Hollender, H. (1978):   Physiological
     and morphological changes in cells of the lateral
     geniculate nucleus in monoculary-deprived and reverse-
     sutured cats.   J. Comp. Neurol. 177: 145-158.

17. Jacobs, G.H. (1969):   Transneuronal changes in the
     lateral geniculate nucleus of the squirrel monkey
     Saimiri sciurens).   J. Comp. Neurol. 135: 81-84.

18. Klosovski, B.N. and Kosmarskaja, E.N. (1961):
     Dejatelnoje i tormoznoe sostojanie mozga.   Megdiz,
     Moskva.

19. Kononova, E.P. (1926):   Anatomija i fizilogija
     zatiločnih dolej.   Moskva.

20. Kononova, E.P. (1949):   In: Citoarhitektonika kori
     boljšogo mozga celovjeka.   Megdiz, Moskva.

21. Kuppfer, C. and Palmer, P. (1964):   Lateral genicu-
     late nucleus: histological changes following afferent
     degeneration and visual deprivation.   Exp. Neurol. 9:
     400-409.

22. LeVay, S. and Ferster, D. (1977): Proportion of interneurons in the cat's lateral geniculate nucleus. Relay cell classes in the lateral geniculate nucleus of the cat and the effects of visual deprivation. J. Comp. Neurol. 172: 563-584.

23. LeVay, S. and Ferster, D. (1979): Proportion of interneurons in the cat's lateral geniculate nucleus. Brain Res. 164: 304-308.

24. Levine, G.Z. (1957): Embionaljno razvitie naruzhnogo kolenachnogo tela u reptilii i mlekopitajushchich. Arch anat. gistol. i embriol. 34: 83-90.

25. Lyubimov, N.N. (1968): Mnogokanalnaja organizacia afferentnogo provedenia v analizatornih sistemah golovnogo mozga. Doct. dissert.

26. Lyubimov, N.N., Radulovački, M., Kado, R.T. and Adey, R. (1967): Zritelnie i sluhovie analizatori. Simposium Moskva.

27. Minkowsky, M.A. (1913): Experimentelle Untersuchungen über die Beziehungen der Grosshirnrinde und der Netzhaut zu den Primären optischen Zenter, besonders zum Corpus geniculatum externum. Arbeit aus dem Hirnanatomischen Institut, Zürich, 7: 255-362.

28. Newman, B.L. and Monroe, B.C. (1963): Electron microscopy of neuroglia and long term neuronal degeneration in the lateral geniculate body. J. Anat. Res. 145: 266-]67.

29. Obuhova, G.P. (1961): Sinapsi naruzhnogo kolenchastogo tela. Morfologia mezhneuronalnih svjazej. Moskva-Lenjingrad.

30. Peters, A. and Palay, S.L. (1966): The morphology of laminae A and A_1 of the dorsal nucleus of the lateral geniculate body of the cat. J. Anat. 100: 451-486.

31. Polyk, S. (1957): The Vertebrate Visual System. University of Chicago Press, Chicago.

32. Sarkisov, S.A. (1949): Citoarhitektonika kori bolshogo mozga. Moskva.

33. Scheibel, M.E. and Scheibel, A.B. (1966): Patterns of organization in specific and non-specific thalamic fields. In: The Thalamus. D.P. Purpura and N.D. Yohr (eds.), pp. 13-46, Columbia Univ. Press, New York.

34.   Smith, J.M., O'Leary, J.L., Harris, A.B. and Gay, A.J.
       (1964): Ultrastructural features of the lateral gen-
       iculate nucleus of the cat. J. Comp. Neurol. 123:
       357-387.

35.   Stipanović, L., Zvarikin, V.P. and Lyubimov, N.N.
       (1973): Effect of total unilateral deafferentation
       on the development of corpus geniculatum laterale of
       the dog. Acta med. iug. 27: 79-92.

36.   Szentàgothai, J. (1963): The structure of the synapse
       in the lateral geniculate body. Acta Anat. 55: 166-
       185.

37.   Wiesel, T.N. and Hubel, D.H. (1963): Effects of visual
       deprivation on morphology and physiology of cell in
       the cat's lateral geniculate body. J. Neurophys. 26:
       987-993.

38.   Wiesel, T.N. and Hubel, D.H. (1965): Projections
       from lateral geniculate. Extent of recovery from the
       effects of visual deprivation in kittens. J. Neurophys.
       28: 1060-1072.

ANATOMO-HISTOLOGICAL STUDIES OF CAUDATE NUCLEUS -

VISUAL SYSTEM RELATIONS

Lj. M. Rakić, Lidigja Lazarević,
N.N. Lyubimov and J.J. Ivanuš

Brain Research Laboratory
Institute for Biological and
Medical Research of Montenegro
Kotor, Yugoslavia

Brain Institute
Academy of Medical Sciences
Moscow, USSR

The caudate nucleus has an inhibitory role (3, 14) in the integrational function of the central nervous system. Electrophysiological and behavioral signs of inhibition might be produced by low frequency stimulation of the caudate and blocked by natural or artificial arousing stimuli (4). The inhibitory effects integrate into the basic neural processes of excitation and inhibition which interact with the impulses reaching the central nervous system trhough the various sensory modalities (5, 8), and especially the visual system (2, 15). Inhibitory electrographical responses in the electroencephalogram elicited by caudate stimulation ("caudate spindle") are blocked by the stimulation of the visual system. Similar relationships were observed in the other electrophysiological parameters as well (1, 15).

Behavioral experiments pointed out that the blocking of "caudate spindle" produced by the simultaneous stimulation of the caudate nucleus and the visual system relay sites, especially the lateral geniculate body, eliminates inhibitory effects in the conditioned reflex behavior (1, 15).

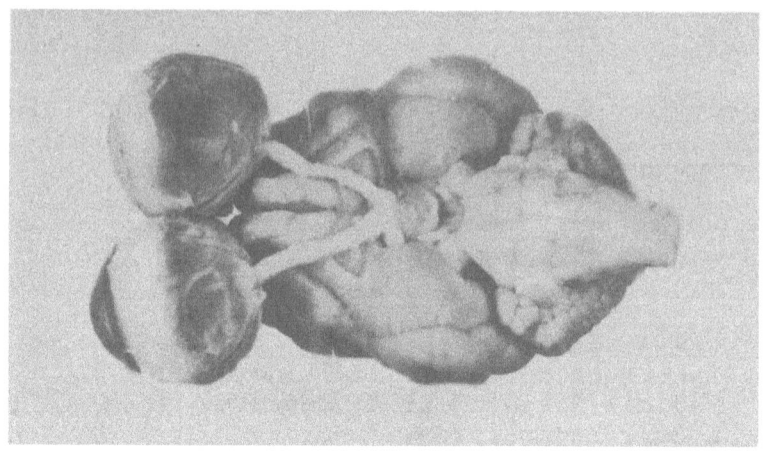

Fig. 1.   The base of the brain with the sectioned left
         optic tract.

It is known that the striatum receives fibers from
two main sources, the cerebral cortex (9, 19, 20) and the
thalamus, (7) as well as the smaller projections from the
midbrain (13) and the contralateral cortex (6, 10).   The
connections between striatum and visual system were
studied by evaluating the sequelae of unilateral visual
deafferentation in the caudate nucleus.

MATERIAL AND METHODS

Unilateral optic tract excision (1-2 mm in length)
was performed under Nembutal^R anesthesia and aseptic
conditions on 14 adult cats (Fig. 1).   Intact animals
served as controls.

Experimental animals (with the sectioned optic tract)
were divided into two groups with different survival times
(a) 10 days (early changes), (b) 18 months (late changes).
The planimetric circumference and shape of cerebral
structure were measured in experimental and control
animals using the Kononova method (11).

The animals were perfused through the aorta with
saline and 10% neutral formalin under Nembutal anesthesia.
The brains were immersed in formalin for 2-4 weeks and then
embedded in paraffin.   Coronal sections of the brains were

cut at 20 μm and one of five serial sections was stained
with Van der Loos' modified technique of Nissl for the
investigation of gross and micro cytoarchitecture of the
caudate nucleus (18). The pattern of the caudate nucleus
degeneration was determined on frozen sections stained with
the Nauta-Gygax method (12). Quantitative analyses of the
neurons and glial cells were made on three different levels
(rostral, medial, caudal) of the caudate nucleus in experimental and control animals.

The evaluation included cell counts in the macroscopically visible and nonvisible deformed regions of the
caudate nucleus. A comparison of both areas was made to
the contralateral caudate nucleus where the optic tract
was intact. Moreover, quantitative cellular analysis of
the caudate nucleus from each side of the brain was
carried out since left and right asymmetry was described
in corresponding left right cortical and subcortical
structures (16, 17).

RESULTS AND DISCUSSION

A. Early changes were manifested by marked focal
degeneration of the caudate nucleus on the side of the
sectioned optic tract while changes on the contralateral
side were slight or not present (Figs. 2, 3). In serial
sections, there were many degenerative fibers in the
head of the caudate nucleus but only a few in the tail.
These changes were absent in the body and on the narrow
tissue strip between the head and the body of the caudate
nucleus (Fig. 4).

B. Late changes. Grossly visible, deformed foci of
the caudate nucleus were seen on the side of the optic
tract transection (Fig. 5). Moreover on gross histological inspection, the caudate nucleus appeared to
contain fewer neurons and a greater number of glial cells
on the affected than on the contralateral side (Figs. 6,
7). Quantitative cellular analysis confirmed the gross
impression. The glia-neuron ratio was considerably
higher, due to a marked increase in the number of glial
cells and a decrease in the neurons (Table 1). A lower
ratio was found in the surrounding area (Table 1) and in
the unaffected regions (Table 2). The controls showed
an equal number of neurons and glial cells in the caudate
nucleus on either side of the brain (Table 3).

The grossly and macroscopically altered structures
of the caudate nucleus on the ipsilateral in contrast to
the contralateral side of the segmental optic tract

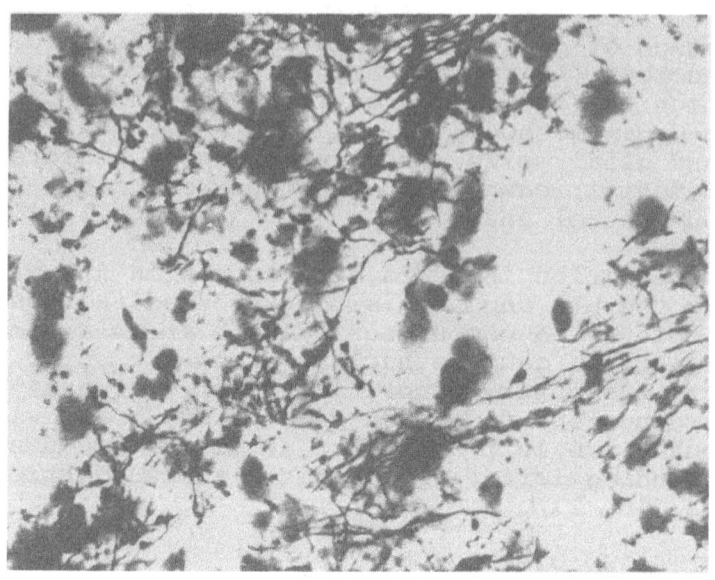

Fig. 2. Nauta-Gygax staining of the caudate nucleus of
         the contralateral side of the optic tract

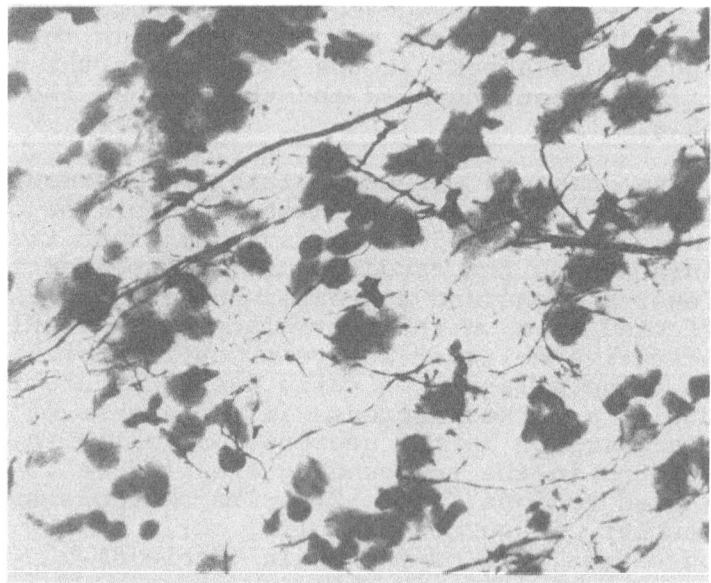

Fig. 3. Nauta-Gygax staining of the caudate nucleus on
         the side of the optic tract section.

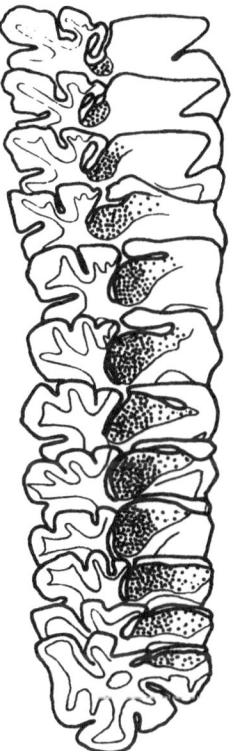

Fig. 4. Reconstruction of degenerated patches in the
         caudate nucleus on the side of the sectioned
         optic tract.

Fig. 5.   Reconstruction of gross caudate nucleus section.
           The arrows indicate the macroscopically visible
           deformations.

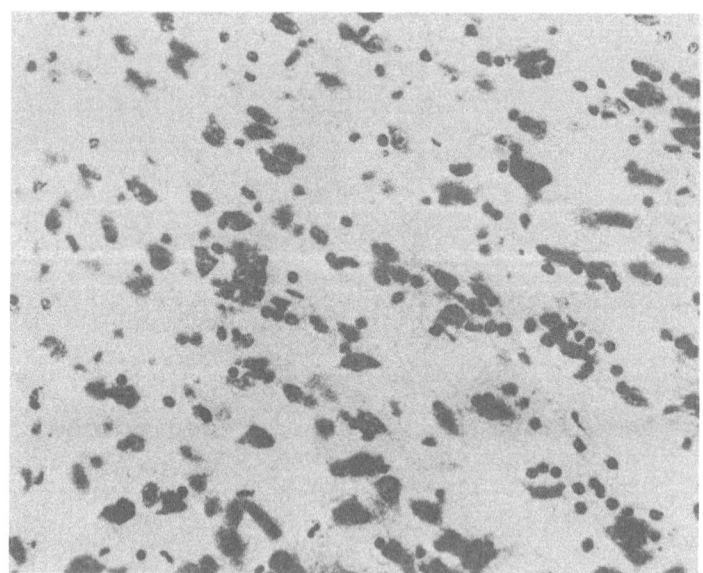

Fig. 6.   Nissl staining of the caudate nucleus on the
           side of the sectioned optic tract.

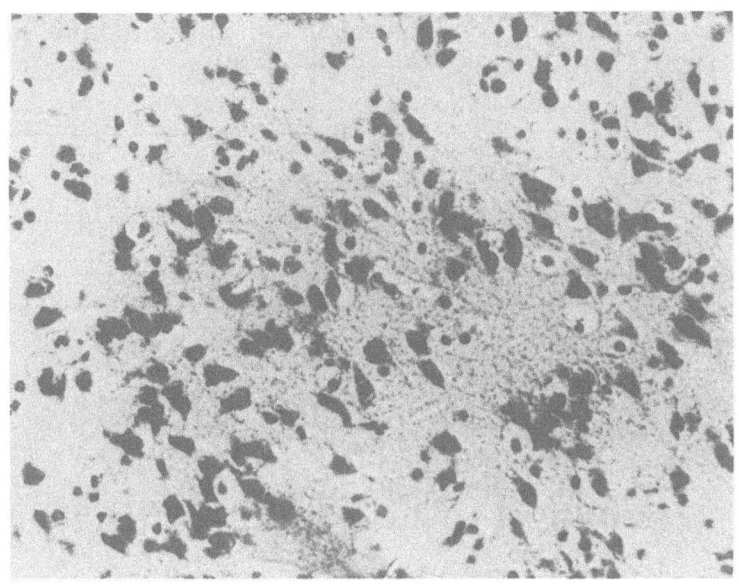

Fig. 7.   Nissl staining of the caudate nucleus contra-
          lateral to the sectioned optic tract.

excision indicate that the optic tract projections are
abundant on the ipsilateral side but few terminate on the
opposite side of the brain.

     In support of the observation that the optic tract
fibers extend from one side and terminate on the other
side is the higher ratio of glia to neuronal cells
(Index 1.7) found in the caudate nucleus contralateral
to the optic tract transection than the ratio of these
cells (Index 1.0) found in the controls.

     The regional quantitative analysis of caudate
nuclei serially sectioned demonstrates that the unilateral transection of the optic tract leads to focal
rather than diffuse degeneration, which might reflect the
specific mode of the optic tract fiber termination.

     These findings together with the described physio-
logical properties of the visual-striate complex (2, 4)
support the participation of the caudate nucleus in the
sensory integrational processes (1, 2, 5).

Table 1. Cellular composition of the caudate nucleus
          after optic tract section.

Cell Type	Left			Right			
	a	b	c	a	b	c	d
Neurons	36			44			
	42	35		54	41		
	28			27			
A.			3.9			1.7	2.3
Glia	188			72			
	122	137		66	71		
	100			75			
Neurons	76			41			
	51	64		48	46		
	66			60			
B.			1.9			1.5	1
Glia	81			68			
	100	98		83	72		
	112			79			

A. = Macroscopically visible deformed areas.
B. = Areas surrounding the deformed structures.
    a = Number of cells in each defined area.
    b = Mean cell number
    c = Glia - neuronal cellular ratio.
    d = The left to right ratio of the total number
        of cells

Table 2. The cellular composition of the caudate nucleus after optic tract section in grossly unaffected regions.

Cell Type	Left			Right			
	a	b	c	a	b	c	d
	56			57			
Neurons	64	58		63	58		
	53			53			
			1.8			1.2	1.5
	93			67			
Glia	108	104		62	69		
	110			78			

* Symbols same as in Table 1.

Table 3.   Number of cells in the caudate nucleus of
           intact cats.

Cell Type	Left	Mean	Right	Mean
	60		48	
Neurons	40	45	45	47
	47		45	
	47		45	
Glia cells	30	39	45	45
	40		45	

REFERENCES

1. Buchwald, N.A., Hull, C.D. and Trachtenberg, N.C.
    (1967): Concomitant behavioral and neural
    inhibition and disinhibition in response to
    subcortical stimulation. Exp. Brain Res. 4: 58-72.

2. Buchwald, N.A., Rakić, Lj., Wyers, E.J., Hull, C.
    and Heuser, G. (1962): Integration of visual
    impulses and the caudate loop. Exp. Neurol. 5:
    1-20.

3. Buchwald, N.A., Wyers, E.J., Lauprecht, C.W. and
    Heuser, G. (1961): The "caudate spindle". IV
    A behavioral index of caudate-induced inhibition
    Electroencephalog. clin. Neurophysiol. 531-537.

4. Buchwald, N.A., Wyers, E.J., Okuwa, T. and Heuser,
    G. (1961): The "caudate spindle". I. Electro-
    physiological properties. Electroencephalog. clin.
    Neurophysiol. 13: 509-518.

5. Buchwald, N.A., Wyers, E.J. and Rakić, Lj. (1962):
    The caudate nucleus and inhibition. Integration
    of sensory inputs and caudate inhibition. Proc.
    int. Union Physiol. Sci. Leiden (Abstract) 2: 1102.

6. Carman, J.B., Cowan, W.M. and Powel, T.P.S. (1963):
    The organization of the cortico-striate connections
    in the rabbit. Brain 86: 525-562.

7. Droogleever-Fortuyn, J. (1953): Anatomical basis
    of cortico-subcortical interrelationships.
    Third International EEG Congress, Symposia,
    pp. 149-162.

8. Ervin, F.R. and Mark, V.H. (1961): Thalamic
    organization of pain in man. Fed. Proc. 20: 345.

9. Kemp, J.M. and Powel, T.P.S. (1970): The cortico-
    striate projection in the monkey. Brain 93:
    525-546.

10. Kemp, J.M. and Powel, T.P.S. (1971): The site of
    termination of afferent fibers in the caudate
    nucleus. Phil. Trans. R. Soc., London, B., 262:
    413-427.

11. Kononova, E.P. (1949): In: Citoarhitektonika kori
     bolshogo mozga cheloveka. Megdiz, Moskva.

12. Nauta, W.J.H. and Gygax, P.A. (1954): Silver im-
     pregnation degenerating axons in the central nervous
     system: a modified technic. Stain Technol. 29:
     91-93.

13. Nauta, W.J.H. and Mehler, W.R. (1969): Fiber
     connections of the basal ganglia. In: Psychotropic
     Drugs and Dysfunction of the Basal Ganglia. G. E.
     Crane and R. Gardner (eds.), pp. 68-74. Public
     Health Service Publ., Washington, D.C.

14. Rakic, Lj., (1966): Cortical inhibition and sub-
     cortical inhibitory influence. In: Impact of Basic
     Sciences on Medicine. B. Shapiro and M. Pryves
     (eds.), Acad. Press, New York - London, pp. 298-305.

15. Rakic, Lj. (1978): Systems regulating behavior.
     Acad. Sci. and Art Kosovo Publ. Prishtine
     pp. 1-120.

16. Sarkisov, S.A. (1948): Nekotorie osobenosti
     stroenia nejronalnih svjazej kori bolshogo mozga.
     AMN SSSR, Moscow.

17. Stipanovic, L., Zvarikin, V.P. and Lyubimov, N.N.
     (1973): Influence of total unilateral deafferenta-
     tion on development. Acta med. Iugosl. 27: 79-92.

18. Van der Loos, H. (1956): Une combination de deux
     vicilles methodes histologiques pour le systeme
     nerveux centrale. Monatschrift fur Psychiatrie
     und Neurologie, 132: 330-334.

19. Webster, K.E. (1961): Cortico-striate inter-
     relations in the albino rat. J. Anat. 95: 532-544.

20. Webster, K.E. (1965): The cortico-striate
     projection in the cat. J. Anat. 99: 329-337.

DEVELOPMENT OF THE MEDULLO-SPINAL CEREBROSPINAL

FLUID-CONTACTING NEURONS

B. Vigh, I. Vigh-Teichmann and
R. Olsson

2nd Department of Anatomy
Medical University Budapest, Hungary, and
Zoological Institute of the University
Stockholm, Sweden

The neurons which directly contact cerebrospinal fluid have been shown around the central canal of the medulla oblongata and of the spinal cord of vetebrates (14). Their perikarya and axons do not differ from "ordinary" nerve cells, but their dendrites characteristically penetrate the ependyma and directly contact the cerebrospinal fluid (CSF). The dendrites, which are covered with many straight-standing stereocilia and solitary kinocilia, terminate freely in the central canal (16, 17, 19).

As early as 1909, Held described that some neuronal processes terminate freely in the central canal (4). The bipolar neuroblasts were shown to extend one process to the ependyma and the other one to the myotomes, in the embryos of the pig, lizard and lamprey (4). Hence, the early neuroblasts of motor neuron cells have features similar to the CSF-contacting neurons that were first described by Tretjakoff (12, 13). However, Edinger (3) and Stendall (11) described a population of ependymal neurons which were similar to the primary sensory cells of the spinal cord and also sent processes into the lumen. Subsequently, Kolmer (5) and Agduhr (1) found primary sensory cells around the central canal in more than 200 vertebrate species. These neurons were also present in adults but were dissimilar to the neuroblasts. According to Agduhr (1), the ependymal processes of most

neuroblasts in the embryos resemble those of the primary
sensory cells. However, during development the neuroblast
migrates away from the central canal lumen and forms
various types of neurons found in the spinal cord.   Only
a portion of the remaining neuroblasts develop into
intraependymal and subependymal primary sensory cells
(1).   This concept is consistent with the phylogenetic
development of the neural tube from a simple sensory
epithelium of the primitive deuterostomian chordates
and vertebrates (11).

Ultrastructurally, the terminal dendrites of the
CSF-contacting nerve cells were found to be similar to
those of some mechanoreceptors in the inner ear and in
the lateral line organ.   It is possible that the CSF-
contacting neurons may function as a mechanoreceptor
and perhaps sense the intensity of the CSF flow.   These
receptors could maintain the CSF flow as it passes down
the central canal to the terminal filum and then into
the subarachnoid space and connective tissue (2, 7).
Kolmer postulated a functional correlation between
Reissner's fibers and these terminals.   The fibers
originate from the condensed secretory material of the
subcommissural organ and terminate freely in the central
canal (5, 6).   It was further suggested that there is a
mechanoreceptive function for the intraependymal cells,
but that the specific stimulation was secondary to the
dislocation of the Reissner's fiber, which in turn
excited the terminals of the sensory cells.

The observations from the scanning electron microscope that the Reissner's fiber sometimes touches the
cilia of CSF-contacting neurons indicates a possible
functional relationship analogous to that between the
tectorial membrane and the hair cells of the organ of
Corti.   This view is strengthened by the structural
similarity between Reissner's fiber and the tectorial
membrane (15).   Since the Reissner's fiber leaves the
central canal at the caudal opening of the terminal
filum, it may facilitate the outflow and clearing of
the CSF (9, 10).

In regard to the mechanical role of the Reissner's
fiber, the movement of the CSF has to be considered.
The ependymal ciliary motion propels the CSF downward,
and therefore the fluctuations of the CSF flow and/or
of the ciliary motion may affect the fiber position.
Such a dislocation may then mechanically stimulate the
receptor endings of the CSF-contacting neurons.   However, this present hypothesis requires verification.

In our present work, we studied the ultrastructural relation of this problem in the lamprey (Lampetra fluviatilis). We found that the fine structure of its spinal CSF-contacting neurons was similar to that of CSF-contacting neurons of higher verbetrates (Fig. 1). In the lamprey, the stereocilia are rather short, but the presence of the many striated rootlet fibers underlines the importance of the solitary kinocilium motility of the CSF-contacting dendrite terminal (Fig. 2).

The perikarya of the CSF-contacting neurons contain granular vesicles of up to 2000 Å in diameter, partly resembling in size the peptidergic neurosecretory granules (Fig. 3). It is noteworthy that two kinds of perikarya were demonstrated in higher vertebrates, one being AChE-positive and the other showing induced monoamine fluorescence (16). These findings may suggest a double mechanism, a cholinergic or peptidergic and a monoaminergic mediated transmission of the stimuli received by the CSF-contacting neurons.

To the question where is the information being transmitted to, the following experiments were performed.

Using silver impregnation method, the CSF-contacting neurons in the turtle were shown to converge at a fiber bundle on both sides of the central canal and then run to the surface of the spinal cord. This fiber bundle, the "centro-superficial tract," extends to the surface of the cord where the axons terminate on the basal lamina (20). The axon terminals which are of considerable size were also found in Lampetra fluviatilis (Fig. 4) and in Lampetra japonica (8). The superficial axons form a terminal zone ventrolaterally at both sides of the spinal cord. The axons pierce the membrana limitans gliae superficialis formed by ependymal-glial endfeet and directly contact the space surrounding the cord. In the lamprey, the ependymal endfeet are attached to the basal lamina by exceptionally well developed structures (Fig. 5). The axon terminals on the surface of the spinal cord are similar to those of the known neurosecretory release sites, such as the urophysis, neurohypophysis, median eminence, epiphysial pulvinal and vascular organ of the terminal lamina (18). Therefore, they probably represent a release site for substances produced in the CSF-contacting neuronal perikarya. If these bioactive materials act through the external CSF (e.g., on vessels), they could regulate the blood flow

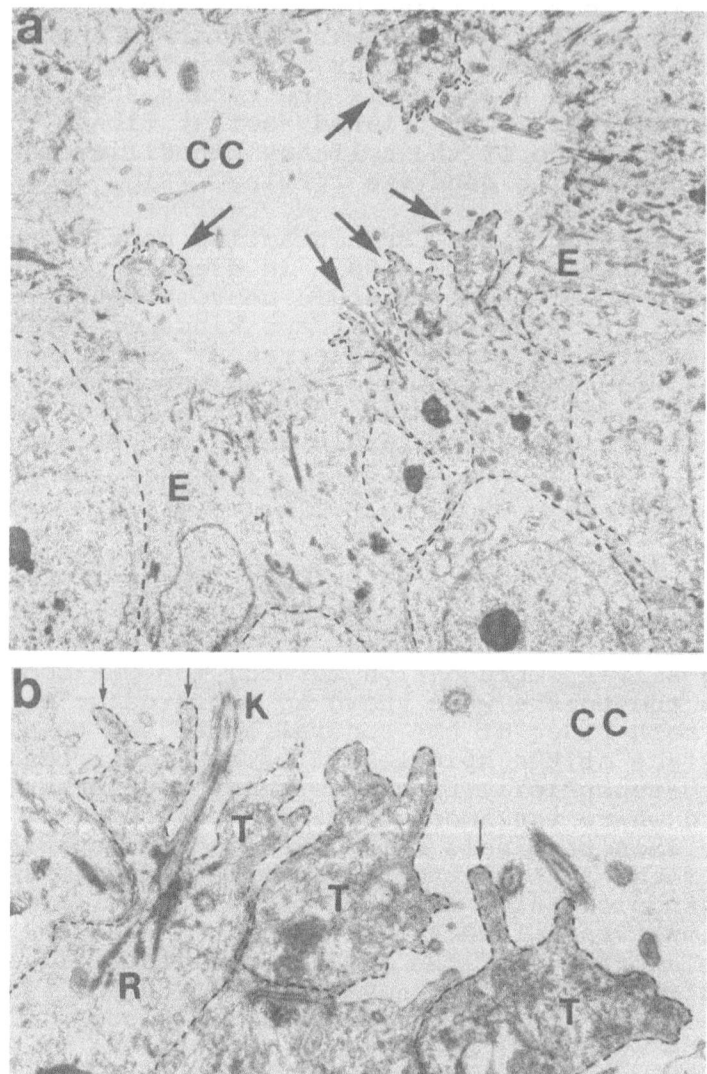

Fig. 1.(a-b). CSF-contacting neurons (plasmalemma dotted) around the central canal (CC) of the spinal cord in the lamprey, Lampetra fluviatilis. a) Arrows indicate the CSF-contacting dendrite terminals. E ependyma. x 3,400.
b) Three CSF-contacting dendrite terminals (T, plasmalemma dotted) with stereocilia (arrows) and kinocilium (K). R striated rootlet fiber. x 15,700. (Reproduced at 20% reduction.)

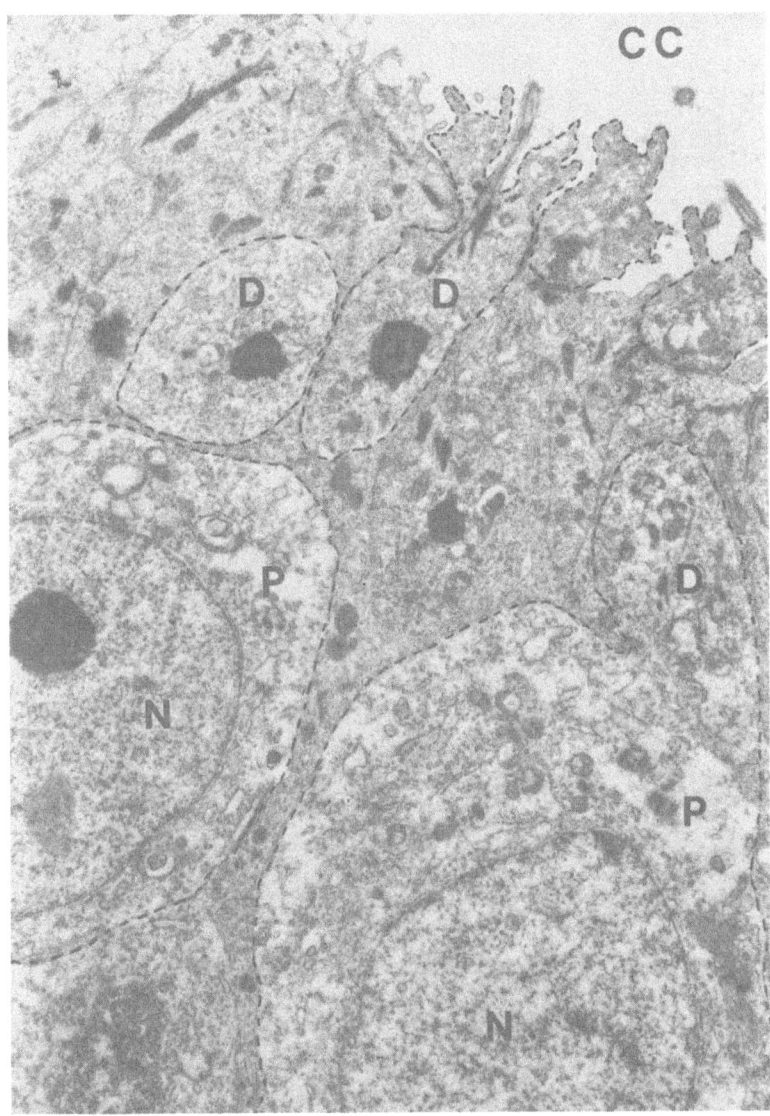

Fig. 2. Perikarya (P) of CSF-contacting neurons (plasma-lemma dotted) in Lampetra fluviatilis. CC central canal, D dendrites (dotted), N nucleus. x 7,650. (Reproduced at 10% reduction.)

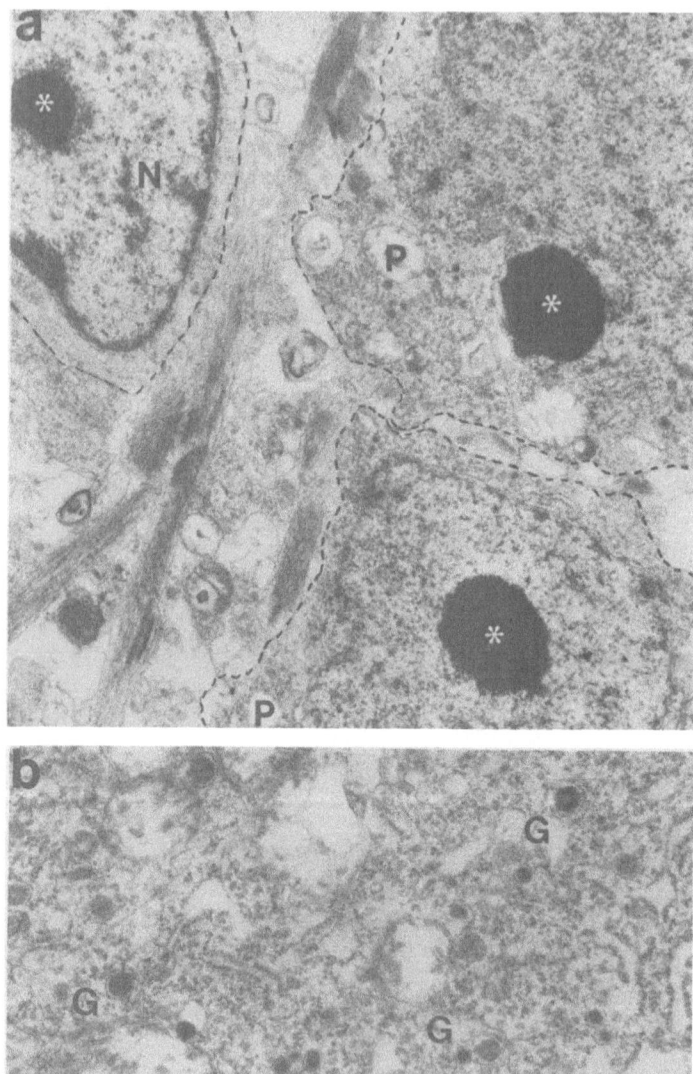

Fig. 3.(a-b). Perikarya around the central canal in the
lamprey. a) On the right two neuronal perikarya (P). On
the left ependymal perikaryon. Nucleoli (asterisks). N
nucleus of the glial cell. x 10,200. b) Detail of perikaryon of CSF-contacting neuron containing granular vesicles (G). x 25,560.
(Reduced 20% for reproduction.)

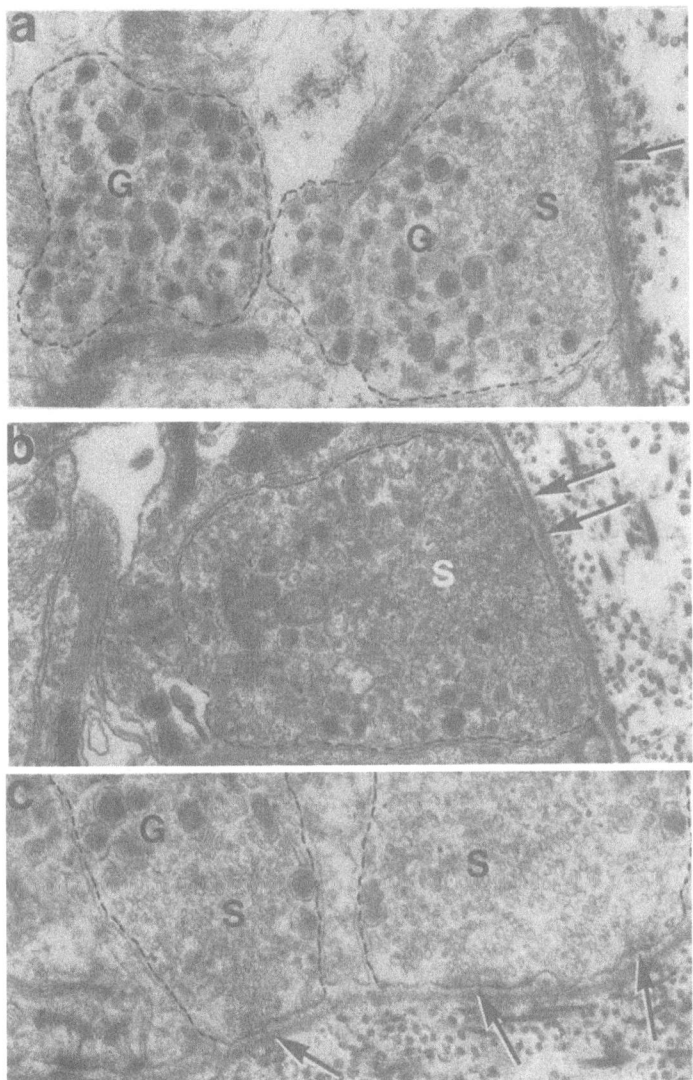

Fig. 4.(a-c). Axons (plasmalemma dotted) of CSF-contacting neurons terminate on basal lamina (arrows) of the spinal cord in the lamprey. G granular vesicles, S synaptic vesicles. a) On the left cross section of the axon below the external surface of spinal cord. x 25,560. b,c) Accumulation of synaptic vesicles and dense projections near the basal lamina (arrows). x 25,560, x 37,200.
(Reduced 15% for reproduction.)

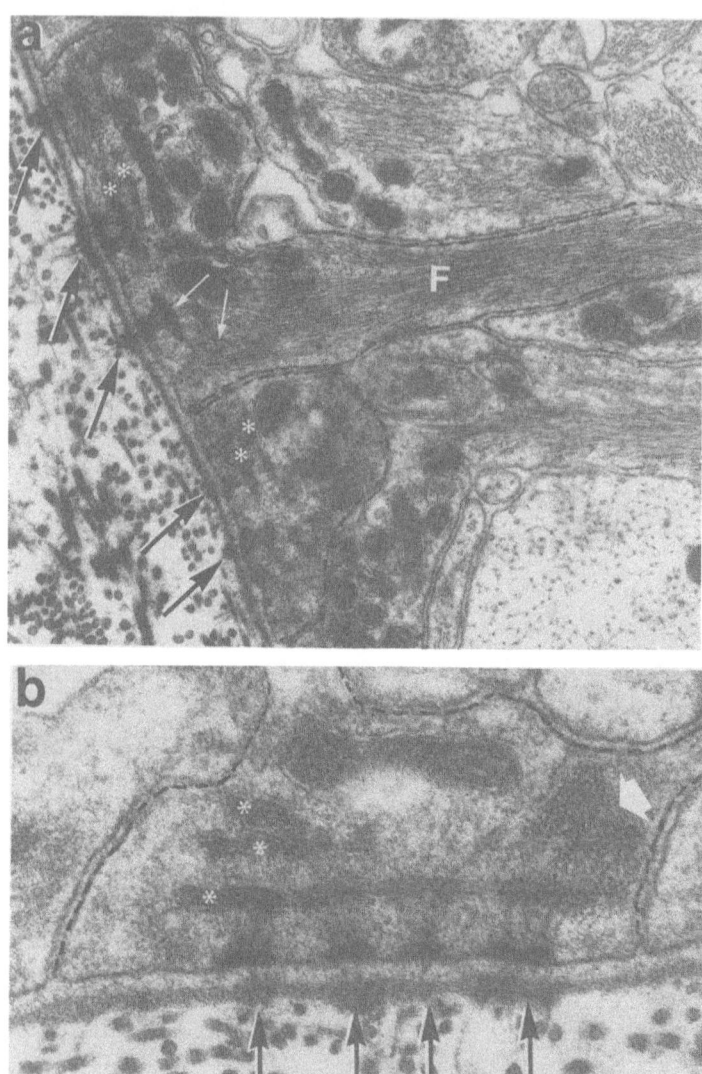

Fig. 5.(a-b). Basal processes of ependymal cells (plasma-lemma dotted) in the spinal cord of the lamprey. Note thickening of basal lamina (black arrows) opposite to hemidesmosomes of the ependymal endfeet. Several rows of filaments bound to hemidesmosomes (asterisks). a) White arrows show connection between longitudinally running filaments (F) of the ependymal process and hemidesmosome. x 25,560. b) Note cross sections of filament bundles (big arrow). High magnification, x 64,800.
(Reduced 20% for reproduction.)

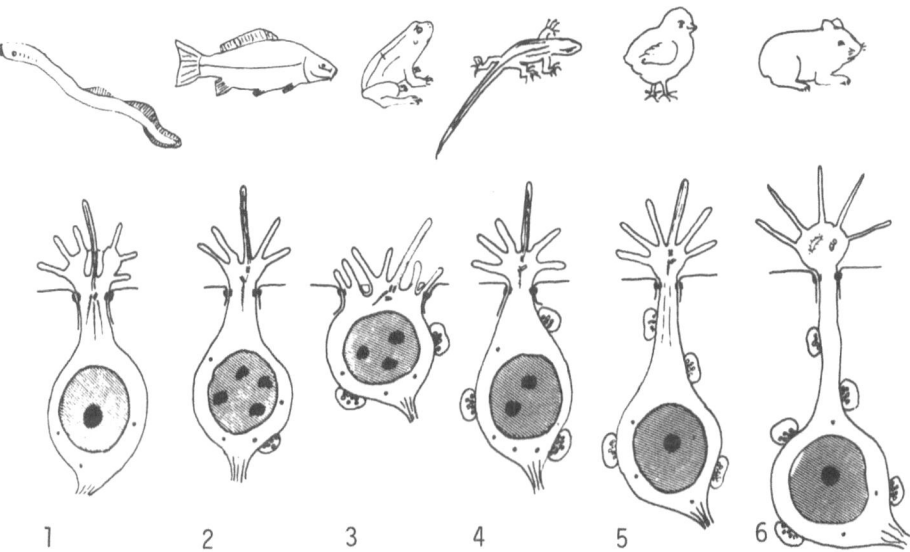

Fig. 6. Graphic sketch of the spinal CSF-contacting neurons in the lamprey (1) and in various vertebrates (2-6). The CSF-contacting dendrite terminals of the cells are charac-terized by numerous sterocilia and solitary cilia. Affer-entations (neuronal synapses) are more numerous in higher vertebrates.
(Reduced 10% for reproduction.)

of the spinal cord and secondarily the characteristics
of the CSF.

SUMMARY

The ultrastructure of the medullo-spinal CSF-contacting neurons studied in the lamprey (Lampetra fluviatilis) is similar to that of the CSF-contacting nerve
cells of higher vertebrates (Fig. 6). The CSF-contacting
dendrite terminals of the lamprey have well developed
solitary kinocilia and less well-developed stereocilia,
and therefore may serve as receptors to sense the flow
of the CSF. A sensory role for the CSF-contacting
neurons in conjunction with the Reissner's fibers is
proposed. The neurosecretory axon terminals of the
CSF-contacting neurons may release a bioactive substance
into the external CSF at the ventrolateral surface of the
spinal cord. thus transmitting the information sensed by
the CSF-contacting dendrites from the central canal.

REFERENCES

1. Agduhr, E. (1922): Über ein zentrales Sinnesorgan
    (?) bei den Vertebraten. Z. Anat. Entwickl.-
    Gesch. 66: 223-360.

2. Bradbury, M.W.B., Davson, H., Lathem, W. (1964):
    A flow of cerebrospinal fluid along the central
    canal of the spinal cord of the rabbit. J. Physiol.
    London 172: 16-17.

3. Edinger, L. (1906): Einiges vom "Gehirn" des
    Amphioxus. Anat. Anz. 28: 417-428.

4. Held, H. (1909): Die Entwicklung des Nervengewebes
    bei den Wirbeltieren. Leipzig.

5. Kolmer, W. (1921): Das "Sagittalorgan" der
    Wirbeltiere. Z. Anat. Entwickl.-Gesch. 60:
    652-717.

6. Kolmer, W. (1931): Über das Sagittalorgan, ein
    zentrales Sinnesorgan der Wirbeltiere, insbesondere
    beim Affen. Z. Zellforsch. 13: 1236-248.

7. Nakayama, Y. (1976): The openings of the central
    canal in the filum terminale internum of some
    mammals. J. Neurocytol. 5: 531-544.

8. Ochi, J. and Hosoya, Y. (1975): Monoamine neurons in the brain and spinal cord of the lamprey, Lampetra japonica. A fluorescence and electronmicroscopic study. In: Proc. 10th Intern. Confer. Anat. E. Yamada (ed.), p. 141, Tokyo.

9. Olsson, R. (1955): Structure and development of Reissner's fibre in the caudal end of Amphioxus and some lower vertebrates. Acta Zool. Stockholm 36: 167.

10. Olsson, R. (1969): Phylogeny of the ventricle system. In: Zirkumventrikuläre Organe und Liquor, G. Sterba, (ed.), pp. 291-305, VEB Fischer, Jena.

11. Stendell, W. (1914): Zur Histologie des Rückenmarkes von Amphioxus. Anat. Anz. 46: 258-267.

12. Tretjakoff, D. (1909): Das Nervensystem von Ammocoetes. II. Gehirn. Arch. Mikr. Anat. 74: 636-779.

13. Tretjakoff, D. (1913): Die zentralen Sinnesorgane bei Petromyzon. Arch. Mikr. Anat. 83: 68-117.

14. Vigh, B., Teichmann, I. and Aros, B. (1969): Das Paraventrikularorgan und das Liquorkontakt-neuronensystem. Anat. Anz. Suppl. 125: 683-688.

15. Vigh, B., Teichmann, I. and Aros, B. (1976): Hogan látjuk ma a neuroszekréciós sejtet? In Hungarian: "A today concept on the neurosecretory cell." In: A Biológia Aktuális Problémái. pp. 101-151.

16. Vigh, B. and Vigh-Teichmann, I. (1971): Structure of the medullo-spinal liquor contacting neuronal system. Acta Biol. Acad. Sci. Hung. 22: 227-243.

17. Vigh, B. and Vigh-Teichmann, I. (1973): Comparative ultrastructure of the CSF contacting neurons. Int. Review Cytol. 35: 189-251.

18. Vigh-Teichmann, I. and Vigh, B. (1979): A comparison of epithalamic, hypothalamic and spinal neurosecretory terminals. Acta Biol. Acad. Sci. Hung. 30: in press.

19. Vigh, B., Vigh-Teichmann, I., Koritsanszky, S.
     and Aros, B. (1970): Ultrastruktur der Liquor-
     kontaktneurone des Rückenmarkes von Reptilien.
     Z. Zellforsch. 109: 180-194.

20. Vigh, B., Vigh-Teichmann, I. and Aros, B. (1977):
     Special dendritic and axonal endings formed by the
     cerebrospinal fluid contacting neurons of the
     spinal cord. Cell Tiss. Res. 183: 541-552.

PHYLOGENETIC ASPECTS OF THE SENSORY NEURONS OF
THE WALL OF THE DIENCEPHALON

I. Vigh-Teichmann, B. Vigh, P. Rohlich and
R. Olsson

Second Department of Anatomy, Medical University
Budapest, Hungary, and
Zoological Institute of the University
Stockholm, Sweden

Physiological investigations suggest the presence
of light receptors, thermoreceptors, as well as sodium
and glucose receptors in the wall of the diencephalon
(22). Morphological data likewise indicate the occurrence
of sensory cells in the hypothalamus (12); for instance,
the so-called liquor-contacting neurons (13, 14, 19) and
the coronet cells of the vascular sac of fishes (16). The
photoreceptor cells of the retina and the similarly built
pinealocytes develop from the diencephalon as well.

In this paper, we deal with the cerebrospinal fluid
(CSF)-contacting neurons and the pinealocytes. Our main
aim was to characterize cells as to whether they are
sensory at all and to what type of receptor they may
belong.

CSF-CONTACTING NEURONS

The CSF-contacting neurons have neuronal elements
which are in direct contact with the cerebrospinal fluid
(CSF). All the other nerve cells respect the CSF-brain
barrier formed by the ependyma. Therefore, it is peculiar
that the CSF-contacting neurons penetrate the glial barrier to contact the CSF directly. This particular histological feature suggests a special function for these
cells.

Three kinds of CSF-contacting nerve cells are distinguished, by contacting the CSF either with one of
their dendrites (I) or with their axon (II) or even with
their cell bodies (III).

I.   The CSF-contacting neurons whose dendrites
enter the ventricle are the most likely to be sensory
neurons since dendrites are generally considered to be
the receptor part of the neuron (9).   It is interesting
that these dendrites form bulb-like terminal enlargements
in the CSF.   Further, the dendrite terminals contain
solitary cilia.   The presence of ciliated terminals is
characteristic of many known sensory cells.   The cilia of
the CSF-contacting dendrites differ from those of ependymal cells by their shortness and the absence of the
central pair of tubules (cilium of type 9 x 2 + 0).   We
found CSF-contacting neurons forming such ciliated dendrite endings in several hypothalamic nuclei: the parvocellular and magnocellular preoptic nucleus of lower
vertebrates, the magnocellular paraventricular nucleus of
reptiles, the anterior periventricular nucleus of fishes,
amphibians and reptiles, the paraventricular organ and
the infundibular nucleus from fishes up to birds, the
nucleus lateralis tuberis and the vascular sac of fishes
(14, 18, 19, 21-23).

In conclusion, we observed such ciliated CSF-contacting neurons in all classes of submammalian vertebrates.
CSF-contacting neurons were also described in Cyclostomata by light and scanning electron microscopy (6, 10).
Recently, we investigated the transmission electron
microscopy of the parvocellular preoptic nucleus in the
lamprey, Lampetra fluviatilis (Cyclostomata).   We found a
large number of CSF-contacting neurons in the periventricular gray matter.   They are situated intraependymally,
subependymally and farther away from the ventricle.   Subependymal CSF-contacting neurons may occur closely, side
by side (Fig. 1).   In general, the CSF-contacting dendrite terminals are relatively large (diameter up to 6 μm)
compared to those of the corresponding area in the newt
(23).   Their solitary cilia are short, provided with basal
bodies and multiple rootlet fibers, and belong to type
9 x 2 + 0, 8 x 2 + 1 (Fig. 2a, b, c).   The ciliary shaft
appears twisted and consists of three parts (Fig. 2b).
The initial part (a) has an electron-dense plasmalemma.
The transitional zone (b) is slightly tapered and shows a
strongly electron-dense area inside the circle of tubules
(Fig. 2b).   The distal part (c) of the ciliary shaft has
a wavy plasmalemma surface (Fig. 2b).   Thus, the CSF-contacting dendrites of the lamprey resemble those of other
species (23, Fig. 3a).

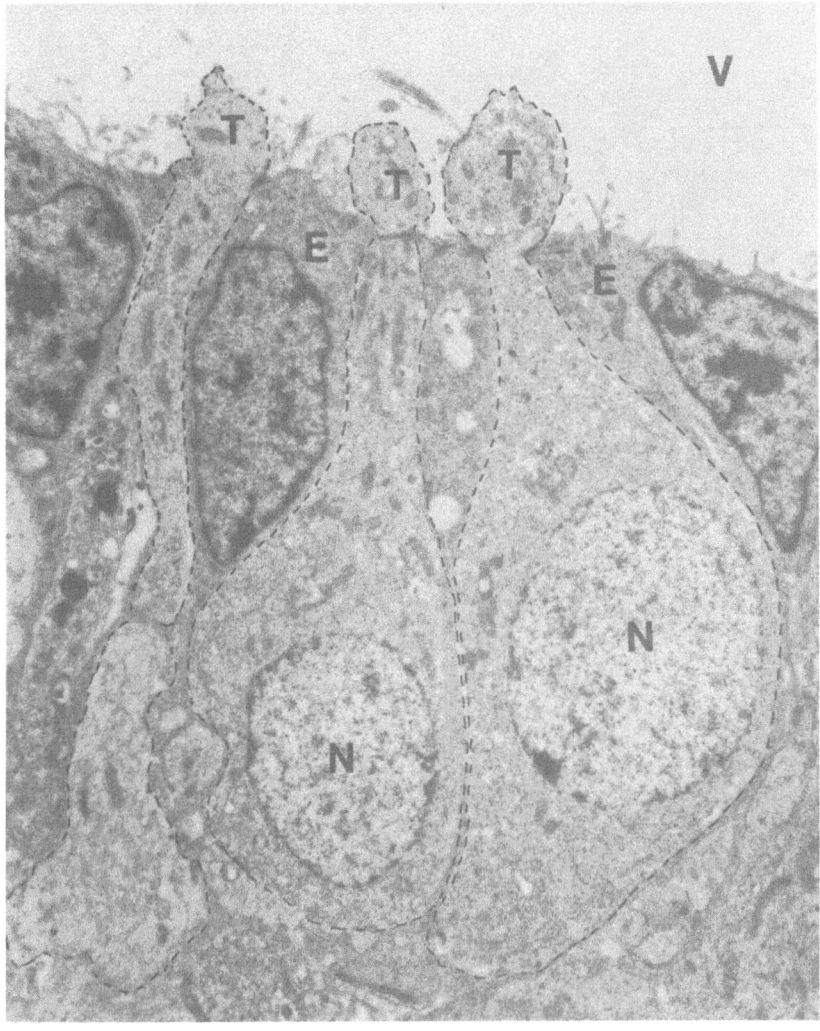

Fig. 1. Subependymal CSF-contacting neurons (plasmalemma dotted) of the parvocellular preoptic nucleus of lamprey form bulb-like dendrite terminals (T) in 3rd ventricle (V). E ependyma, N nucleus. x 4.800.
(Reduced 10% for reproduction.)

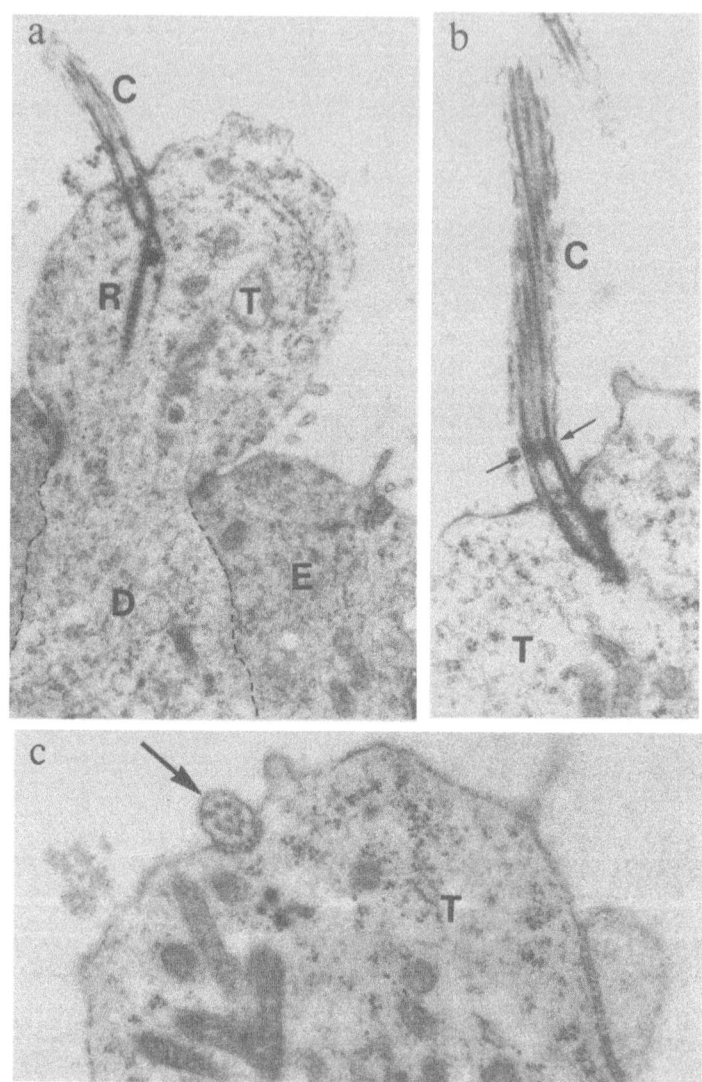

Fig. 2.(a-c). Details of CSF-contacting dendrite terminals of the preoptic nucleus of the lamprey. a) Short solitary 9 x 2 + 0 cilium (C) provided with basal bodies and rootlet fibers (R). x 15.720. b) The ciliary shaft consists of initial part with electron-dense plasmalemma, of slightly tapered transitional zone (arrows) and of sometimes twisted, distal part with wavy plasmalemma. Arrows indicate the electron-dense area inside transitional zone. x 22.800. c) 8 x 2 + 1 cilium (arrow) of the CSF-contacting dendrite. x 25.560
(Reduced 15% for reproduction.)

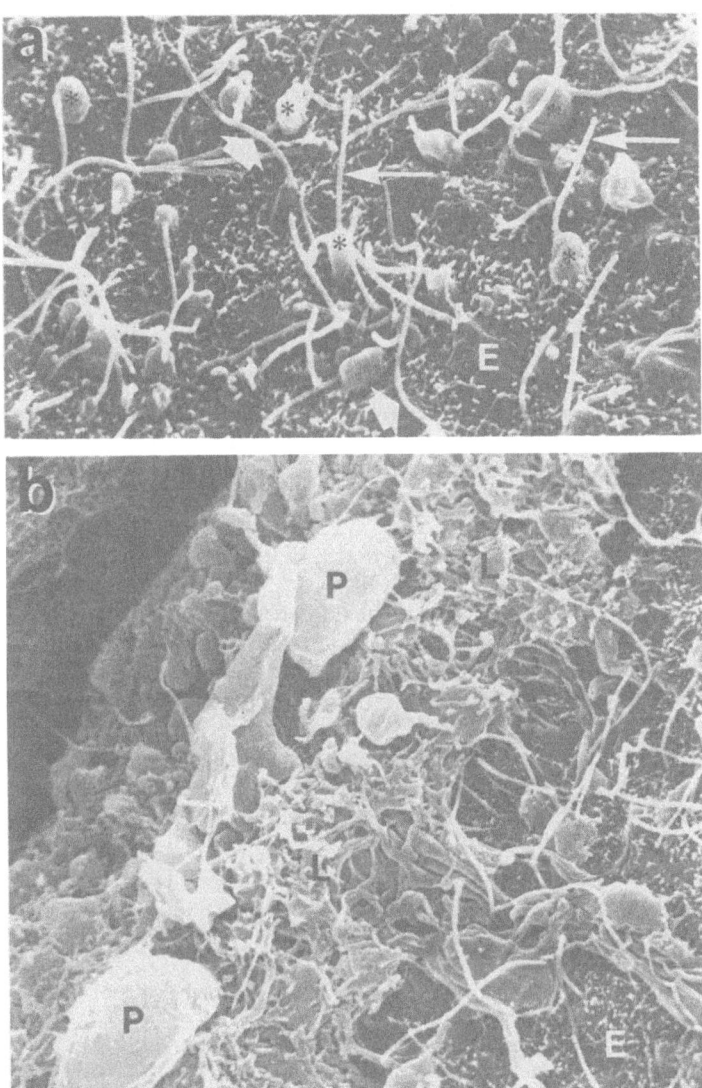

Fig. 3.(a-b). Scanning electron microscopy of the ventricular surface of the infundibular nucleus (a) and the paraventricular organ (b) in the newt. a) At asterisks, bulblike CSF-contacting dendrite terminals. Their cilia (arrows) are shorter than ependymal kinocilia (large arrows). E ependyma. x 7.200. b) Round-shaped CSF-contacting neuronal cell bodies (P) on top of the labyrinthic supraependymal layer (L) of the CSF-contacting dendrites. E ependyma. x 5.500.
(Reduced 20% for reproduction.)

Nerve cells of similar appearance were also reported in the lancelet (7) and the starfish (3). These data strengthen our view that the CSF-contacting neurons are phylogenetically old elements derived perhaps from the superficial sensory epithelim of primitive deuterostomia (11). In higher deuterostomia (chordates, vertebrates), the neuronal epithelium forms a neural tube of which the sensory terminals become inverted to the inner surface of the ventricle, resulting in the formation of CSF-contactin neurons.

II. The periventricular CSF-contacting neurons con-tact the CSF with their axons, which may be in either the internal or the external CSF space. The axons contacting the external CSF space belong to the neurosecretory cells, and they represent neurohormonal release sites of the brain (organon vasculosum laminae terminalis, median eminence, neurohypophysis, subpineal region, medullo-spinal neurosecretory areas, urophysis). The other type of CSF-contacting axon enters the ventricle. These intra-ventricular axons may terminate on the apical surface of the ependyma, on the ciliated CSF-contacting dendrites or on the intraventricular cell bodies representing the third kind of CSF-contacting neurons (19, 20, 23).

III. The CSF-contacting neuronal cell bodies are of various shapes: bipolar, pseudounipolar, multipolar (Fig. 3b). They give rise to ramifying processes (Fig. 4) which contribute to the intraventricular plexus of nerve fibers, already detected in monkeys in 1930 (5). Some of the processes are dendrites, others axons. The latter are thin, straight and exhibit bouton-like enlargements that terminate on the ependyma and on CSF-contacting neuronal elements (18, 19, 20, 23).

PINEALOCYTES

The pinealocytes of lower vertebrates resemble CSF-contacting neurons by their photoreceptor terminals pro-truding into the pineal recess, a diverticle of the third ventricle (15). In mammals, the lumen of the pineal organ disappears, but we found that the mammalian pinealocytes are structurally analogous to the photoreceptor cells, and not to the glial cells as thought by earlier authors (17). Also, the mammalian pinealocyte disposes of an axon with synaptic ribbons and of a rudimentary "inner segment". The "outer segment" is represented by a 9 x 2 + 0 cilium which lacks membrane specialization (17). It is an open question whether the pinealocytes of mammals can percept

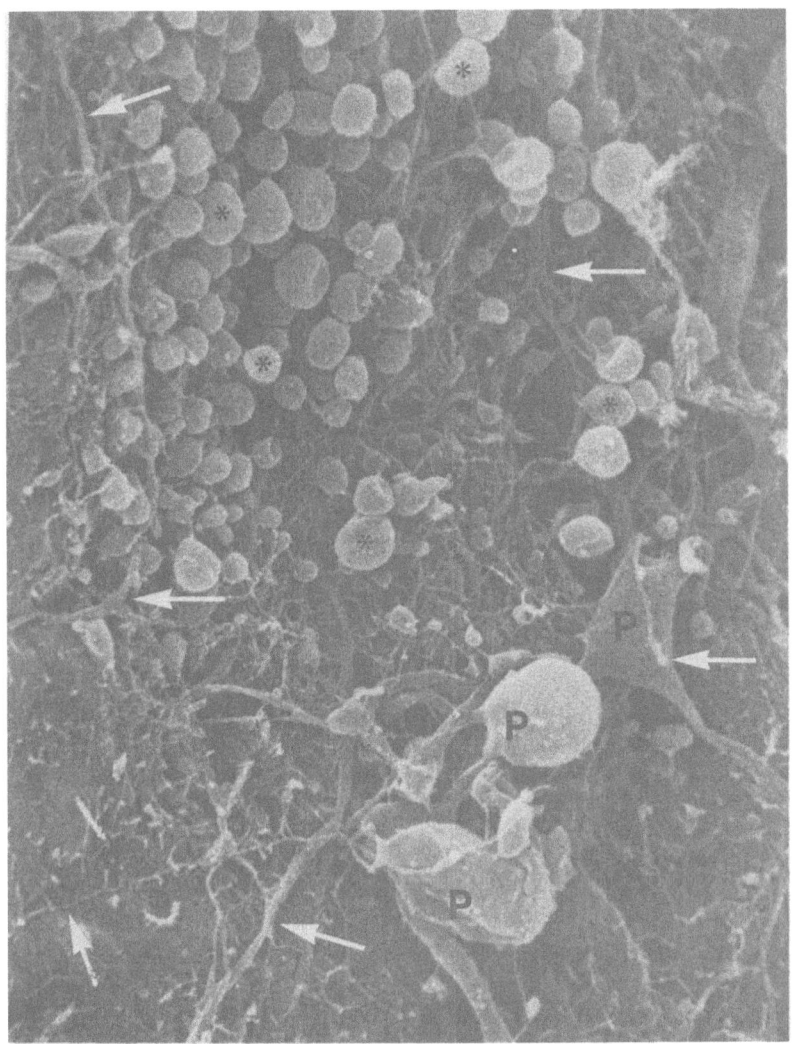

Fig. 4. Multipolar CSF-contacting neurons (P) ramify near the CSF-contacting dendrite terminals (asterisks) of the paraventricular organ in the carp. Arrows indicate the thick branching supraependymal dendrites and the thin axons. The latter form varicosities and contact the intra-ventricular neuronal elements. x 4.800. (Reduced 10% for reproduction.)

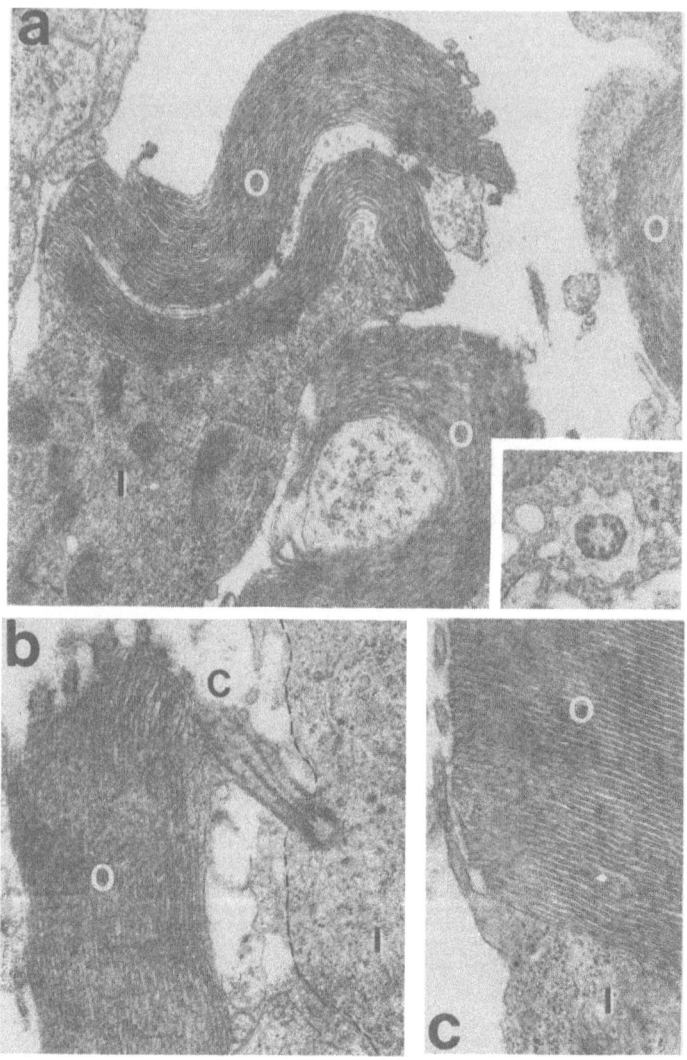

Fig. 5.(a-c). Details of the photoreceptor terminals of th
pineal organ of the goldfish (a, b) and of the frontal
organ of the frog (c). a) Inner segment (I) and outer segments (O) of the pinealocytes of the pineal stalk. x 27.12
Inset: Cross section of 9 x 2 + 0 cilium of the inner segment. x 22.800. b) Connecting piece of the cilium (C) aris
ing from the inner segment (I) of the pinealocyte in the
terminal part of the pineal organ. Numerous photoreceptor
membranes of the outer segment (O). x 25.560 c) Part of
inner (I) and outer (O) segments of the photoreceptor cell
of the frontal organ. x 32.400.
(Reduced 20% for reproduction.)

light as is supposed for the pinealocytes of lower vertebrates (2). It is known that the photoreception in the retina is related to the presence of the visual pigment rhodopsin, which has been demonstrated by immunocytochemical method (1, 4, 8). In our present paper, we examined the presence of rhodopsin in the sensory cells of the pineal complex (pineal organ, parapineal organ, frontal organ) of various vertebrates, and in the hypothalamus of the newt by immunohistochemistry. The unlabeled antibody peroxidase method was employed (4).

Semithin sections of material (Carassius auratus, Triturus vulgaris, Bombina bombina, Rana esculenta, Pseudemys scripta elegans, Lacerta viridis et agilis, white leghorn chickens, 12-day-old rats), fixed by cardiac perfusion with 4% glutaraldehyde in cacodylate buffer, were freed of araldite in sodium methoxide. Goat antibodies specific to the frog rhodopsin (a gift of Dr. Papermaster, Yale University, New Haven, Conn.) were coupled to the antigoat peroxidase complex. Antisera dilution and incubation time were: antirhodopsin 1:1000, second antiserum 1:200, 1:400, for 24 or 40 h at 4°C and 90 min at 22°C, respectively. The specificity of the method was tested.

The photoreceptor terminals of the stalk and the terminal vesicle of the pineal organ (Fig. 5a, b) and of the parapineal organ of the goldfish (Fig. 6a) were strongly positive with the immunohistochemical antirhodopsin reaction. The reaction product seemed to be located mainly in the outer segments of the pinealocytes (Fig. 6b, c, d). In the amphibians, the photoreceptor terminals of the frontal organ (Fig. 5c) and of the pineal organ were also reactive (Fig. 6e, f, g), mainly in the outer segments (Fig. 6g). It is noteworthy that in the frog we observed pinealocytes also in the entrance to the pineal recess and even on top of the third ventricle (Fig. 6f, g). The outer segments of these cells were also immunoreactive. Although the frog pinealocytes are relatively small, their antirhodopsin reaction was stronger than that of the large photoreceptor cells of the retina (Fig. 6h). In the reptiles, chickens and rats, the antirhodopsin reaction was rather weak around the lumina of the pineal organ. Furthermore, no reaction product was found in the ciliated CSF-contacting dendrite terminals of the paraventricular organ and infundibular nucleus of the newt (and in the subcommissural organ of the newt and the turtle).

Fig. 6a-h.   Details of the parapineal and the pineal
             organs and of the retina in the goldfish
             (a-d) and frog (e-h).
             a)   Semithin section of the parapineal organ
             (encircled by dotted line) above, and the
             stalk of the pineal organ (dotted) below.
             Toluidine blue azure II. X 245
             b)   Positive antirhodopsin reaction in
             photoreceptor terminals of the parapineal
             organ (dotted, above) and the pineal stalk
             (dotted, below). X 245
             c)   Antirhodopsin reaction in outer segments
             of the parapineal organ (dotted). X 450
             d)   Two immunoreactive outer segments of the
             pinealocytes in the pineal stalk (border
             dotted). X 1100
             e)   Antirhodopsin-reactive photoreceptor
             terminals of the pinealocytes around lumen
             (asterisk) of the pineal organ. X 245
             f)   Immunoreactive photoreceptor terminals
             of pinealocytes on top of the third ventricle
             (V) near the pineal recess. X 110
             g)   Pineal organ and recess (asterisk) with
             higher magnification. Antirhodopsin reaction
             in the outer segments of the photoreceptor
             terminals (at arrows). X 450
             h)   Antirhodopsin reaction in the outer
             segments of the retinal photoreceptor cells.
             X 1100 (Reduced 10% for reproduction)

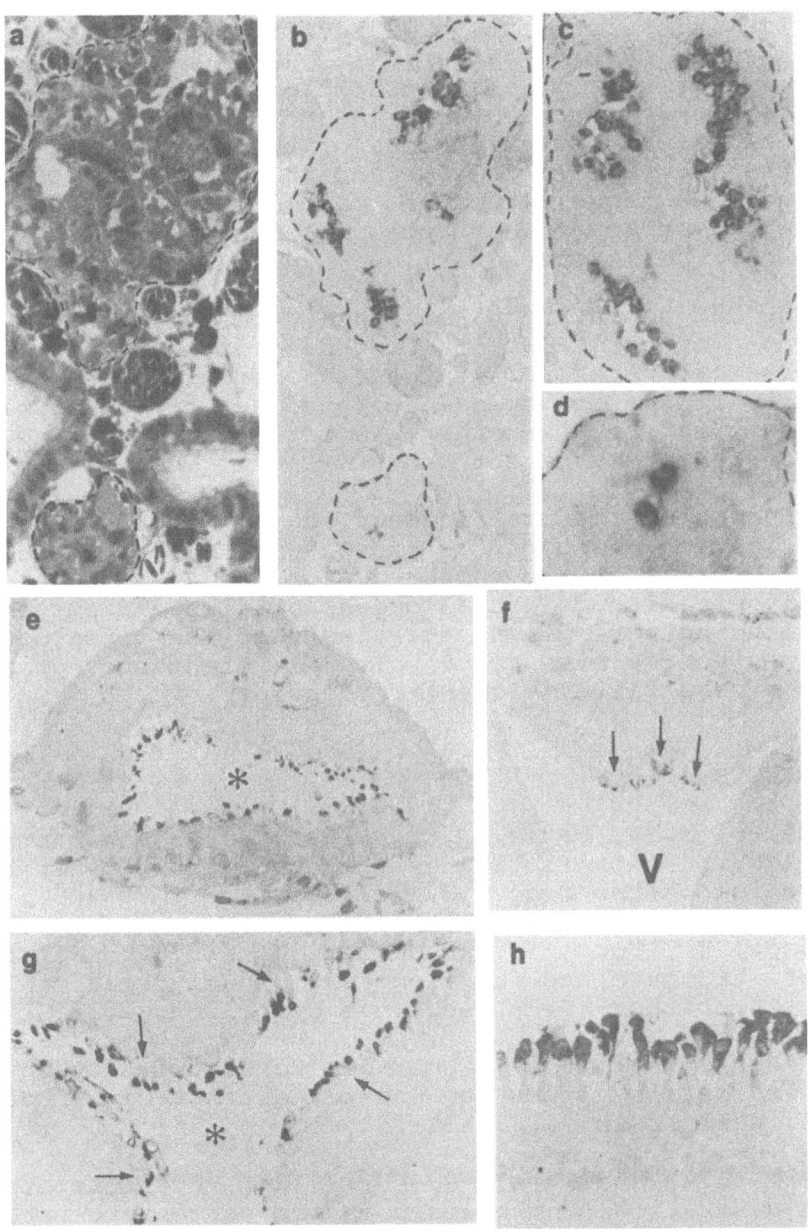

SUMMARY

   This paper describes the ultrastructure of the parvocellular preoptic nucleus of the lamprey. Its periventricular neurons form ciliated CSF-contacting dendrite
terminals which are similar to those of higher vertebrate
and resemble sensory elements. Light microscopically, the
ciliated CSF-contacting dendrites of the paraventricular
organ and infundibular nucleus of the newt do not react
with the immunohistochemical antirhodopsin reaction, suggesting that these structures may not be light receptors
but represent some other type of sensory elements (sodium-
glucose-, thermoreceptors?).

   The CSF-contacting axons innervate the ependyma and
terminate on the CSF-contacting dendrites and neuronal
cell bodies. Presumably, these axons influence the
activity of the ependyma and of the sensory dendrites.
It is not yet known whether the intraventricular neurons
are CSF receptors or not.

   Concerning the pinealocytes, our immunohistochemical
investigation revealed the presence of rhodopsin in the
outer segments of the photoreceptor cells of the parapineal, pineal and frontal organs of fish and amphibians.
This finding strengthens the view that the pinealocytes
of lower vertebrates are involved in light perception by
means of the visual pigment rhodopsin.

REFERENCES

1. Dewey, M.M., Davis, P.K., Blasif, J.K. and Barr, L.
    (1969): Localization of rhodopsin antibody in the
    retina of the frog. J. Mol. Biol. 39: 395-405.

2. Dodt, E., Ueck, M. and Oksche, A. (1971): Relations
    of structure and function: The pineal organ of lower
    vertebrates. Proc. Purkinje Symp. Brno 253-278.

3. Hehn, G. v. (1970): Über den Feinbau des hyponeur-
    alen Nervensystems des Seesternes. Asterial Rubens L.
    Z. Zellforsch. 105: 137-154.

4. Jan, L.Y. and Revel, J.-P. (1974): Ultrastructural
    localization of rhodopsin in the vertebrate retina.
    J. Cell Biol. 62: 257-273.

5. Kolmer, W. (1930): Über einen supraependymalen
    Nervenplexus in den Hirnventrikeln der Affen.
    Z. Anat. Entwickl.-Gesch. 93: 182-187.

6. Konstantinova. M. (1973): Monoamines in the liquor-
    contacting nerve cells in the hypothalamus of the
    lamprey, Lampetra fluviatilis L. Z. Zellforsch. 144:
    549-557.

7. Meves, A. (1973): Elektronenmikroskopische Unter-
    suchungen über die Zytoarchitektur des Gehirns von
    Branchiostoma lanceolatum. Z. Zellforsch. 139:
    511-532.

8. Papermaster, D.S., Schneider, B.G., Zorn, M.A. and
    Kraehenbuhl, J.P. (1978): Immunocytochemical local-
    ization of opsin in outer segments and Golgi zones
    of frog photoreceptor cells. An electron microscope
    analysis of cross-linked albumin-embedded retinas.
    J. Cell Biol. 77: 196-210.

9. Ramon y Cajal, S. (1935): Die Neuronenlehre. In:
    Handbuch der Neurologie 1, O. Bunke and O. Foerster
    (eds.), pp. 887-982, Springer, Berlin.

10. Shioda, S., Honma, Y., Yoshie, S. and Hosoya, Y.
    (1977): Scanning electron microscopy of the third
    ventricular wall in the lamprey, Lampetra japonica.
    Arch. histol. jap. 40: 41-49.

11. Stendell, W. (1914): Zur Histologie des Rücken-
    markes von Amphioxus. Anat. Anz. 46: 258-267.

12. Tretjakoff, D. (1909): Das Nervensystem von Ammo-
    coetes. II. Gehirn. Arch. mikr. Anat. 74: 636-779.

13. Vigh, B., Teichmann, I. and Aros, B. (1969): Das
    Paraventrikularorgan und das Liquorkontaktneuronen-
    system. Anat. Anz. Suppl. 125: 683-688.

14. Vigh, B. and Vigh-Teichmann, I. (1973): Comparative
    ultrastructure of the CSF-contacting neurons. Int.
    Review Cytol. 35: 189-251.

15. Vigh, B. and Vigh-Teichmann, I. (1974): Vergleich
    der Ultrastruktur der Liquorkontaktneurone und
    Pinealozyten. Anat. Anz. Suppl. 68: 433-443.

16. Vigh, B. and Vigh-Teichmann, I. (1977): Studies on
     the vascular sac and related structures. Nova Acta
     Leopoldina Suppl. 9: 97-102.

17. Vigh, B., Vigh-Teichmann, I. and Aros, B. (1975):
     Comparative ultrastructure of cerebrospinal fluid-
     contacting neurons and pinealocytes. Cell Tiss.
     Res. 158: 409-424.

18. Vigh-Teichmann, I. (1971): A hypothalamikus peri-
     ventricularis szürkeállómány és a liquor cerebro-
     spinalis kapcsolatának összehasonlitó morfólogiai
     vizsgálata. (Comparative morphological study of
     the relation between hypothalamic periventricular
     gray substance and cerebrospinal fluid. In Hungar-
     ian) Cand. Med. Sci. Thesis, Hung. Acad. Sci.,
     Budapest.

19. Vigh-Teichmann, I. and Vigh, B. (1974): The infun-
     dibular cerebrospinal fluid-contacting neurons.
     Advances Anat. Embryol. Cell Biol. 50: 2.

20. Vigh-Teichmann, I. and Vigh, B. (1977): Zilien-
     tragende Perikaryen im Diencephalon. Verh. Anat.
     Ges. 71: 989-995.

21. Vigh-Teichmann, I., Vigh, B. and Aros, B. (1976):
     Cerebrospinal fluid-contacting neurons, ciliated
     perikarya and "peptidergic" synapses in the magno-
     cellular preoptic nucleus of teleostean fishes.
     Cell Tiss. Res. 165: 397-413.

22. Vigh-Teichmann, I., Vigh, B. and Aros, B. (1976):
     Ciliated neurons and different types of synapses
     in anterior hypothalamic nuclei of reptiles. Cell
     Tiss. Res. 174: 139-160.

23. Vigh-Teichmann, I., Vigh, B., Aros, B., Jennes, L.,
     Sikora, K. and Kovács, J. (1979): Scanning and
     transmission electron microscopy of intraventricular
     dendrite terminals of hypothalamic cerebrospinal
     fluid-contacting neurons in Triturus vulgaris.
     Z. mikr.-anat. Forsch. 93: 4, in press.

K. Abe, Laboratory of Neuropathology and Neuroanatomical
   Sciences, National Institute of Neurological and
   Communicative Disorders and Stroke, National Insti-
   tutes of Health, Bethesda, Maryland 20205, U.S.A.

T. Abe, Laboratory of Neuropathology and Neuroanatomical
   Sciences, National Institute of Neurological and
   Communicative Disorders and Stroke, National Insti-
   tutes of Health, Bethesda, Maryland 20205, U.S.A.

R.K. Andjus, Institute of Physiology, Faculty of Science,
   University of Belgrade, 1100 Belgrade, Yugoslavia.

M.E. Bitterman, Békésy Laboratory of Neurobiology,
   University of Hawaii, Honolulu, Hawaii 96822.

G.A. Buznikov, N.K.Kol'tsov Institute of Developmental
   Biology of the Academy of Sciences of USSR, Moscow,
   USSR.

V. Cvejić, Laboratory of Neurochemistry, Institute of
   Biochemistry, Faculty of Medicine, Belgrade, Yugoslavia.

C. Di Benedetta, Istituto di Fisiologia Umana, Università,
   di Bari, Italy.

N.H. Diemer, Institutes of Neuropathology and Medical
   Physiology, University of Copenhagen, Denmark.

B.M. Djuričić, Laboratory of Neurochemistry, Institute
   of Biochemistry, Faculty of Medicine, Belgrade,
   Yugoslavia.

E. Dux, Institute of Biophysics, Hungarian Academy of
   Sciences, Szeged, Hungary.

T. Fujimoto, Department of Neurosurgery, Tokyo Medical
    and Dental University, Tokyo, Japan.

I.V. Gannushkina, Institute of Neurology, Academy of
    Medical Science, Moscow, USSR.

J. Halsey, Department of Pathology, University of Alabama
    Medical Center, Birmingham, Alabama 35294, U.S.A.

K.-A. Hossmann, Max-Planck-Institut für Hirnforschung,
    Forschungsstelle für Hirnreislaufforschung, Köln
    (Merheim), West Germany.

Y. Inaba, Department of Neurosurgery, Tokyo Medical and
    Dental University, Tokyo, Japan.

J.J. Ivanuš, Department of Molecular Biology, Boris
    Kidrich Institute, Belgrade, Yugoslavia.

F. Joó, Laboratory of Molecular Biology, Institute of
    Biophysics, Biological Research Center, Szeged,
    Hungary.

D. Kanazir, Serbian Academy of Sciences, Belgrade, Yugo-
    slavia.

S. Kanazir, Institute for Biological Research, Belgrade,
    Yugoslavia.

I. Karnushina, Laboratory of Molecular Biology, Insti-
    tute of Biophysics, Biological Research Center,
    Szeged, Hungary.

I. Klatzo, Lboratory of Neuropathology and Neuroana-
    tomical Sciences, National Institute of Neurological
    and Communicative Disorders and Stroke, National
    Institutes of Health, Bethesda, Maryland 20205,
    U.S.A.

D. Kostić, Laboratory for Neurochemistry, Institute of
    Biochemistry, Faculty of Medicine, University of
    Belgrade, Yugoslavia.

A.G.B. Kovách, Experimental Research Department, Semmel-
    weis Medical University, Budapest, Hungary.

L. Lazarević, Brain Research Laboratory, Institute for
    Biological and Medical Research of Montenegro,
    Kotor, Yugoslavia.

M. Levental, Department of Neurochemistry, Institute for Biological Research of Montenegro, Kotor, Yugoslavia.

W.D. Lust, Laboratory of Neurochemistry, National Institute of Neurological and Communicative Disorders and Stroke, National Institutes of Health, Bethesda, Maryland 20205, U.S.A.

B. Manukhin, N.K.Kol'tsov Institute of Developmental Biology of the Adademy of Sciences of USSR, Moscow, USSR.

D. Mićić, Institute of Biochemistry, Faculty of Medicine, Belgrade, Yugoslavia.

R. Mileusnić, Institute of Biochemistry, Faculty of Medicine, Belgrade, Yugoslavia.

R. Miline, Institute of Histology and Embryology, Faculty of Medicine, Novi Sad, Yugoslavia.

B.B. Mršulja, Laboratory of Neurochemistry, Institute of Biochemistry, Faculty of Medicine, Belgrade, Yugoslavia.

B.J. Mršulja, Division of Neurophysiology and Neurochemistry, Institute for Biological Research, Belgrade, Yugoslavia.

M.B. Novaković, Department of Molecular Biology, Boris Kidrich Institute, Belgrade, Yugoslavia.

V. Pantić, Institute of Histology and Embryology, Faculty of Veterinary Sciences, Belgrade, Yugoslavia.

S. Petrović, Department of Molecular Biology, Boris Kidrich Institute, Belgrade, Yugoslavia.

Lj.M. Rakić, Department of Neurochemistry, Institute for Biological Research, Belgrade, Yugoslavia.

R. Rodnight, Department of Biochemistry, Institute of Psychiatry, British Postgraduate Medical Federation, London University, London SE5 8AF, United Kingdom.

W. Roggendorf, Institut für Neuropathologie, Klinikum Steglitz, Freie Universität, Berlin, Federal Republic of Germany.

S.P.R. Rose, Biology Department, The Open University,
    Molton Keynes, UK.

M. Rusić, Department of Neurochemistry, Institute for
    Biological Research, Belgrade, Yugoslavia.

O.Z. Sellinger, Laboratory of Neurochemistry, Mental
    Health Research Institute, University of Michigan
    Medical Center, Ann Arbor, Michigan 48109, U.S.A.

M. Spatz, Laboratory of Neuropathology and Neuroanatomi-
    cal Sciences, National Institute of Neurological
    and Communicative Disorders and Stroke, National
    Institutes of Health, Bethesda, Maryland 20205,
    U.S.A.

V. Šušić, Institute of Physiology, Faculty of Medicine,
    University of Belgrade, Yugoslavia.

I. Tóth, Laboratory of Molecular Biology, Institute of
    Biophysics, Biological Research Center, Szeged,
    Hungary.

A. Vernadakis, Department of Psychiatry and Pharmacology,
    University of Colorado School of Medicine, Denver,
    Colorado 80260, U.S.A.

B. Vigh, 2nd Department of Anatomy, Medical University,
    Budapest, Hungary.

I. Vigh-Teichman, 2nd Department of Anatomy, Medical
    University, Budapest, Hungary.

A. Vranešević, Laboratory for Neurochemistry, Institute
    of Biochemistry, Faculty of Medicine, University
    of Belgrade, Yugoslavia.

E. Westergaard, Anatomy Department C, University of
    Copenhagen, Denmark.

T. Yanagihara, Department of Neurology, Mayo Clinic
    and Mayo Medical School, Rochester, Minn. 55901,
    U.S.A.